THE
POPULIST REVOLT

THE POPULIST REVOLT

*A History of the Farmers' Alliance
and the People's Party*

by

JOHN D. HICKS

A Bison Book
University of Nebraska Press
1961

Originally published by the University of Minnesota Press
Copyright 1931 by the University of Minnesota
Copyright renewed 1959 by John D. Hicks
All rights reserved
Library of Congress catalog card number 61–7237
International Standard Book Number 0–8032–5085–1
Manufactured in the United States of America

First Bison Book printing February, 1961
Most recent printing shown by first digit below:

8   9   10

TO MY GRADUATE STUDENTS

PREFACE

THE rôle of the farmer in American history has always been prominent, but it was only as the West wore out and cheap lands were no longer abundant that well-developed agrarian movements began to appear. First on the list came the Granger movement, which held together long enough to advance materially the cause of governmental regulation of the railroads in the interest of the farmers. Then came the Greenback party, which voiced the sentiments of the debtor West lustily enough to make the currency question a leading issue in American politics for the rest of the century. Then, coincident with the annihilation of the frontier, came the Farmers' Alliance and the People's party with longer and more complicated programs for ameliorating the evils from which agriculture was suffering. As a result of these agitations many of the remedies for which the farmers clamored were at length adopted, but it seemed as though for every problem solved a new one came to view. Even in the fat years of the early twentieth century farm relief as a political problem was rarely forgotten, and now it promises to become a hardy perennial.

The Grangers and the Greenbackers have already found their historians, and the time has come when a more adequate account can be written of the Farmers' Alliance and the People's party than has hitherto appeared. The pioneer work on this subject, Frank L. McVey's *Populist Movement*, was published in 1896 in the midst of the hot prejudice that

the rise of Populism had engendered and while the outlines
of the movement were yet indistinct. Twenty years later,
when Fred E. Haynes wrote his *Third Party Movements
since the Civil War*, the old anti-Populist rancor had somewhat died down, and the work of the Populists could more
nearly be assessed for what it was worth. But Haynes lacked
monographic material on which to rely, and in the case of a
movement so widespread and so many-sided as Populism the
work of a single investigator was bound to be inadequate.
Since the time when Haynes wrote, books and articles dealing with various phases and segments of the Populist movement have multiplied amazingly, and for this reason, if for
no other, the time is ripe for another general treatise on the
subject. The lapse of a few more years should have made it
possible, also, to get a better focus on the actual accomplish-
ments of the Populists and to review even their wildest
activities with complete good nature.

It is a pleasure to acknowledge the great indebtedness
I owe to the long list of investigators whose researches into
Populist history have preceded or accompanied my own.
Such names as Solon J. Buck, Francis B. Simkins, Alex M.
Arnett, John D. Barnhart, Raymond C. Miller, Hallie Farmer,
Herman C. Nixon, and Ernest D. Stewart appear with great
frequency and for good reason throughout the footnotes. It
is only fair, also, to point out that the graduate students who
have sat in my seminars at the University of Nebraska during the past eight years have had a large share in the making
of this book. Their names and their particular contributions
appear in the bibliography. From Professor Thomas M.
Raysor, chairman of the Department of English at the University of Nebraska, from Mrs. Margaret S. Harding, managing editor of the University of Minnesota Press, and from
Miss Livia Appel, assistant editor, have come many valuable
suggestions. The attendants at many libraries have shown
me unfailing courtesy, but special mention should be made

of Mrs. Clarence S. Paine of the Nebraska State Historical Society, who has suffered many things in the interest of Populism, not only from myself but also from my students. Nor can I forget the help I have received in the preparation of the manuscript from my secretary, Miss Florence G. Beers, who patiently reduced my handwriting to type, and from my wife, who came to my aid in the dull business of proofreading.

JOHN D. HICKS

Lincoln, Nebraska
July, 1931.

CONTENTS

THE
POPULIST REVOLT

For this is a moral entertainment, good people, planned to show you that yesterday makes to-day and that they both make to-morrow, and so the world spins around the sun.

A Certain Rich Man

CHAPTER ONE

THE FRONTIER BACKGROUND

THE growth of the United States west of the Alleghanies during the past fifty years is due not so much to free institutions, or climate, or the fertility of the soil, as to railways. If the institutions and climate and soil had not been favorable to the development of commonwealths, railways would not have been constructed; but if railways had not been invented, the freedom and natural advantages of our western states would have beckoned to human immigration and industry in vain. Civilization would have crept slowly on, in a toilsome march over the immense spaces that lie between the Appalachian ranges and the Pacific Ocean; and what we now style the Great West would be, except in the valley of the Mississippi, an unknown and unproductive wilderness.[1]

To the nineteenth century American who penned these words it probably never occurred that the rapid expansion of the West was perhaps not an unmixed good. The land, he assumed, was there to be conquered, and the sooner the conquest could be completed the better. Whatever helped to further that end was wise and good and right. Whatever held it back was an obstacle in the way of "progress." The railways had doubtless accelerated the conquest of the continent quite as much as he believed, and they were therefore

[1] Sidney Dillon, "The West and the Railroads," *North American Review*, 152:443.

of the very essence of all that was best. The suggestion of a foreigner, Lord Bryce, that perhaps the American West had grown too fast and that if development had been slower " it might have moved upon better lines," [2] would have filled him with amazement and disgust.

But Bryce was doubtless right. Certainly when the railways assumed command of the westward movement, they pushed it forward with unseemly haste. Never deliberate, the advance of the frontier now became a headlong rush. It took a century and a half for the frontier to reach the Appalachians; it took a good half century for it to move from the Appalachians to the Mississippi, even though in this second West the rivers generally solved the problem of transportation by flowing the right way; but it took only another half century to annihilate the frontier altogether, in spite of the vast mountain and desert spaces that blocked the way in the trans-Mississippi West. Under railway leadership the population came in too rapidly to permit of thoughtful and deliberate readjustments. A society at once so new and so numerous was immediately confronted with problems that it could not comprehend, much less hope to solve. Flight to a new frontier could no longer avail, for the era of free lands was over. The various agrarian movements, particularly the Alliance and the Populist revolts, were but the inevitable attempts of a bewildered people to find relief from a state of economic distress made certain by the unprecedented size and suddenness of their assault upon the West and by the finality with which they had conquered it. [3]

The building of the transcontinental railways marked the beginning of the end for the last American frontier. [4] These lines were projected before the Civil War, and most

[2] James Bryce, *The Study of American History*, 35.
[3] Frederick J. Turner, *The Frontier in American History*, 219–221.
[4] Frederic L. Paxson, "The Pacific Railroads and the Disappearance of the Frontier in America," *Annual Report of the American Historical Association, 1907*, 1:105–118.

of the available routes were surveyed during the decade of the fifties,[5] but the impending struggle between North and South prevented immediate action. Only when the war was on and the objections of "state righters" and strict constructionists were no longer heard in Congress was the nationalistic North free to go ahead with what was essentially a nationalistic policy. Then one Pacific railway bill after another was enacted into law. The Union Pacific received its original charter in 1862, the Northern Pacific in 1864, the Atlantic and Pacific, whose charter rights were later enjoyed mainly by the Santa Fe, in 1866, and the Texas Pacific in 1871.[6]

These laws carried the practice of governmental assistance to railways a step further than it had ever gone before. Grants of government land to states in aid of internal improvements had been common since 1850, and great quantities of land thus given to the states had been handed on to the railways. But the Pacific railways received their lands directly from the national government. The Union Pacific and the Central Pacific of California each obtained a cash loan of from sixteen to forty-eight thousand dollars per mile of track constructed, in addition to alternate sections of land for a distance of twenty miles into the interior on each side of their right of way. The other roads got no cash, but they were given forty sections of land per mile of track in each of the territories and twenty sections in each of the states. In the meantime the policy of land grants to the states for railway purposes enabled some of the newly created western commonwealths to vie with the national government in the generosity of their gifts. Altogether no less than 129,000,000

[5] Reports of Explorations and Surveys to Ascertain the Most Practicable and Economical Route for a Railroad from the Mississippi River to the Pacific Ocean, *House Executive Document* No. 91, 33 Congress, Session 2.
[6] *United States Statutes at Large*, 12:489; 13:356, 365; 14:292; 16:573. The Texas Pacific land grant was later forfeited. See John B. Sanborn, *Congressional Grants of Land in Aid of Railways*, 125.

acres of the public lands were distributed, directly or indirectly, to the railways.[7]

The period of active railway expansion which followed the Civil War and which was induced, at least in part, by these extraordinary gifts, forms a familiar chapter in national history. During the seventies and the early eighties most of the great roads of the trans-Mississippi West mapped out their through routes, while numerous lesser and more local roads filled in the gaps, especially in the region just west of the Mississippi River. Following the junction of the Union Pacific and the Central Pacific in 1869, one great railway system after another approached completion. Within a few years the Northern Pacific, the Southern Pacific, the Kansas Pacific, the Missouri Pacific, the Santa Fe, the Burlington, the Rock Island, the Great Northern, all became household names. It was due to them that western Iowa and western Minnesota, Dakota Territory, Kansas, and Nebraska, to say nothing of the farther West, became habitable and inhabited.[8] Through their efforts the American citizen was introduced to another frontier, a region aptly named by Hamlin Garland the "Middle Border."

It was quite as much the business of the roads to people this West as to penetrate it. Only by so doing could they hope to build up the thriving local business upon which their future prosperity would depend. Handsome dividends could come only from local freight rates, and local freight rates could be paid only by actual settlers. The transcontinental lines had, of course, a certain secure through business, but no one was sanguine enough to believe that this business alone, divided up as it must be among four or five competitors, would supply much revenue. The new frontier must be advertised and settled — the sooner the better. Naturally enough, this duty was assumed in the main by the land-grant roads, each of which speedily organized a land department

[7] Benjamin H. Hibbard, *A History of the Public Land Policies*, 264.
[8] Hallie Farmer, "The Economic Background of Frontier Populism," *Mississippi Valley Historical Review*, 10:407.

and began to boom its wares. The land departments, ably managed, made it their chief concern to collect and disseminate by every known method such information about the territory adjacent to their roads as might induce people to come there to live.[9]

This task was not altogether an easy one, in spite of the normal proneness of the American farmer to go West, for the old fiction of the Great American Desert was not yet fully discredited.[10] Many people knew better, but for the multitudes the tradition that the country to the west of the Missouri River was good only for the Indians and of no value to the whites was still sound doctrine. Political necessities rather than the pressure of population had forced the organization of Kansas and Nebraska, and the sparseness of their settlements in the sixties testified eloquently to the common belief in the worthlessness of their soil and climate.[11] The average man who had heard of these and neighboring regions was afraid of them. He might concede that agriculture would have a chance for a few miles beyond the Missouri River, but he solemnly warned the prospective emigrant that to adventure farther was to enter a region where "nothing would grow." Some placed the dead line as close as ten miles to the west of the river; the most optimistic made it less than a hundred; there was general agreement that agricultural settlement beyond that distance was impossible.[12]

One of the first tasks of the advertisers, therefore, was to break down this old prejudice about western lands and climate. In the motley array of pamphlets and brochures that they sent forth, these points were dwelt upon with consid-

[9] Addison E. Sheldon, "Early Railroad Development of Nebraska," *Nebraska History and Record of Pioneer Days*, 7:16–17.

[10] *Fourth Annual Report of the President and Secretary of the Nebraska State Board of Agriculture*, 1873, p. 11.

[11] Frederic L. Paxson, *The Last American Frontier*, 137.

[12] Ralph C. Morris, "The Notion of a Great American Desert East of the Rockies," *Mississippi Valley Historical Review*, 12:197–199; C. J. Ernst, "The Railroad as a Creator of Wealth in the Development of a Community or District," *Nebraska History and Record of Pioneer Days*, 7:19. Mr. Ernst was in charge of the land department of the Burlington from 1876 to 1924.

erable solicitude. "Follow prairie dogs and Mormons and
you will find good land," was the attractive slogan unearthed
by a Burlington pamphleteer.[13] The Union Pacific called
attention to its lands in the Platte Valley, a "flowery mea-
dow of great fertility clothed in nutritious grasses, and watered by numerous streams."[14] As settlers came in, the
records of their successes were sent back. By 1877 the Burlington could testify that crop yields in Nebraska were amazingly large. "Many fields of properly cultivated wheat have
yielded *over thirty bushels of grain per acre, and many fields
of corn over seventy.*"[15] Pamphlets were published giving
in detail the experiences of early settlers — men who had
come to the West with nothing and in a few short years
had arrived at a comfortable prosperity.[16]

Nor did the people of the new states and territories stop
with the mere contribution of these personal details. As
desirous of securing new settlers as ever the railways could
be, they paid out public funds for the work of state boards
of immigration and of agriculture, whose chief duty it was
to advertise the West.[17] One state publication nailed forever
the myth about the lack of rainfall west of the Missouri.
It published the records kept by one Dr. A. L. Child of
Plattsmouth, Nebraska, to show that the average rainfall
in Nebraska, as he had observed it from 1866 to 1879, was
32.29 inches, as compared with 34.13 inches in Illinois.[18]
What further evidence could one desire?

[13] *Views and Descriptions of Burlington and Missouri River Railroad Lands,
with Important Information Concerning How to Select and Purchase Farms in
Iowa and Nebraska, on Ten Years Credit*, opposite plate 14.
[14] *Guide to the Union Pacific Railroad Lands*, 1870, p. 7.
[15] *B and M Railroad — 750,000 Acres of the Best Lands for Sale, Southern
Iowa and Southeastern Nebraska*, 6.
[16] J. T. Allan, *Nebraska and Its Settlers*, 7.
[17] Farmer, in the *Mississippi Valley Historical Review*, 10:409. The states,
moreover, had for sale lands of their own, given them by the national government in aid of education. These "school lands" were usually offered on very
easy terms.
[18] *The State of Nebraska and Its Resources*, 10.

As the illusion of the Great American Desert gradually disappeared, the tremendous attractiveness of this new West became more and more apparent. The terms offered by the railways were clearly designed to bring in settlers rather than to produce a revenue. The prices asked were uniformly low. The average price at which the Burlington sold its entire Nebraska grant was $5.14 an acre, but it disposed of enormous tracts of land at as low a price as twenty-five cents an acre. The usual government minimum of $1.25 an acre and "double minimum" of $2.50 an acre were prices very commonly asked and obtained.[19]

The railway companies, moreover, granted credit of the most liberal kind. The Burlington, for example, advertised in an elaborate booklet two plans of payment, "short credit" and "long credit." By "short credit" was meant a three-year plan of payment, under which one-third of the principal fell due each year and the interest rate was ten per cent. Long credit was designed to meet the needs of the great majority, who had little or no cash. Small annual payments, the first two of which amounted only to the interest at six per cent, were to extend over a period of ten years. Settlers were generally advised to buy on this plan, since it would leave them free to use what money they had or could earn for the purchase of implements and the improvement of their farms.[20] Liberal discounts were usually obtainable for those who could pay cash or buy on short credit,[21] but bargains were always available for the man whose resources were slight. The Burlington, in order to enable prospective settlers to examine land in advance of purchase, sold land-

[19] Ernst, in *Nebraska History*, 7:21; *Guide to the Union Pacific Railroad Lands*, 24; *Views and Descriptions of Burlington and Missouri River Railroad Lands*, opposite plate 4; *Nebraska and Its Resources*, 31–32.
[20] *Views and Descriptions of Burlington and Missouri River Railroad Lands*, opposite plates 5, 6, 18.
[21] *North Platte Lands of the Burlington and Missouri River Railroad Company in Nebraska*, 12.

exploring round-trip tickets at three-fifths the usual fare and gave with each ticket a rebate certificate entitling the holder to a handsome refund in case he made a purchase. The same road accorded to actual settlers advantages not open to speculators. Passenger fares for entire families and necessary freight charges on their movable property were refunded to those who moved upon their land within two years' time. Newcomers who would cultivate a certain portion of their land were given a "premium for improvement," consisting of a fifteen or twenty per cent discount on their purchase price. "Better terms than these have never been offered to purchasers of land and probably never will be," [22] was a Burlington boast, but the other roads were never far behind, if behind at all, in generosity to the prospective settler. Railroad lands were not quite "free lands," but they were nearly that. [23]

Free lands, however, did exist. Under the operation of the Homestead Act of 1862 it was possible to get land from the government for no more than a few dollars in fees. Since the railways were primarily interested in settlers rather than in sales, they advertised these homestead opportunities to the limit. Scarcely a pamphlet or booklet failed to set forth in detail facts concerning the amount and character of the government lands and the methods by which they might be secured. According to the original law, any person who was over twenty-one, the head of a family, and a citizen of the United States, or an alien who had filed his intention of becoming a citizen, could acquire one hundred and sixty acres of government land by living on it for five years. [24] This was

[22] B and M Railroad Lands, 27–28; Ernst, in Nebraska History, 7:20; Views and Descriptions of Burlington and Missouri River Railroad Lands, opposite plate 5.

[23] See also James B. Hedges, "Colonization Work of the Northern Pacific Railroad," Mississippi Valley Historical Review, 13:321, and Helen Anderson, "The Influence of Railway Advertising upon the Settlement of Nebraska," master's thesis, University of Nebraska.

[24] United States Statutes at Large, 12:392.

qualified, however, with reference to the alternate sections that lay within the limits of the railway land grants. Here one could obtain a homestead of only eighty acres, since, it was argued, the nearness of the railway would at least double the value of the land.[25] Free lands, however, rarely lasted long in a region opened up by a railroad. Often settlers seeking homesteads preceded the railway, and except in the arid or semi-arid regions they were never far behind it. In 1877 a Burlington pamphlet pointed out that " to-day there is but a small quantity of government land, lying within reasonable distance from railroads and fit for agricultural purposes, open to settlement. The government lands contiguous to this line of road were all eagerly taken up four or five years ago." Those remaining were from twenty to forty miles from the tracks.[26] A Nebraska state publication asserted in 1879 that homestead lands still existed within the borders of the state but admitted that " those who want these will have to go considerable west." [27]

Numerous among those who sought homesteads were Union veterans of the Civil War and their widows and dependent children, for whom the government made the acquisition of land especially easy. Gifts of land to veterans of a war had been an American tradition, but the Homestead Act of 1862 virtually extended this privilege to all citizens. Therefore, in order to help the soldier or his dependents more than others, Congress provided by law in 1870 that soldiers might take one hundred and sixty acre homesteads from the government land within the railway land grants, whereas others could take only eighty acres.[28] This was a valuable privilege, for the lands near the railroads had a decidedly higher sale price than other lands and were dis-

[25] Sheldon, in *Nebraska History,* 7:16.
[26] *B and M Railroad Lands,* 24.
[27] *Nebraska and Its Resources,* 31.
[28] *United States Statutes at Large,* 16:321.

posed of more easily.[29] By a law enacted in 1872 Congress did even more for the soldiers, by offering homesteads to them or their widows or minor children sooner than ordinary citizens could obtain them. Whatever time, up to four years, that a veteran had spent in the service he might subtract from the five years necessary to prove up on his homestead. Wounded or disabled veterans, discharged before their term of enlistment had expired, might deduct from the five years their entire term of enlistment, and the unmarried widows and minor orphans of veterans who died in service had a similar privilege.[30] Soldiers also had the privilege of locating their claims through an attorney, whereas others must file in person. These privileges, which unquestionably were responsible for a large influx of settlement into the West, were widely heralded by the railways.[31]

Homesteading, however, was not the only way of acquiring land. The right of preemption had been exercised long before the passage of the Homestead Act, and it was not then discontinued. The status of preemptor was obtained by six months' residence upon a quarter section, the amount of land that, according to the law, he was privileged to acquire. After the six months, however, he must pay for his holding at the rate of $1.25 per acre, except in the case of sections retained by the government in railway land-grant areas, where the price was doubled. Homesteaders who did not wish to wait five years to acquire title to their land were permitted after six months' residence to change to the preemption right and pay out on their claim at the same rates.[32] This right of commutation, little used in the early years of the Homestead Act, became exceedingly popular later on, and the details concerning it and the original right of pre-

---

[29] Hibbard, *Public Land Policies*, 130.
[30] *United States Statutes at Large*, 17:49, 333.
[31] *Views and Decriptions of Burlington and Missouri River Railroad Lands*, opposite plate 15; *B and M Railroad Lands*, 24; Sheldon, in *Nebraska History*, 7:17; Hibbard, *Public Land Policies*, 131.
[32] Hibbard, *op. cit.*, 158, 168, 245, 386.

emption were invariably set forth fully in the advertisements
sent out by the railways.[33]

Under the Timber Culture Act of 1873 and its various
subsequent amendments one more right to public land was
conferred. The purpose of these measures was to encourage
the planting of trees on the western plains. Those who lived
up to the elaborate requirements of the law and, in practice,
more who did not, might receive title to a maximum of one
hundred and sixty acres of land at the expiration of a period
of eight years. This act was from the first interpreted by the
settler as a chance to increase the amount of his holding
rather than an obligation to plant trees.[34] The three rights,
homestead, preemption, and timber culture, could be, and
frequently were, all exercised by the same individual. A Burlington pamphlet told precisely how this could be done. The
prospective settler should first take a timber claim and a
preemption, paying up for the latter after six months. He
could then take a homestead and thus secure a total of four
hundred and eighty acres of land "at a cash outlay of about
50¢ per acre."[35]

With such opportunities as these to portray, it was an
easy task for the advertiser to make a most effective appeal
to residents of the states to the east who were not wholly
contented with their lot. Chief reliance was placed upon
pamphlets, but long and obviously inspired articles in such
papers as the *Western Rural* and the *Chicago Times* also
extolled the virtues of the new frontier. Pamphlets were
distributed locally for redistribution by the settlers who had
already answered the "call of the West." Wherever the railway company had agents the brochures were scattered
broadcast. In Europe they were distributed by steamship

[33] See, for example, *Views and Descriptions of Burlington and Missouri
River Railroad Lands*, opposite plate 3; *The Broken Bow Country in Central
and Western Nebraska and How to Get There*, 3; and *Eastern Colorado. A Brief
Description of the New Lands Now Being Opened Up*, 11–12.
[34] Hibbard, *Public Land Policies*, 414–415.
[35] *Eastern Colorado*, 11–12.

officials as well as by official railway representatives, and
they were placed in the steerage of Atlantic steamers bound
for America.[36] There was no considerable region without its
special pleader. The Santa Fe advertised as the "best thing
in the West" two and a half million acres of land in southwestern Kansas and offered statistics to prove that with an
investment of eight thousand dollars in the cattle business
one might reasonably expect an annual return of eleven
thousand dollars. The Chicago, Milwaukee, and St. Paul described Dakota as a "land of promise," in which every "ce-
real, vegetable, and flower grown in this latitude in the United
States" would thrive.[37] Burlington and Union Pacific pamphlets sounded ominous warnings against going too far either
to the north or to the south and advocated that newcomers
keep in the latitude to which they were accustomed. For
Nebraska they claimed the advantage of a more central loca-
tion, superior railway facilities, and a soil and climate which
guaranteed that every agricultural pursuit would be as re-
plete with pleasure as with profit.[38]

The chance of acquiring wealth in a few short years was
an almost irresistible inducement. The new lands were described as "cheaper, more easily tilled, better adapted to
cultivating and harvesting by machinery, more productive,
and increasing in value much faster than any of the Eastern
states." Here was a man who had made a fortune of ten
thousand dollars in seven years; there, an invalid whose
move to the West had netted him health as well as wealth.[39]
"Settling on the prairie which is ready for the plow is different from plunging into a region covered with timber," wrote
one observer. "Nature seems to have provided protection

[36] *The Broken Bow Country*, 16; Anderson, "Railway Advertising," 24–29.
[37] Farmer, in the *Mississippi Valley Historical Review*, 10:407–408.
[38] *Views and Descriptions of Burlington and Missouri River Railroad Lands*,
opposite plate 13; *B and M Railroad Lands*, 3; *Guide to Union Pacific Railroad Lands*, 13–14.
[39] *Views and Descriptions of Burlington and Missouri River Railroad Lands*,
opposite plate 13; *B and M Railroad Lands*, 9–10.

and food for man and beast; all that is required is diligent labor and economy to ensure an early reward."[40] "I have never, before coming here, taken much stock in the stories of wonderfully cheap homes in the West," wrote another, "but I find people here . . . living almost in affluence on their prairie farms, who I know left New York State only a few years ago with next to nothing."[41] The prospect of wealth, however, was not the only inducement held forth. Few arguments that might conceivably lure an extra settler to the West were neglected. One pamphleteer even saw fit to point out that males far outnumbered females in the western states. "Accordingly, when a daughter of the East is once beyond the Missouri she rarely recrosses it except on a bridal tour."[42]

Naturally enough the appeal for settlers was not confined to residents of the United States alone. Foreign immigration became an early concern of all the western land-grant railroads. The Burlington land department was headed by a man who counted as one of his greatest assets the ability to "speak, read, and write two languages and . . . understand and make myself understood by two or three other nationalities."[43] All the advertising roads caused pamphlets to be prepared in the various foreign languages for distribution in Europe. Some of them, possibly all of them, sent open or secret immigration agents abroad. The Chicago, Milwaukee, and St. Paul had a resident agent at Liverpool to direct the work of subordinates, who scattered throughout the continent in their search for prospects. To poverty-stricken peasants they portrayed America as a land of riches and plenty. To the politically oppressed they revealed the American West as still the "home of the free." To manufac-

[40] *Guide to the Lands of the Union Pacific Railway Company in Nebraska,* 1883, p. 2.
[41] *B and M Railroad Lands,* 14.
[42] *Views and Descriptions of Burlington and Missouri River Railroad Lands,* opposite plate 19.
[43] Ernst, in *Nebraska History,* 7:20.

turers and artisans who might possibly be persuaded to come
together and establish factories in the western cities, they offered the most generous aid. "Emigrants were given every
possible assistance. They were carefully guided, guarded
from the agents of rival companies, and passed from the
hands of the European agent to the New York agent, from
the New York agent to the Chicago agent, and by this agent
were consigned to the local agents for distribution." [44]

An example of foreign advertising in its most active form
occurred after 1873, when it was learned that large numbers
of Germans, long resident in Russia, had been virtually
ordered out of the country. Since the time of Catherine the
Great these German-Russians had enjoyed special privileges
of a most extraordinary sort; not only were they permitted
to use their own language and to maintain their own political and religious institutions without Russian interference,
but in deference to their pacifist principles they were even
exempted from the usual requirements with regard to military service. These privileges the czar now annulled, giving
his German subjects the alternative of conforming to Russian laws within ten years or leaving the country. The opportunity of luring the prospective exiles to America was of
course quickly seized upon by the railway advertisers. According to Ernst, the head of the Burlington land department, every land-grant road soon had its representatives in
Russia who sought purchasers for their lands with great
success.

Ernst's colorful narrative of the keen competition among
the roads for these emigrants, although written fifty years
later, is doubtless essentially correct:

I met them several times by the trainload, and on one occasion swiped a whole trainload from the two Kansas roads, each
of which had a special train waiting their arrival at Atchison,

[44] Farmer, in the *Mississippi Valley Historical Review*, 10:409; Hedges, *ibid.*,
13:322–325.

but I stole the whole bunch, except less than a dozen unmarried young men, and carried them all by special train, free, to Lincoln, Nebraska. Those were certainly strenuous days for settling up our prairie states.

Suggestive of another reason for the numerous foreign-language publications of the railways is the following statement from the same favored observer:

We located large settlements, not only of the German colonists from Southern Russia, but many other settlements of Germans, Swedes, Norwegians, Danes, Hollanders, Bohemians, and Polanders, most of them coming from other states where land was too expensive to enable them to acquire farms of their own.

Ernst placed the cost of the advertising done by the Burlington alone at $969,500.35. A Union Pacific official placed the "expense of selling" the lands of his road in Nebraska at $855,414.92. Judging from these figures, the total sum expended by all the roads must have amounted to many millions.[45]

It was not until well along in the seventies that settlers began to pour into the new West in satisfactory numbers. The few who came first were not entirely contented with their lot. In addition to the usual hard times of the frontier the early venturers were confronted with grasshopper plagues of such appalling severity that many of the fainter-hearted gave up and returned to their eastern homes.[46] The period of hard times following the panic of 1873, however, caused considerable unemployment and unrest in the eastern portion of the United States and created a class who were not unwilling to listen to the blandishments of the railway advertisers.[47] By the late seventies a tremendous westward migration had set in, the magnitude of which was fully attested by

[45] Ernst, in *Nebraska History*, 7:20–21; letter from Edson Rich to Helen Anderson, April 6, 1926, in the possession of the Nebraska Historical Society.
[46] Ernst, in *Nebraska History*, 7:21.
[47] Farmer, in the *Mississippi Valley Historical Review*, 10:406; Hedges, *ibid.*, 13:327–329.

the census reports of 1880. In the decade of the seventies Kansas had advanced from a third of a million to a million inhabitants; Nebraska, from a quarter of a million to a half million; Dakota Territory, from fourteen thousand to nearly ten times that number. The neighboring states of Iowa and Minnesota, in the western portions of which there was still much vacant land in 1870, were similarly affected, although to a smaller degree. In this decade Iowa added nearly half a million to her population and jumped from twentieth to eleventh place among the states with respect to number of inhabitants. Minnesota gained nearly three hundred thousand, an increase of about seventy-five per cent over the figures for 1870.[48]

Flushed with these successes, the railway advertisers redoubled their efforts. "Has there ever been progress equal to that of this wonderful state?" wrote a Nebraska apologist, who added to the dazzling figures on population increase others, even more dazzling, with respect to the number of acres lately brought under cultivation. With an enthusiasm typical of the period the same writer proclaimed: "The truth is, a poor man may live richer, and a rich man easier in this portion of Nebraska than in most other parts of the world's domain. A farmer with but little means can come into this section now, and by . . . following the dictates of prudence and economy soon become a very nabob."[49] The experiences of farmers who attained wealth in a few short

[48] *Twelfth Census of the United States*, 1900, *Population*, Vol. 1, Part 1, pp. 2–4. Following are the exact figures:

INCREASE OF POPULATION IN WESTERN STATES, 1870–80

State	1870	1880	Percentage of Increase
Kansas	364,399	996,096	173.4
Nebraska	122,993	452,402	267.8
Dakota Territory	14,181	135,177	853.2
Iowa	1,194,020	1,624,615	36.1
Minnesota	439,706	780,773	77.6

[49] *The Broken Bow Country*, 4, 8.

years were available in apparently unending numbers, and bulky pamphlets consisting almost wholly of such testimony were sent forth in prodigious numbers. The high prices obtainable for wheat and corn during these years had much to do with these sudden fortunes. Instances are actually on record of farmers who paid for their land from the profits of a single season. No wonder the advertisers featured "the profit, the clear money profit that can be made." [50]

Railway advertising was supplemented more and more effectively by state, local, and individual activities of the same nature. The biennial report of the Kansas State Board of Agriculture upheld the honor of the sunflower state by claims so extravagant that one might readily infer, as a later board said, that the mere fact of coming to Kansas meant "a life of ease, perpetual June weather . . . milk and honey." [51] State boards of immigration imitated in every possible particular the methods already made familiar by the railroads. Town councils and city chambers of commerce, real estate men and speculators, newspaper editors and church societies, all joined in. Information that might deter settlers from coming was rigorously suppressed. Everyone wanted more people to pay taxes and build up the country. More people would insure better schools, better roads, better public buildings. More people would increase the demand for land and boom land prices. More people would provide a market for innumerable town lots platted, but not yet built upon, in all the budding cities of the plains. [52]

[50] *Ibid.*, 8; Raymond C. Miller, "The Economic Background of Populism in Kansas," *Mississippi Valley Historical Review*, 11:470, 482; Ernst, in *Nebraska History*, 7:21. Typical pamphlets are *Corn Is King, Nebraska and Its Settlers*, and *Central Western Nebraska and the Experiences of Its Stockgrowers*, all published by the Union Pacific Railroad Company during the eighties. On the prices of wheat and corn see C. Wood Davis, "Why the Farmer Is Not Prosperous," *Forum*, 9:237–238.
[51] Quoted by Miller, in the *Mississippi Valley Historical Review*, 11:471–472.
[52] Farmer, in the *Mississippi Valley Historical Review*, 10:406, 409 note; Ernst, in *Nebraska History*, 7:213. The *Nebraska State Journal*, April 6, 1887, describes a circular of this kind sent out with the compliments of the First National Bank of Lincoln, Nebraska.

And the people came. The frontier expansion reflected by
the census of 1880 gained such momentum in the succeeding
years that by the middle eighties it had attained the full proportions of a boom. According to Senator Peffer, who did
not exaggerate greatly, "a territory larger than that of the
thirteen original states was populated in half a dozen
years."[53] Whereas the movement of the seventies had resulted mainly in the settlement of the eastern third of the
trans-Missouri states, that of the eighties pushed beyond into the central portions. In Kansas, Nebraska, and Dakota
Territory each new legislature created new counties in this
central region literally by the dozen, and new villages, new
towns, and new cities sprang up by the hundreds.[54] Each
little town became a boom town, foolishly optimistic about
its future and ready enough to believe the rhetorical absurdi-
ties of its pamphleteers.[55]

Indeed, settlers pushed on into the arid zone of western
Kansas and Nebraska and even farther. Eastern Colorado,
where the "only crop for years was bankrupts," now became
the scene of a lively boom. Here land that had an elevation
of thirty-five hundred feet was seized upon as eagerly as land
along the Missouri River where the elevation was only seven
hundred and fifty feet. The Colorado boomers claimed that
in a single year fifty thousand people had entered the eastern
portion of the state. These newcomers shared with the settlers in western Kansas and Nebraska a conviction, duly
fostered by the advertisers, that the rain belt had lately
moved so far to the westward that the high table-lands of
the plains could now no longer properly be classified as arid.[56]
The fact was that in the eight years preceding 1887 there had

[53] William A. Peffer, "The Farmer's Defensive Movement," *Forum*, 8:464.
[54] Farmer, in the *Mississippi Valley Historical Review*, 10:410; Miller, *ibid.*,
11:481.
[55] For a fair sample see *The Broken Bow Country*, 11.
[56] *Eastern Colorado*, 4–5; Charles M. Harger, "New Era in the Middle West,"
Harper's New Monthly Magazine, 97: 276; Robert W. Furnas, *Nebraska, Her
Resources and Advantages*, 20.

been an unusual amount of rainfall in this area, and the belief was common that the plowing of ground, the planting of trees, and the introduction of irrigation were the causes: "These vast extents of plowed land not only create a rainfall by their evaporation, but invite rains by their contrasts of temperature."[57]

Whatever the causes, people were soon persuaded that the change had taken place, and they seemed not to question its permanence. Even a professor of natural science at the University of Nebraska appears to have shared the common delusion. He seemed to think that ultimately rainfall in the western part of the state would equal the rainfall in the eastern portions. One Colorado enthusiast affected to complain that "so much rain now falls in the eastern portion of the arid lands of Colorado that it is no longer fit for a winter range for cattle." In general, however, the range country seemed better off for the abundance of rain. Grass-producing summers and mild winters combined to make the profits of cattle-raising mount to great heights. Eastern and foreign capital bent on speculative profits flowed into the "cow country," overstocked the range, pushed cattle into areas where the grazing was poor, and prepared the way for an appalling disaster.[58]

It is difficult to state what peaks were reached by the populations of the states affected by the boom, for the census came after the boom had collapsed and many thousands of settlers had gone back East. But even the census figures record a net population increase of a million and a half in Kansas, Nebraska, and Dakota. The neighboring states to the east and some western states, notably Colorado and Washington, also showed the effects of the boom. Iowa

[57] Dillon, in the *North American Review*, 152:445; Miller, in the *Mississippi Valley Historical Review*, 11:477, 486.

[58] *Eastern Colorado*, 15; John D. Barnhart, "Rainfall and the Populist Party in Nebraska," *American Political Science Review*, 19:532; Ernest S. Osgood, *The Day of the Cattleman*, ch. 4.

was apparently affected adversely by the competition of the frontier lands, for the percentage of population increase was less in the eighties than in any preceding decade and less than the average for the country at large. Most of the other near-by states showed enormous increases.[59]

This rapid movement of population to the West was accompanied by an equally extraordinary movement of capital in the same direction. Not much of this capital, however, was brought in by the settlers. Especially in the earlier years of the migration the men who went West were poor, and sometimes, in spite of the warning of the railway advertisers that disappointment, homesickness, and discontent might be the lot of those who came entirely destitute of means, they were penniless. Even those who had some savings to depend upon rarely had enough to tide them through the first hard year, for it took several hundred dollars at the very least to pay the cost of living until a crop could be harvested, to erect farm buildings and fences, and to buy seed grain and machinery. Those who had no money when they came and were unable to obtain it from interested friends or relatives often worked as farm hands until they had saved the necessary minimum upon which to make a start. Almost all the newcomers were potential borrowers. Loans upon "the possible products of wild, unfenced, uncleared, or un-

[59] The exact figures are as follows:

INCREASE OF POPULATION IN WESTERN STATES, 1889-90

State	1880	1890	Percentage of Increase
Kansas	996,096	1,427,096	43.3
Nebraska	452,402	1,058,910	134.1
North and South Dakota	135,177	511,527	278.4
Iowa	1,624,615	1,911,896	17.7
Minnesota	780,773	1,301,826	66.7
Washington	75,116	349,390	365.1
Colorado	194,327	412,198	112.1

The average increase for the country at large was twenty-six per cent. Despite the fact that it furnished the Populists with their presidential candidate in 1892, Iowa was never a third-party stronghold. The table is computed from the *Twelfth Census of the United States*, 1900, Vol. 1, *Population*, Part 1, pp. xx, 2-4.

broken lands" were indeed hard to get, but when the pioneers had built their houses and barns and had broken up a few acres of land, they might confidently expect such help. In fact, outside capital by the millions of dollars was necessary if these men were to succeed.[60]

It chanced that the money needed by the western farmers could be supplied without much difficulty by eastern investors, who, as the hard times of the seventies gave way to the prosperity of the eighties, found their savings increasing by leaps and bounds. The man who had only a small sum to invest was especially attracted by western mortgages on which he could easily obtain high interest rates — from six to eight per cent on real estate and from ten to eighteen per cent on chattels. Such an investment seemed the more attractive because of the well-advertised rise in western land values. The competition of free lands, always before this time a potent cause of the low price of land, was now nearly at an end. Moreover, crop yields in the West over a period of years had averaged high, prices were good, and collections were easily made. The mortgage notes themselves, "gorgeous with gold and green ink," looked the part of stability, and the idea spread throughout the East that savings placed in this class of investments were as safe as they were remunerative. Small wonder that money descended like a flood upon those who made it their business to place loans in the West![61]

Banks were numerous throughout the West, and they did

[60] *Pioneer Press*, January 19, 1884; *Nebraska and Its Settlers*, 3, 7; *Guide to Union Pacific Railroad Lands*, 22; Daniel R. Goodloe, "Western Farm Mortgages," *Forum*, 10:351; Paul R. Fossum, *The Agrarian Movement in North Dakota*, 52. It should be noted, however, that many of the settlers who came in the eighties were better off than the earlier pioneers. Many from near-by states who had "sold out and moved west" had the proceeds of their sale with which to begin again.
[61] Charles S. Gleed, "The True Significance of the Western Unrest," *Forum*, 16:258; J. Willis Gleed, "Western Mortgages," *Forum*, 9:105, and "Western Lands and Mortgages," *Forum*, 11:468; Farmer, in the *Mississippi Valley Historical Review*, 10:411; Miller, *ibid.*, 11:474. See also Seth K. Humphrey, *Following the Prairie Frontier* (The University of Minnesota Press, 1931), 95 ff., for a first-hand account.

a flourishing business, but much of the work of lending the savings of the East to the borrowers of the West devolved upon mortgage companies organized for the purpose. These companies were of all sorts and sizes. The larger ones had central offices in such places as St. Paul, Omaha, Kansas City, St. Joseph, and Denver, with local agents and field men distributed throughout the surrounding territory. There were also many smaller companies and many individuals who acted as loan agents on their own responsibility. Estimates of the number of corporations engaged in the mortgage business in the two states of Kansas and Nebraska vary from one hundred and thirty-seven to about two hundred. The mortgage company, of course, made its money from the margin between the relatively low interest rates that the eastern investor would accept and the high interest rates that the western borrowers were willing to pay. There were also commissions for securing the money and bonuses for the more successful agents. The large company cared little what its local representatives did, so long as it could count on an interest return from the investment of from six to eight per cent on a five-year loan. If the local agents could get more, they were welcome to the difference between what they got and what the company required of them. Many "wild cat" organizations, managed by men who were innocent of any proper knowledge of their business but guilty of a determination to make all the money they could out of it, rose and flourished. According to one observer, the local agents of some of these corporations were mere "clerks who did not know a sandhill pasture from a bottomland garden," from whom anybody could borrow; "indeed, they beseeched the borrowers to place money on their farms — the more mortgages, the more fees." [62]

So great became the craze for western mortgages that

[62] J. W. Gleed, in the *Forum*, 9:94–96; W. F. Mappin, "Farm Mortgages and the Small Farmer," *Political Science Quarterly*, 4:438; Charles M. Harger, "The Farm Mortgage of To-day," *Review of Reviews*, 33:572; Herman C. Nixon,

these companies, however prodigal with their funds they might be, seldom lacked money to invest. The manager of one of them declared that during many months of 1886 and 1887 he was unable to get mortgages enough to supply the eastern demand. "My desk was piled high every morning with hundreds of letters, each enclosing a draft and asking me to send a farm mortgage from Kansas or Nebraska." [63] According to William Allen White, agents in Kansas with a plethora of money on their hands drove about the country in buggies, soliciting patronage and freely placing loans on real estate up to its full valuation, pointing in justification to the steadily mounting price of land. The companies tended almost without exception to lend too much on each farm. It was not their own capital that was at stake but the capital of distant investors, and the more they lent the more they made for their own profits. [64]

It was inevitable that this avalanche of credit, which far outran the real needs of the situation, should tempt the new West to extravagance, overinvestment, and speculation. Farmers could rarely resist the funds proffered them. With bumper crops, high prices, and rising land values, it appeared, indeed, the part of wisdom to borrow money for enlarging holdings, improving breeds of stock, and purchasing the latest and best machinery. The federal census figures of 1890 relating to real estate mortgages furnish convincing evidence of what went on. They show that Kansas, Nebraska, North and South Dakota, and Minnesota, in spite of the comparative poverty of their inhabitants, ranked well toward the top of the list of states in the amount of per capita mortgage debt. The ratio of the real estate mortgage debt to the true value of all taxed real estate was higher in

"The Economic Basis of the Populist Movement in Iowa," *Iowa Journal of History and Politics*, 21:379.
[63] Harger, in *Harper's New Monthly Magazine*, 97:277.
[64] Miller, in the *Mississippi Valley Historical Review*, 11:475; Herbert L. Glynn, "The Urban Real Estate Boom in Nebraska during the Eighties," *Nebraska Law Bulletin*, 6:464–467.

Kansas than in any other state except New York, and none
of the other states mentioned ranked lower than ninth. In
Kansas, also, where the amount of speculation was greatest
and the figures were always most extreme, the new mortgage
debt incurred per capita in the year 1885 was more than
double and in 1887 more than treble, the amount of 1880;
but some of the other frontier states were not far behind.
In Kansas and North Dakota there was in 1890 a mortgage
for every two persons, and in Nebraska, South Dakota, and
Minnesota, one for every three persons — more than one to a
family in all five states. In certain counties where seventy-
five per cent of the farms occupied by their owners were
mortgaged and where the total mortgage debt was about
three-fourths of the true valuation of all the land taxed, it is
not an unfair inference that most of the mortgaged farms
were mortgaged literally for all they were worth. The census
statistics show, in short, that Minnesota, the Dakotas, and
Nebraska were all appreciably better off than Kansas, approximately in the order named, but that throughout this
region mortgages had multiplied in number and amount far
beyond any reasonable limits.[65]

It is not surprising that land values rose to unprecedented
heights. In a block of six counties in southeastern Nebraska
the price of land doubled in the years 1881 to 1887, achieving
by the latter year a price of about $17.50 an acre. This may
be regarded as a typical nonspeculative gain, inasmuch as
that region was as little affected by the boom as any and
experienced no material decline in land prices in the period of
hard times that followed.[66] Extravagant figures, however,
became a commonplace, particularly in or near the boom
towns. Near Clifton, Kansas, a quarter section of land, once

[65] *Eleventh Census of the United States,* 1890, Vol. 12, *Real Estate Mortgages,* 17, 37, 116, 158, 160–161.
[66] Unpublished statistics compiled by Professor J. O. Rankin, formerly of the
Department of Rural Economics, University of Nebraska. Professor Rankin's
statistics are based upon the average sale price per acre of warranty deed farm
land transfers in Clay, Polk, Saline, York, Fillmore, and Platte counties.

thought worthless, brought $6,000. A farm near Abilene in the same state that had cost $6.25 an acre in 1867 sold twenty years later for $270 an acre. Wild and unimproved land in eastern Colorado was held at from $3 to $10 an acre and when slightly cultivated, at from $8 to $20 an acre. If newspaper advertisements are to be trusted, increases in value of from 400 to 600 per cent from 1881 to 1887 were by no means unusual.[67]

Speculation reached its peak in the numerous booms that broke forth in most of the western towns during the spring and summer of 1887. Omaha, Lincoln, Kansas City, Atchison, Topeka, and Wichita were among the most conspicuous victims of the boom fever, but there was scarcely a village of any consequence that escaped it entirely. Money to lend on city lots seemed to be quite as plentiful as money to lend on farm lands, and with almost unlimited funds available speculation ran riot. Each community acted as if convinced that it was destined to become a future metropolis and that those who made investments in its real estate could not conceivably lose.[68]

The Wichita boom was perhaps worse than any other. Colonel M. M. Murdock, editor of the *Eagle*, was credited with having consciously started it by the injudicious advertising of his home town through newspaper channels, but precisely how the boom did get started would be hard to determine. By February, 1887, however, Wichita, according to one observer, had become "the liveliest city in Kansas. The streets are full of people, the hotels and boarding houses overrun with them, and every train brings in additional scores to swell the boom . . . everybody is talking real estate and shouting for Wichita." Prices of lots rose to unbelievable heights, and speculators made money right and left.

[67] C. S. Walker, " The Farmers' Alliance," *Andover Review*, 14:131; Farmer, in the *Mississippi Valley Historical Review*, 10:145; Miller, *ibid.*, 11:470, 474, 482.
[68] Farmer, in the *Mississippi Valley Historical Review*, 10:414; Glynn, in the *Nebraska Law Bulletin*, 6:468–481.

A clerk who put his $200 savings into a lot sold it two months later for $2,000. A barber who dabbled in real estate made $7,000. Real estate agents, many of whom made much larger fortunes, swarmed over the place by the hundreds; they were so numerous that the city derived a considerable revenue from the license fees they had to pay. In the five months from January to May, 1887, forty-two sections of land were platted into lots and sold, and the total amount involved in real estate exchanges reached $35,000,000.[69]

What happened in Wichita happened elsewhere on a less extravagant scale. Atchison, Kansas, had a boom but evidently not so great a one as Wichita. An Atchison booster, who could see the sinister side of the Wichita boom and could write sarcastically about it, insisted with perfect naïveté, "If Atchison . . . had Wichita's boom, it would be permanent and substantial prosperity."[70] In Kansas City the extent of the boom may be gauged by the fact that in April of 1887 real estate transfers involved a total of $13,000,000 — as much as in the entire year of 1885; while in a single day in May Topeka ran up its sales to $272,644. In Lincoln, Nebraska, an enthusiastic editor pointed with pride to the fact that on a single day in March $202,650 was "planted in Lincoln real estate by gentlemen who knew exactly what they were about." A minister of the same city lifted his voice in protest against the way in which church members were vying with "the rash speculators and the gamblers of the city" in buying options on corner lots.[71] In Omaha, at a time when new divisions far on the outskirts of the city were being sold out at absurdly high prices, deep resentment was expressed against "including this city among the places which are having a 'boom.' So far as Omaha is

[69] *Atchison Champion,* quoted in the *Nebraska State Journal,* March 5, 1887.
[70] *Ibid.*
[71] *Nebraska State Journal,* March 2, 5, 1887; Farmer, in the *Mississippi Valley Historical Review,* 10:414; Glynn, in the *Nebraska Law Bulletin,* 6:468–481.

concerned, there is no 'boom' here. On the contrary, it is
nothing but legitimate growth." [72] This failure to appreciate
the situation was fairly general throughout the areas of
speculation. "There is nothing strange or peculiar about
the great activity in real estate in all the trade centers in the
northwest this winter," ran one explanation. "It is not a
sudden impulse in Lincoln or Omaha or Hastings or Atchison
or Wichita or Kansas City. It is simply the effect of the
exhaustion of the public lands and the prosperity of the
trans-Missouri region." [73]

The example of the larger towns was successfully imitated
by the smaller towns. The refrain adopted by a real estate
firm in one Nebraska village shows how well the lesson had
been learned:

> Beatrice is not dead or dying,
> Real estate is simply flying,
> He who buys to-day is wise,
> For Beatrice dirt is on the rise.[74]

Advertisers of Hastings, Nebraska, who had just platted a
forty-acre tract in town lots, urged all enterprising citizens
to make their selections early "and double your money in
30 days." Throughout the boom-stricken area, whether in
Kansas, Nebraska, or the Dakotas, new towns appeared with
magical rapidity. "Nothing causes the Nebraska farmer
more dismay," one editor declared, "than to return from
town after spending a few hours there, and find that his farm
has been converted into a thriving city with street cars and
electric lights during his absence. But such things will occur
now and then and should be regarded with comparative
calmness." [75]

The needs, real or fancied, of these boom towns and vil-

[72] *Omaha Daily Bee*, May 5, June 1, 1887.
[73] *Nebraska State Journal*, March 5, 1887.
[74] *Ibid.*, April 24, 1887. The boom in the smaller cities and towns of
Nebraska is well described by Glynn in the *Nebraska Law Bulletin*, 7:228–243.
[75] *Nebraska State Journal*, July 5, October 2, 1887; Farmer, in the *Mississippi
Valley Historical Review*, 10:415; Miller, *ibid.*, 11:474.

lages opened up still other fields for the investment of eastern capital in the west. Municipal bonds were voted with the most reckless abandon in order to provide courthouses and jails, schoolhouses, waterworks and electric light plants, sewer systems, paved streets, and better roads and bridges. "Do not be afraid of going into debt," wrote the editor of the Kansas *Belle Plaine News*.

Spend money for the city's betterment as free as water. But judiciously. Too much cannot be spent this year if properly applied. Let the bugaboo of high taxes be nursed by old women. Do all you can for Belle Plaine regardless of money, and let the increase of population and wealth take care of the taxes.[76]

This was the sort of advice that the boom territory liked, and great quantities of bonds were voted in aid of private as well as public enterprises. Aid to private enterprises usually went to railroads, but assistance was given even to "cheese factories and . . . packing plants," for the success of such local businesses could be counted upon to help along the boom. Fifteen Kansas towns, with a pretentiousness utterly unjustified by the facts, installed street-car systems. Strangely enough the prodigal bond issues necessary to finance these various ventures found a ready market in the East. Whereas the small investor tended to put his money into a real estate mortgage, the large investor was more likely to buy municipal bonds. So popular were these securities that they were not only floated in unreasonable quantities but frauds of the most transparent nature were also easily perpetrated. Given the artificial stimulus of so much outside capital to spend, the western towns grew with mushroom rapidity.[77]

Perhaps railway building was regarded as more essential to the purposes of the boomers than any other activity.

[76] Quoted by Miller, *op. cit.*, 473.
[77] Farmer, in the *Mississippi Valley Historical Review*, 10:410-415. See also James E. Boyle, *The Financial History of Kansas*, chs. 6-7.

Towns and cities in the West were judged in importance somewhat in accordance with the number of railway lines converging upon them. Each sizeable village aspired, therefore, to become a railway center and stood ready to bond itself heavily in order to grant favors to prospective roads. In fact, the mania for railroads became so acute that a class of promoters developed who made it their business to project and construct lines, not because they were needed or could hope to pay dividends for long but because the gifts that the counties, cities, and even the states were willing to shower upon them insured for the promoters a handsome initial profit. Once the road was built, they lost no time in unloading its obligations upon others. "The roads are poor," declared one writer, "while the builders are rich." Small wonder that in most of the frontier states railway building during the decade of the eighties was far in excess of the needs of the situation. In Kansas, for example, there were 3,102 miles of track in 1880 and 8,810 miles in 1890. In Nebraska the mileage grew in that decade from 1,634 to 5,407 and in Dakota from 399 to 4,726. Kansas had a mile of track for every nine and a third square miles of land and for every one thousand inhabitants. Practically all the roads, however bountiful the gifts that had made them possible, issued stocks and bonds to a figure far beyond their real value and strove heroically to pay interest and dividends in spite of the excessive competition for business. And yet the stocks and bonds of these western roads were purchased by eastern investors with as little discrimination and as great avidity as any other type of western securities.[78]

That the boom could not continue indefinitely, sensible people should have foreseen, but few of them apparently had any premonition of what was to come. The process of

[78] Henry V. Poor (ed.), *Manual of the Railroads of the United States, 1890,* vi; John R. Dodge, "The Discontent of the Farmer," *Century,* 21:452; *Farmers' Alliance* (Lincoln), September 22, 1890; Farmer, in the *Mississippi Valley*

deflation began in 1887. In Kansas the end came with dramatic suddenness; elsewhere it arrived more slowly but with disheartening certainty. The immediate cause of the slump was the lack of rainfall during the season of 1887 and, with a few exceptions, each succeeding season for a period of ten years. With the year 1886 the era of abundant rainfall on the frontier ended more or less abruptly, and the attractive stories that had circulated concerning the westward migration of the rain belt were soon demonstrated to be false. About eighteen or twenty inches of rainfall annually coming at the proper time of year is regarded by agriculturists as sufficient to produce a good crop. For a considerable period prior to 1887 these conditions had been fulfilled throughout the eastern and central portions of Kansas, Nebraska, and Dakota, and even well into the western regions. But from 1887 to 1897 there were only two years in which the central and western areas had enough rainfall to insure a full crop, and for five seasons out of the ten they had practically no crops at all. To the distress that resulted from lack of moisture were added heavy losses due to the hot winds that persisted throughout the summer months and burned the already suffering crops to a crisp. "Week after week," wrote H. W. Foght in his *History of the Loup River Region,* "the hot burning sun glared down from a cloudless steel-blue sky. The dread hot winds blew in from the south. Day after day they continued. All fodder, small grain, and corn were cut short. Where farming had been carried on extensively rather than intensively the yield amounted to preciously near nothing. The careful expert got some returns for his work, though small." Other losses came from the chinch bugs, whose dep-

Historical Review, 10:412–413; Miller, *ibid.,* 11:470–471. *Appletons' Annual Cyclopaedia, 1888,* 459, gives the following figures for Kansas:

	Main Track	Side Track	Total
January 1, 1888	8,799	899	9,698
January 1, 1885	4,064	489	4,553
Increase in three years..............	4,735	410	5,145

redations are worst in a period of drought, and from early frosts, especially in Dakota.[79]

The summer of 1887, giving evidence of impending crop failure, called a halt to the boom. As the hot weeks wore on the number of real estate transfers and the prices paid for land and lots declined precipitately. The boom towns were no less affected than was the country, for their future depended quite as completely upon a continuing agricultural prosperity. It was in these towns, indeed, that the collapse was most complete, for with popular confidence shaken, there was a rush to sell out, and this speedily brought prices to a minimum. Eastern investors, learning of the turn of events in the West, no longer clamored for western securities; while countless numbers of real estate men, mortgage vendors, railway promoters, and bankers went out of business altogether, many of them hopelessly bankrupt. The cattle industry on the northwestern plains was all but destroyed. The hot summer of 1886 had left the range in poor condition, and the winter of 1886–87 had been merciless. When spring came at last only a few pitiful remnants of the great herds remained. Speculative live-stock companies lost all they had invested; "cattle barons" and "bovine kings" ceased to exist; the ranges were almost stripped of cattle in a vain effort to satisfy the demands of creditors. Hard times settled down upon the whole frontier, not to be shaken off for a decade.[80]

[79] Foght is quoted by Barnhart in the *American Political Science Review*, 19:534. See also Farmer, in the *Mississippi Valley Historical Review*, 10:416–418; Miller, *ibid.*, 11:486–487; Glynn, in the *Nebraska Law Bulletin*, 7:244–248.
[80] *Omaha Daily Bee*, June 27, 1887; *Nebraska State Journal*, July 26, 1887; Harger, in *Harpers' New Monthly Magazine*, 97:277, and in the *Review of Reviews*, 33:572; J. W. Gleed, in the *Forum*, 11:468; C. S. Gleed, in the *Forum*, 16:258; Farmer, in the *Mississippi Valley Historical Review*, 10:416; Miller, *ibid.*, 11:475; Osgood, *Day of the Cattleman*, ch. 7. In the cattle country, as in Kansas, the end of prosperity came in 1887, but in much of Nebraska the first really heart-breaking season was 1890. "The efforts of the Kansas newspapermen to show that the corn crop in that state is unparalleled are quite painful," wrote a Nebraska editor in 1887. "For corn . . . come to Nebraska." *Nebraska State Journal*, October 2, 1887.

Convinced by bitter experience that they had pushed too
far into the arid West, people who had moved hopefully
into western Kansas, Nebraska, Dakota, or beyond before
1887 now began to retrace their steps. Within a few years
whole districts in this region were almost totally depopulated
except for the older cattlemen who had been there before the
boom began and did not depend for success upon a heavy
rainfall. Covered wagons, sometimes bearing such legends as
"Going back to the wife's folks," or "In God we trusted, in
Kansas we busted," streamed toward the East. Fully half
the people of western Kansas left the country between 1888
and 1892. Twenty well-built towns in that part of the state
were reputed to have been left without a single inhabitant,
and in one of them an opera house worth thirty thousand
dollars, a schoolhouse worth twenty thousand dollars, and a
number of fine business houses were abandoned. In the single
year of 1891 no less than eighteen thousand prairie schooners
crossed from the Nebraska to the Iowa side of the Missouri
River in full retreat from the hopeless hard times. Twenty-
six counties in South Dakota lost thirty thousand people
during the nineties. Disappointed pioneers handed over their
farms to the loan companies by which they had been mortgaged or abandoned them outright. Some of the more coura-
geous may have headed for Oklahoma, where other frontier
lands were being opened to settlement, but doubtless the
rank and file had had enough of pioneering.[81]

In the central portion of the frontier states it was harder
for the real farmer to get away. He had suffered acutely
from the drouth — perhaps more acutely than would have
been necessary had he known how to farm in a region of
scanty but usually adequate rainfall. His investment was

[81] Harger, in *Harper's New Monthly Magazine*, 97:277; Barnhart, in the
American Political Science Review, 19:539; Farmer, in the *Mississippi Valley
Historical Review*, 10:406, 420–422; Miller, *ibid.*, 11:477, 486–487; *Chicago
Herald*, quoted by *Greensboro Daily Record*, July 19, 1892. See also "Some
Lost Towns in Kansas," *Kansas Historical Collections*, 12:1–65, and Humphrey,
Following the Prairie Frontier, 153 ff.

apt to be greater and less a purely speculative enterprise than the investment of the farmer in the extreme west. He could and did have some hope as to the future, for he still believed that in this region enough rainfall to insure a crop was the normal thing. He was seriously injured by the boom, however, for he had usually yielded to the temptation to overmortgage his land, and he now found himself ever on the verge of foreclosure proceedings. Frequently he was a newcomer only lately attracted to the region by the boom and not yet fully established. He was distressed by the falling price of real estate and hurt even worse by the high taxes that he had so lightly shouldered in the boom period. But in most instances he had staked too much to lose it all, and he was still solvent. When the boom broke, many farmers in this section left, but no such proportion of the total population as in the western counties — probably not over one out of every eight. The great majority of the actual farmers stayed on and suffered, forming a discontented class ideally prepared for the doctrines of political and economic revolt, which agitators were quick to introduce.[82]

The farmers in the eastern portions of the frontier states, where the normal rainfall was well above the minimum required to produce a crop, did not suffer appreciably more from the drought than their neighbors in such states as Iowa and Missouri. They never failed entirely to gather a harvest; nor did the collapse of the boom cause them the great distress that came to the residents of the central and western regions. In the counties along the Missouri River the land had been well settled before the boom began, and there was little room outside of the cities for enormous population increases. When the boom came, land prices rose, mortgages multiplied, and many who speculated on the high-priced land lost, but the real farmers of the eastern coun-

[82] Barnhart, in the *American Political Science Review*, 19:538; Miller, in the *Mississippi Valley Historical Review*, 11:483–485.

ties did not need to flee the country. Indeed, there is some evidence that refugees from farther west halted here in their flight; at any rate, the population in the strictly rural portions of this region actually increased during the period of hard times.[83]

Throughout the frontier region the exodus from the boom towns was appalling. Statistics are not available for all the frontier states, but for Kansas the biennial report of the state board of agriculture furnishes convincing proof of what occurred. In the two years covered by the report some sixteen cities in the eastern half of the state lost a total of more than forty-five thousand people. Leavenworth lost fifteen thousand; Wichita, thirteen thousand; and some of the smaller towns were almost wiped out. Although the figures are not available, it is clear that much the same thing happened in Nebraska and the Dakotas. Census statistics for the decade of the nineties are obviously of no great value in estimating the immediate effect of the collapse of the boom, for generally the most striking loss of population was before 1890, and there was a compensating increase after 1897, when prosperity returned. Furthermore, in nearly every western community that had been hard hit by the collapse of the boom, census statistics in 1890 were deliberately falsified to conceal as completely as possible what had happened. But it is perhaps worth noting that, according to census statistics, from one-third to one-half of the counties in Kansas, Nebraska, and South Dakota had a smaller population in 1900 than in 1890.[84]

Aside from the fact that they may have carried the seeds of agricultural discontent back with them, the later fortunes of those who left the frontier in the period of hard times are of no interest here. The history of the West was made not

[83] Miller, *op. cit.*, 479–480.
[84] *Twelfth Census of the United States*, 1900, Vol. 1, *Population*, Part 1, pp. 19, 29, 33; *Seventh Biennial Report of the Kansas State Board of Agriculture*, 1889–90, pp. vi–vii; Miller, in the *Mississippi Valley Historical Review*, 11:477, 480.

by those who moved out but by those who stayed on. The withdrawal of practically all the newcomers from the arid region in the extreme western part of the frontier states left there a class accustomed to the environment and not inordinately dissatisfied. The flight of the floating and speculating element from the boom towns and cities placed control in the hands of the more substantial and conservative citizens. Nor did the eastern counties have the necessary economic basis for radicalism. But in the rural portions of the central regions, where farmers stayed and struggled with failing crops and low prices, with unyielding debts and relentless taxes, where they fought a battle, now successful, now unavailing, to retain the land they had bought and to redeem the high hopes with which they had come to the West — in this region unrest and discontent prevailed, and the grievances that later found statement in the Populist creed smouldered for a season, finally to break forth in a program of open revolt.

CHAPTER TWO

SOUTHERN AGRICULTURE AFTER THE CIVIL WAR

EVEN before the Civil War, when the South was ruled by the great planters, the South and the West had much in common. In both regions agriculture predominated, and most other interests existed only to serve this one. In both regions large cities were rare, and the problems they produced were of no great consequence. Political combinations between the agricultural South and the agricultural West, founded upon common economic interests, were frequent enough. It is not surprising that some southern statesmen even deluded themselves with the hope that the South and the West could draw together in a joint contest-at-arms with the industrial Northeast.

The effects of the Civil War upon the South were such as to make this community of interest even greater, for the planter aristocracy, which with its power and wealth had most distinguished the South from the West, was gone. The war, indeed, operated to restore the South to essentially the frontier stage of development. The great accumulations of property that ordinarily characterize an older society were conspicuously absent. Those who had put their savings into Confederate securities had nothing to show for their investment. Those who had counted their wealth in slaves were as penniless as the freedmen themselves. Those who had land were but little better off than those who had not, for they were, temporarily at least, without the means to work it. They lacked live stock, machinery, seed, and, most of all, a dependable supply of labor. Thousands of acres of the best land were for sale at from three to five dollars an acre — prices that compared favorably with those in the West. Other thousands of acres were simply abandoned by their owners for what the westerners would have called "squatters" to use at will. In many parts of the South houses and barns were burned, fences and railroads torn up, and public buildings destroyed or rendered unfit for use. Factories that had flourished in war time were gone and their machines carried away. In short, there was in the South after 1865 a return to the primitive and an actual equality of conditions for all men seldom met with anywhere except on the frontier, and rarely on so large a scale even there. The social and economic basis for a closer union between the West and South was thus beginning to appear.[1]

Now that the ownership of land was no longer the special privilege of the few, the small farmer soon came to be the predominating type of agriculturist in the South as well as in the West. The lands of the planters, placed on sale at a negligible fraction of their former values, attracted the poor whites in large numbers, the negroes to a lesser extent, and outsiders to a very slight extent. "Never perhaps," wrote a Georgia editor,

was there a rural movement, accomplished without revolution or exodus, that equalled in extent and swiftness the partition of the plantations of the ex-slave holders into small farms. As remarkable as was the eagerness of the negroes — who bought in Georgia alone 6,850 farms in three years — the earth-hunger of the poorer class of the whites, who had been unable under the slave-holding oligarchy to own land, was even more striking.

[1] Walter L. Fleming, *The Sequel of Appomattox*, ch. 1; Benjamin B. Kendrick, "Agrarian Discontent in the South: 1880–1900," *Annual Report of the American Historical Association, 1920*, 267–272.

38   THE POPULIST REVOLT

The former overseer, for example, now purchased land and
began to exploit for himself the talent for management he
had developed in slavery times. The small farmer who had
owned land adjacent to the great plantations enlarged his
holdings. Even the "poor white" was tempted to acquire a
land title. This assault upon the lands once held by the
planters constituted an interesting parallel to the rapid acquisition by home-seekers of the cheap farm lands of the
West.[2]

It must not be supposed, however, that the typical small
farmer of the South was necessarily a landowner. He might
be, and far more probably was, a tenant. The negro farmer,
indeed, was almost certainly a tenant, who farmed his land
according to the "share system" or, as it was more frequently called in the South, the "cropping system." This
system of tenantry, which developed out of the experience
of Reconstruction years, had certain peculiarities due to the
circumstances that called it into being. The negro, it was
found, resented the supervision and the gang-labor methods
of the slavery era, and yet he required some spur other than
the memory of wages spent or the hope of wages deferred to
keep him at work. The share system in some one of its
numerous manifestations seemed the best answer to this
necessity. This system ordinarily consisted of the planter
giving to each tenant a separate plot of ground and a portion of whatever crop he raised upon it — one-half the crop
or less, if the planter furnished the tools, seed, and mules;
and more in case the freedman could furnish all these himself. The negro now had a direct interest in what he pro-

[2] Henry W. Grady, "Cotton and Its Kingdom," *Harper's New Monthly
Magazine*, 63:721-722. Grady goes on to say, "In Mississippi there were in
1867 but 412 farms of less than ten acres, and in 1870, 11,003; only 2,314 of
over ten and less than twenty acres, and in 1870, 8,981; only 16,024 between
twenty and one hundred acres, and in 1870, 38,015. There was thus in this one
state a gain of nearly forty thousand small farms of less than one hundred
acres in about three years. In Georgia the number of small farms sliced off the
big plantations from 1868 to 1873 was 32,824. In Liberty County there were
in 1866 only three farms of less than ten acres, in 1870 there were 616, and 749

duced. If the crop were good, he prospered accordingly; if not, he suffered as much as the planter. As a cropper he was less likely to break his contract and desert his post than as a wage laborer, for if he had worked well he had too much at stake. It was partly by the application of this system and partly by the development of a class of small landed proprietors that the cotton-growing industry got on its feet again. While the whites seemed to show greater interest in securing land of their own than did the negroes, many of the whites also became tenants; and, conversely, while the negroes were usually content with their status as croppers, not a few of them became landowners. Cash tenants were not altogether unknown and in some regions were fairly common.[3]

Whether tenants or proprietors, the post-war farmers were almost without exception in a chronic condition of abject poverty. Those who purchased land could do so usually only by means of deferred payments, which they met with difficulty. Those who were tenants were at best barely able to live on what they could make. Having started with nothing at the close of the war, both classes of farmers found it next to impossible to accumulate enough reserve to carry them from one harvest to the next, and they had therefore to resort constantly to borrowing in one form or another.[4]

The western farmer who was similarly in need of funds could usually secure advances from banks or from loan companies, but the southern farmer was in a far less favored position. Southern banking had been paralyzed by the war,

farms between ten and twenty acres. This splitting of the old plantations into farms went on rapidly all over the South." See also Francis B. Simkins, *The Tillman Movement in South Carolina,* 8–9; R. P. Brooks, *The Agrarian Revolution in Georgia, 1865–1912,* ch. 3.

[3] Matthew B. Hammond, *The Cotton Industry: An Essay in American Economic History,* 130–134; Brooks, *Agrarian Revolution in Georgia,* pp. 44–48, ch. 4; Francis B. Simkins, "The Problems of South Carolina Agriculture after the Civil War," *North Carolina Historical Review,* 7:46–77.

[4] Hammond, *Cotton Industry,* 128, 130–131; J. G. de Roulhac Hamilton, *North Carolina since 1860,* 222.

and for many years thereafter it existed only on the feeblest possible footing. Local capital for investment in the banking business, as for investment in any other, was exceedingly scarce, and outside capital required more certain profits and greater security than the farm lands of the impoverished South seemed to promise. Such banks as were founded loaned money to some extent to the planters and even more to the merchants, since both planters and merchants were able to give some security, but not often to the small farmer, whose only real asset lay in his growing or unplanted crop. As late as 1895 the southern banking system was still regarded as highly imperfect and wholly incapable of handling the total credit load of the South. In that year there were only four hundred and seventeen national banks in the ten states of the cotton belt, and more than half of them were in Texas. State and private banking institutions were correspondingly rare. One hundred and twenty-three counties in the state of Georgia were without any banking facilities whatever.[5]

The need for credit, however, did not long remain unsupplied, for the merchant quickly shouldered the responsibility that the banker shirked. This he did not by lending money but by advancing the goods that the farmer needed, in return for a lien upon his growing crop. Precedent for this practice existed in the relations between the planters and factors of the old South, when it was common enough for the former to obtain from the latter advances upon the crops soon to be marketed. These loans were not at all deplored by the factors, for the interest rate was high, and the agreement insured that a crop so mortgaged must be marketed through the factor who supplied the loan. By advancing a reasonable sum, rarely in excess of ten dollars a bale on a conservative estimate of the growing cotton, the factor insured himself against competition for whatever profits might come from

[5] Hammond, *Cotton Industry*, 160–161.

handling the business of any planter who was forced to secure a loan in this way.[6]

After the war there was some disposition to revive this system, and the planters actually borrowed considerable sums from such of the factors as were able to re-establish their businesses. Southern cotton was in great demand, and the factor who found himself short of funds to lend had little difficulty in securing them from the North or from Europe. To facilitate matters the legislatures in several states passed "lien laws," which legalized the whole procedure. These laws permitted the mortgaging of prospective crops and gave those who loaned money on them first claim upon the proceeds of the harvest. Once the lien laws were passed, and, indeed, often in anticipation of them, the system was extended to cover the accounts of farmers with the owners of country stores and those of tenants with their landlords. To the poverty-stricken South this method of finance appeared to be a great boon, for credit was essential and any certain means of securing it seemed desirable, whatever the cost. Outside capital could easily be obtained by the factors and the bankers, who in turn could supply money to the planters and the merchants, who could then pass it along in the shape of supplies and merchandise to the tenant farmers and the small proprietors.[7]

In some quarters there was at first a tendency on the part of the planters to seek for themselves the exclusive privilege of mortgaging crops grown upon their own land. If such a system could be enforced the tenants could secure supplies only from, or by leave of, their respective landlords. Obviously this would be greatly to the advantage of the landlords, for by maintaining their own stores they might to their considerable profit monopolize the trade of their ten-

[6] *Ibid.*, 141; Matthew B. Hammond, "The Southern Farmer and the Cotton Question," *Political Science Quarterly*, 12: 460–461.

[7] Hammond, *Cotton Industry*, 141–142; A. M. Arnett, *The Populist Movement in Georgia*, 50–51.

ants, and in any event a veto upon the excessive and unwarranted expenditures of the thriftless was not to be despised. But the small-town merchants, whom the profits of the crop-lien system were making richer and more numerous every year, objected strenuously to any limitation whatever of the full freedom of contract for the tenants as well as for the small proprietors. Doubtless it was due to their influence rather than to that of the tenants themselves that the latter came to enjoy, quite as if they were landowners, the privilege of independent store accounts based upon independent liens. While, according to law, the landlords might retain the technical privilege of granting or withholding consent to such accounts, in actual practice the keenness of the competition for tenants made it necessary for the landlords to be exceedingly cautious about the way in which they exercised their rights. In some cases they even found it expedient to release their share of the tenant's crop in favor of the merchant, lest the latter withhold from the tenant the necessary credit.[8]

The conflict of interest between planters and merchants was but temporary, however, for as time went on the two classes tended more and more to become one. Planters were drawn into the store business by the necessity or the desirability of being able to supply their tenants; storekeepers frequently became landowners by taking over the farms of those who were indebted to them or by direct purchase at the prevailing low prices. The typical landowner thus came to be also a merchant, and, since there was a strong tendency on the part of the landlord-merchant to move to town, he was likely to be an absentee. As such, he had a distinct advantage over the resident owner in competing for tenants, for nothing was more distasteful to the average tenant, especially the negro, than close supervision, and close supervision

[8] Arnett, op. cit., 51–52; Brooks, Agrarian Revolution in Georgia, 32–33; Hammond, Cotton Industry, 147.

was exceedingly difficult for the absentee owner to give. Tenants deliberately chose to work for the man who would watch them least, and resident farmers sometimes had difficulty in getting tenants at all.[9] The workings of the crop-lien system were about the same in all portions of the South. A written contract was required almost universally, although there were occasional instances in which the merchant relied upon the honor of his customers to fulfill a verbal agreement. The merchant required that in return for an advance of " horses, mules, oxen, necessary provisions, farming tools and implements or money to purchase the same " the contracting farmer should sign over, as a guarantee that the account would be paid, his "entire crop of cotton, cotton-seed, corn, fodder, peas, and potatoes." His personal property, chattels, and real estate, if he had any, might also be included in the mortgage, and in case he should find it impossible to pay his entire indebtedness out of the proceeds of the season's crop, he was legally obligated to continue trading with the merchant who held the lien until the account should be settled in full.[10]

The effect of the crop liens was to establish a condition of peonage throughout the cotton South. The farmer who gave a lien on his crop delivered himself over to the tender mercies of the merchant who held the mortgage. He must submit to the closest scrutiny of all his purchases, and he might buy only what the merchant chose to sell him. He was permitted to trade with no other merchant except for cash, and in most cases his supply of cash was too meager to be worth mentioning. He must pay whatever prices the merchant chose to ask. He must market his crop through the merchant he owed until the entire debt was satisfied, and

[9] Arnett, *Populist Movement in Georgia*, 53; Charles H. Otken, *The Ills of the South; or Related Causes Hostile to the General Prosperity of the Southern People*, 40–43. Cash renters, of course, were far more independent of their landlords than the ordinary cropper. Brooks, *Agrarian Revolution in Georgia*, 68.
[10] Hammond, *Cotton Industry*, 146–150.

only then had he any right to determine the time and method of its disposal. If his crop failed to cancel his debt, as was the case with great regularity, he must remain for another year— perhaps indefinitely — in bondage to the same merchant, or else by removing to a new neighborhood and renting a new farm become a fugitive from the law. Estimates differ, but probably from three-fourths to nine-tenths of the farmers of the cotton South were ensnared to a greater or less degree by the crop-lien system.[11]

The high prices charged by the merchants on credit accounts contributed immeasurably to the distress of the southern farmer. Statistics on this subject are for the most part fragmentary and unreliable, but the consensus of opinion is that the credit purchaser paid from twenty to fifty per cent more for what he bought than he would have paid if he had been able to buy for cash. Instances are on record of flour purchased at three dollars a barrel being sold for ten dollars a barrel and of corn that cost forty cents a bushel being sold at a dollar a bushel. While these cases are doubtless exceptional, the fact that large margins of profit were realized from the credit trade is not open to question. The farmer saw his account at the store mount by leaps and bounds. Try as he might, or quite as often might not, his purchases almost inevitably exceeded the returns from his crop. The thrifty storekeeper was indeed at some pains to see that this was the case. An unpaid account, provided it were not too large, was of some value, for it insured him against the loss of a customer and, often enough, against the loss of a tenant as well.[12]

The charging of high prices on credit accounts was deemed necessary, in part because of the risk involved in such ac-

[11] *Ibid.*, 149, 155; Hammond, in the *Political Science Quarterly*, 12:462; Arnett, *Populist Movement in Georgia*, 57–58; Otken, *Ills of the South*, 76–78.
[12] North Carolina, *Report of the Bureau of Labor Statistics, 1887*, 77, 101; Hammond, *Cotton Industry*, 152–154; Arnett, *Populist Movement in Georgia*, 56–57.

counts and in part because of the obligation it imposed upon the merchant to play the double rôle of storekeeper and banker. The chances of loss were numerous enough. Crops might fail, prices might drop, customers might prove to be impossibly poor farmers, might have sickness in the family, be sent to jail, or even disappear entirely, leaving the merchant to complete their harvests as best he could. Moreover, the amount of capital necessary to carry a whole community for the better part of a year was not inconsiderable, and the right of a merchant to a good interest on his investment seemed justifiable. Frequently enough the merchant had no adequate means of his own to carry his customers and must obtain funds from the bankers at ruinous rates, commonly one and one-half per cent per month. "The road to wealth in the South" was doubtless "merchandising," as one writer declared, but for many it was also the road to bankruptcy.[13]

The credit system contributed also to the one-crop evil, which did more than its full share to insure to the farmer a permanent condition of indebtedness. Cotton almost served the purpose of money, for it was always marketable, it was comparatively imperishable, it could not be consumed by the producer and thus destroyed, as could corn, for example, and it was comparatively easy to handle. The merchant, therefore, wished his customers to raise cotton, and he objected strenuously if they proposed to raise instead such articles as hay, corn, wheat, or potatoes. It was far more expedient, if not more profitable, for the farmer who found himself in need of credit to do what the merchant desired — plant

[13] George K. Holmes, "The Peons of the South," *Annals of the American Academy of Political and Social Science*, 4:265–274; Otken, *Ills of the South*, 74–75; Hammond, *Cotton Industry*, 155–156; Arnett, *Populist Movement in Georgia*, 61. "A tenant whose crop by chance more than suffices to meet his obligations will not infrequently pick enough cotton to discharge his debts to the landlord and the merchant and abandon the remainder, knowing that he can live on the next crop until it is harvested. The merchant who has a lien on his share of the crop pays his taxes, on occasion buries his wife or child, buys him a mule if he needs one, and feeds and clothes him and his family." *Eleventh Census of the United States*, 1890, Vol. 13, *Farms and Homes*, 22–23.

nothing but cotton and buy on each trip to the store " a bale
of Indiana prairie hay, a bag of Richmond meal, a sack of
Milwaukee flour, and a side of Chicago bacon." Even " cabbages shipped from Germany and Irish potatoes from Scot-
land " were among the items that were charged on one
farmer's bill.[14]

Obviously the danger of losing this type of business constituted one of the chief objections that the merchant raised
to every argument savoring of crop diversification. With
cotton the sole crop, all the food supplies consumed by the
farmer, his family, and his live stock had to be purchased at
the store. Should the farmer grow his own food supplies,
however, his purchasing power would be correspondingly
decreased. From the storekeeper's point of view it was far
more profitable for the farmer to purchase every article of
consumption at exorbitant prices and to confine his farming
to the raising of cotton.

Moreover, if the farmer should pasture his land or should
plant crops designed to build up its fertility, the lucrative
trade in fertilizers might suffer. The use of commercial fertilizers, especially in the eastern and older states of the
South, had come to be regarded as an absolute necessity.
A well-informed South Carolinian gave it as his opinion that
" most of our land without the aid of fertilizer would produce about one-fourth of a bale the first year, and in the
course of a few years would not produce over one-eighth or
one-tenth of a bale per acre. Without fertilizer we would
have to quit planting cotton." The trade in fertilizers was
thus of no small proportions. Hammond estimates that from
twelve to thirty-three per cent of the total value of the cotton crop was consumed in this way. In the case of the tenant
farmers, at least, the fertilizer was so applied as to make no
permanent improvement in the soil but merely to stimulate
the growth of the season's crop. Under the one-crop system

[14] North Carolina, *Report of the Bureau of Labor Statistics, 1887*, 76–78;
Progressive Farmer (Raleigh, North Carolina), March 28, 1893; Hammond, in
the *Political Science Quarterly*, 12:463–464.

there was thus as steady and as permanent a market for fertilizers as for groceries.[15]

The merchants, however, were by no means solely responsible for the existence of the one-crop system. It had not originated with them but with the planters before the Civil War, and it was still further encouraged by the high prices that English spinners were willing to pay for American cotton when the war had ended. Furthermore, cotton was the crop that the southern farmer knew best how to raise, and in the case of the freedman it was practically the only crop he knew how to raise. The ignorance of the southern farmer was indeed so complete that most of the propaganda for diversification, so common in the South from Granger times on, was utterly unintelligible to him, if it reached him at all, and doubtless he would have been incapable of acting on such advice even if he had known what it was all about. A typical North Carolina landlord stated that "out of a dozen or more tenants, both white and colored, I have scarcely one who really understands the cultivation of wheat or clover or one who can safely risk the experiment of raising these crops." Such testimony could have been multiplied without limit. The negroes possibly preferred cotton culture to anything else because of the greater opportunity it gave them for the satisfaction of their gregarious instincts. Perhaps, too, their characteristic shiftlessness was less damaging to cotton than it might have been to other crops. There was rarely any resentment among the farmers against the storekeeper's demand that cotton be the exclusive crop. Content with his own ignorance, the southern farmer made little effort to raise the food crops that would have made him independent, and he remained all too willing to buy costly fertilizers in order that he might cultivate his traditional crop of cotton on worn-out ground year after year.[16]

[15] Hammond, *Cotton Industry*, 178–180; Walker, in the *Andover Review*, 14:127.
[16] Hammond, *Cotton Industry*, 121, 139, 151, 184–187; *Progressive Farmer*, March 28, 1893.

The evils of the one-crop system were compounded again and again. When prices went down, the farmer, with a mounting balance against him at the store, saw no way out except to rent more land and raise more cotton. By attempting to farm too much he of course cut down the effectiveness of his work and got a smaller return per acre. He found, moreover, that his expenditures for seed, fertilizers, and supplies had increased as much as the returns from his crop, and his debt at the store might be even more than it had been the year before. But with the lesson still unlearned he sought the next year to raise more bales of cotton rather than to devise means of cutting down his purchases. Could he have produced for himself even the corn and bacon and hay that he bought, he might have freed himself in a short time from the toils of the credit system. Little wonder that intelligent men campaigned earnestly for diversification. "We may join all the farmers' organizations that can be devised," said the *Progressive Farmer,* a leader among southern farm journals, "but hard times will hover around our firesides so long as we buy our meat and bread, hay, fertilizers, and other farm supplies, and attempt to pay for them from the proceeds of one crop." [17]

Of course cotton was not the only crop produced in the South. Corn was grown in every southern state, and in some sections it shared honors more or less equally with cotton. In Texas, for example, tenants were commonly described as farming on the "third and fourth," meaning that they paid one-third of their cotton and one-fourth of their corn as rental. Tobacco was grown extensively in some localities, and such grain crops as wheat, rye, and oats were never wholly ignored, especially in the more northerly and westerly portions of the South. Those farmers who raised foodstuffs were generally in better condition financially than those who

[17] Hammond, *Cotton Industry,* 135–136, 157; *Progressive Farmer,* January 12, 1888; April 23, 1889.

o not, but in spite of their example the hold of cotton upon the ordinary southern farmer remained unbroken. If he did raise corn, he was likely to raise too little of it, and the saying that "the negro will not raise corn" was common. The South, in fact, was not even holding its own in food production, for, considering the increase in population, it was actually farther from being self-sustaining in 1889 than in 1860. During these years cotton culture had increased prodigiously. As Otken put it, "cotton planting has been a mania. The neglected corn field with all its consequences is a part of Southern history."[18]

Reformers who sought to relieve the agricultural distress in the South did not, however, confine their efforts to the preaching of diversification. It was plain enough that the lack of a reasonable system of finance had much to do with the situation, and remedies for this condition were sought after eagerly. One that aroused much hope was the entrance into the South during the early eighties of loan companies not unlike those that were doing such a thriving business in the West. Henry W. Grady, the enthusiastic editor of the *Atlanta Constitution*, thought that he saw in the work of these companies the way of salvation for the hard-pressed southern farmer. The merchants who supplied the South with credit preferred the crop lien as security, and they took real estate only in case of necessity. The loan companies, on the other hand, proposed to lend money generously on land at interest rates as low as seven or eight per cent. This appeared to offer a chance of emancipation from the credit system, at least for those who owned their own farms. But a short trial proved that these loan companies were of comparatively small value. They offered long-time credit, five years being the usual period for which the loan ran, whereas the southern farmer needed short-time credit, loans for a

[18] Otken, *Ills of the South*, ch. 7; Hammond, *Cotton Industry*, 137–138; Clarence Ousley, "A Lesson in Co-operation," *Popular Science Monthly*, 36:822.

few months to carry him through to the end of the crop season. Moreover, the loan companies charged fees and commissions that ran the actual rate of interest far beyond the nominal rate of seven or eight per cent. And for the countless numbers of tenant farmers who owned no property in land, the loan companies had nothing whatever to offer. Clearly the search for better credit facilities for the South could not end here.[19]

It must not be supposed that the South was wholly agricultural. Northern capital, while hesitant enough when it came to backing the southern farmer, had gradually awakened to the opportunities that industrial development in the South seemed to afford. New railway lines had been built, some of which by penetrating heavily forested regions introduced "southern pine" to northern markets, while others by opening up areas rich in coal and iron brought cities like Birmingham, Alabama, into prominence.[20] Still more important, cotton manufacturing, which had existed on a modest scale before and during the Civil War, now grew by leaps and bounds. Observers of the Cotton Exposition held at Atlanta in 1881 and of the New Orleans Exposition in 1884 could predict confidently a great industrial future for the South, since there were nearly twice as many spindles in operation in 1880 as there had been in 1860. This number was doubled again between 1880 and 1885 and was trebled by 1890, while better business methods insured rapidly mounting profits.[21] Tobacco establishments in North Carolina, molasses and sugar refineries in Louisiana, cottonseed oil and oil-cake manufactories in Texas, all added to the new industrialism. Hard-pressed farmers and their families, seeking emancipation from the exactions of the country storekeeper, were easily induced to accept employment in mines,

---

[19] Grady, in *Harper's New Monthly Magazine*, 63:723; Hammond, *Cotton Industry*, 164–165; Arnett, *Populist Movement in Georgia*, 63–64.
[20] Allan Nevins, *The Emergence of Modern America, 1865–1878*, 359.
[21] Holland Thompson, *The New South*, 88–92.

mills, and factories, or on the railroads; and the migration from country to town, so characteristic of the age, went on in the South apace.[22]

At the close of the Reconstruction period political conditions in the South were more favorable to the business man than to the farmer. The struggle for white supremacy had saddled the South with a one-party political system, under which the Democratic machine could count on the unthinking support of a vast majority of the voters. Naturally the men who had led the movement for the overthrow of the negro and carpetbag governments were rewarded with the offices, and naturally, too, since no new and burning issues appeared, they and their friends were continued in office, election after election. These Bourbon leaders associated with the more important people in the growing towns and cities — merchants, bankers, railway men, manufacturers, and the various professional satellites who served these classes — and came to share their points of view. From among these citizens rather than from among the country people came the new politicians, recruited to take the places of such of the older leaders as fell by the way; and from the business men came also the policies that this new, somewhat diluted, Bourbon Democracy chose to follow. Lien laws favorable to the merchants were enacted, and proposed changes in the interest of the poverty-stricken tenants were given scant consideration. Following the well-established Jeffersonian maxim, "the less government the better," railroads and other corporations were rarely scrutinized closely as to their methods and were almost never subjected to any genuine regulation. Formerly the excessive poverty of the South had furnished an excuse for leasing out convicts, chiefly ex-slaves, to work for private individuals and corporations; but this cheap labor supply was still available for those who wished to employ it, even though the finances of

[22] Nevins, *Emergence of Modern America*, 359–360.

the southern states were no longer in so desperate a plight. In short, the post-Reconstruction régime in the South stood ready to guard the welfare of the business man, but to the needs of the farmer it was quite blind.[23]

This situation was aggravated by the existence within the lower South itself of a definite Black Belt, where the negroes far outnumbered the whites. This region was located in the relatively fertile low country that had been the heart of the old plantation system and had always had a preponderance of negroes. When the slaves were freed many of them remained in the vicinity of their old homes, and of those who did migrate not a few came back. Landlordism in the Black Belt was at its worst; here the storekeeper-landowner held his negro tenants in a bondage extremely like slavery. Political rights were of course denied the negroes, and the landlords, supported by the townspeople, controlled nominations and elections to office. In the hill country, where the whites were more numerous than the blacks, the free farmers and the white tenants had the numbers necessary to rule locally if they chose. But in state politics the Black Belt politicians always won, for in all party conventions the black counties were represented not in proportion to the number of actual voters but in proportion to the population. Even when the population of the Black Belt was less than that of the hill counties, the ability of its politicians to manipulate conventions usually more than balanced the difference in numbers. Thus, since the one-party system made a nomination equivalent to an election, a small white minority located in the Black Belt and in the cities was in a position to control the political destinies of a whole state. Naturally the ruling caste used this power to preserve the privileges that it enjoyed.[24]

[23] Arnett, *Populist Movement in Georgia*, 18–23, 32, 82–83; Hamilton, *North Carolina since 1860*, 224–228; Simkins, *Tillman Movement*, ch. 1.
[24] Joseph C. Manning, *The Fadeout of Populism*, 1–3; Brooks, *Agrarian Revolution in Georgia*, ch. 6. See also the maps in Simkins, *Tillman Movement*, opposite page 10, and in Arnett, *Populist Movement in Georgia*, opposite page 184.

Opposition in the white counties to Black Belt control had existed before the Civil War, and after the worst trials of Reconstruction were passed it soon reasserted itself. The Granger movement awakened some echoes in the South and taught the rural whites that if only they could control the state governments they might hope to find legal redress for the economic ills from which they suffered. Popular leaders such as Martin Gary in South Carolina and W. H. Felton in Georgia challenged the Bourbon supremacy, with little success if measured in terms of elections won but with an appreciable result in the way of increased discontent. Given hard enough times and a sufficiently active group of agitators, the rural whites of the South stood ready towards the end of the century to stage a determined revolt against the political leadership they had followed so long.[25]

[25] Simkins, *Tillman Movement,* 16–20; Arnett, *Populist Movement in Georgia,* 36–45.

CHAPTER THREE

THE GRIEVANCES

IN THE spring of 1887 a North Carolina farm journal
stated with rare accuracy what many farmers in all sections of the United States had been thinking for some time.

There is something radically wrong in our industrial system.
There is a screw loose. The wheels have dropped out of balance.
The railroads have never been so prosperous, and yet agriculture languishes. The banks have never done a better or more
profitable business, and yet agriculture languishes. Manufacturing enterprises never made more money or were in a more flourishing condition, and yet agriculture languishes. Towns and
cities flourish and " boom " and grow and " boom," and yet
agriculture languishes. Salaries and fees were never so tempt-
ingly high and desirable, and yet agriculture languishes.[1]

Nor was this situation imputed to America alone. Once
in an unguarded burst of rhetoric a high priest of the Alliance movement pointed out that similar conditions prevailed
in all thickly populated agricultural countries, " high tariff
and low tariff; monarchies, empires, and republics; single
gold standard, silver standard or double standard." [2] It was
true indeed that the blessings of civilization had not fallen
upon all mankind with equal bounty. To the upper and
middle classes more had been given than to the lower; to the
city dweller far more than to his country kinsman. The

[1] *Progressive Farmer* (Raleigh), April 28, 1887.
[2] *Ibid.*, May 13, 1890.

farmer had good reason to believe, as he did believe, that he worked longer hours, under more adverse conditions, and with smaller compensation for his labor than any other man on earth.[3]

For this condition of affairs the farmer did not blame himself. Individual farmers might be lacking in industry and frugality, but farmers as a class were devoted to these virtues. Those who gave up the struggle to win wealth out of the land and went to the cities so generally succeeded in the new environment that a steady migration from farm to city set in. Why should the same man fail as a farmer and succeed as a city laborer? More and more the conviction settled down upon the farmer that he was the victim of " some extrinsic baleful influence." Someone was " walking off with the surplus " that society as a whole was clearly building up and that in part at least should be his.[4] He was accustomed to regard himself as the " bone and sinew of the nation " and as the producer of " the largest share of its wealth." Why should his burdens be " heavier every year and his gains . . . more meager? "[5] Why should he be face to face with a condition of abject servility? Not himself, certainly, but someone else was to blame.[6]

The farmer never doubted that his lack of prosperity was directly traceable to the low prices he received for the commodities he had to sell. The period from 1870 to 1897 was one of steadily declining prices. As one writer put it, the farmer's task had been at the beginning of this era " to make two spears of grass grow where one grew before. He solved that. Now he is struggling hopelessly with the question how to get as much for two spears of grass as he used to get for

---

[3] C. S. Walker, " The Farmers' Movement," *Annals of the American Academy of Political and Social Science,* 4:791; J. M. Rusk, " The Duty of the Hour," *North American Review,* 152:425–430.
[4] Davis, in the *Forum,* 9:231.
[5] Washington Gladden, " The Embattled Farmers," *Forum,* 10:314–315.
[6] R. Means Davis, " The Matter with the Small Farmer," *Forum,* 14:381; Leonidas L. Polk, " The Farmer's Discontent," *North American Review,* 153:9; editorial, " The Farmer and the Carpenter," *Nation,* 50:407.

one."[7] Accurate statistics showing what the farmer really
received for his crops are almost impossible to obtain, but the
figures given by the Department of Agriculture for three
major crops, given in the table below, will at least reveal the
general downward trend of prices.[8]

AVERAGE MARKET PRICES OF THREE CROPS, 1870–1897

YEARS	WHEAT (PER BUSHEL)	CORN (PER BUSHEL)	COTTON (PER POUND)
1870–1873	106.7	43.1	15.1
1874–1877	94.4	40.9	11.1
1878–1881	100.6	43.1	9.5
1882–1885	80.2	39.8	9.1
1886–1889	74.8	35.9	8.3
1890–1893	70.9	41.7	7.8
1894–1897	63.3	29.7	5.8

These prices are subject to certain corrections. They are
as of December 1, whereas the average farmer had to sell
long before that time, often on a glutted market that beat
down the price to a much lower figure They make no
allowance, either, for commissions to dealers, for necessary
warehouse charges, nor for deductions made when the produce could not be regarded as strictly first class. They fail to
show, also, the difference in prices received along the frontier,
where the distance to market was great, and in the eastern
states, where the market was near at hand. In 1889, for example, corn was sold in Kansas for as low a price as ten cents
a bushel and was commonly burned in lieu of coal. In 1890
a farmer in Gosper County, Nebraska, it was said, shot his
hogs because he could neither sell nor give them away.[9]

[7] Henry R. Chamberlain, "The Farmers' Alliance and Other Political Parties," *Chautauquan*, 13:388.
[8] *Yearbook of the United States Department of Agriculture, 1901*, 699, 709,
754. See also Thorstein Veblen, "The Price of Wheat since 1867," *Journal of
Political Economy*, 1:68–103.
[9] *Farmers' Alliance* (Lincoln), August 16, 1890; Elizabeth N. Barr, "The
Populist Uprising," *A Standard History of Kansas and Kansans*, ed. by William
E. Connelly, 2:1138.

So low did the scale of prices drop that in certain sections of the country it was easy enough to prove, statistically at least, that farming was carried on only at an actual loss. It was generally agreed that seven or eight cents of the price received for each pound of cotton went to cover the cost of production; by the later eighties, moreover, many cotton growers were finding it necessary to market their crops for less than they had been getting.[10] The average price per bushel received by northwestern wheat growers dropped as low as from forty-two to forty-eight cents, whereas the cost of raising a bushel of wheat was variously estimated at from forty-five to sixty-seven cents. Statisticians held that it cost about twenty-one cents to produce a bushel of corn, but the western farmer sometimes had to take less than half that sum.[11] Quoth one agitator:

We were told two years ago to go to work and raise a big crop, that was all we needed. We went to work and plowed and planted; the rains fell, the sun shone, nature smiled, and we raised the big crop that they told us to; and what came of it? Eight cent corn, ten cent oats, two cent beef and no price at all for butter and eggs — that's what came of it. Then the politicians said that we suffered from over-production.[12]

Not politicians only but many others who studied the question held that overproduction was the root of the evil. Too many acres were being tilled, with the result that too many bushels of grain, too many bales of cotton, too many tons of hay, too many pounds of beef were being thrown upon the market each year. As the population increased, the number of consumers had advanced correspondingly, but the increase in production had gone on even more rapidly. It was a fact that the per capita output of most commodities had risen with each successive year. The markets of the world were literally broken down. With the supply so far

[10] Hammond, *Cotton Industry*, 169.
[11] *Annual Report of the Railway Commissioner of Minnesota, 1884*, 21.
[12] Barr, in *Kansas and Kansans*, 2:1123, 1150.

in excess of the demand, prices could not possibly be maintained at their former levels.[13]

Those who believed in the overproduction theory argued
that to some extent this condition of affairs was due to the
rapid expansion of the agricultural frontier in the United
States and in the world at large. In the United States the
opportunity to obtain free lands, or lands at a nominal price,
tempted thousands of artisans and laborers to seek their
fortunes in the West. This was true not only in the Northwest, where wheat and corn were the chief products, but also
in the Southwest, where the main reliance was placed on
cotton. The new and fertile lands of the frontier had come
into competition with the old and worn-out lands of the East.
Minnesota and the Dakotas led the nation in the production
of wheat; Kansas and Nebraska, in the production of corn;
Texas, whole regions of which had but lately been opened up,
in the production of cotton; the western ranges, in the production of beef. The eastern farmer had made an effort to
keep pace with the western farmer. He had spent huge sums
on fertilizer and machinery, and he had learned to farm more
scientifically. But his success merely added to the total
output and brought prices tumbling the faster.[14]

Moreover, the revolution in means of transportation that
had been accomplished during the latter half of the nineteenth century had opened up world markets for regions that
had hitherto had small chance to sell their produce. This was
true not only of the American West, which could never have
come into being without an elaborate system of railways over
which to market its crops, but also of distant regions in Russia, India, Australia, Algeria, Canada, Mexico, and the Ar-
gentine, whence, thanks to railways and steamship lines,
harvests of surpassing abundance could now find their way
to the very centers of trade. Such crops as wheat and cotton,

[13] Davis, in the *Forum*, 9:233–241; Dodge, in the *Century*, 21:448.
[14] Peffer, in the *Forum*, 8:466; Rodney Welch, " The Farmer's Changed Condition," *Forum*, 10:699; Hammond, *Cotton Industry*, 135–136, 173–178.

of which the United States had an excess for export, must now often come into competition with these tremendous outpourings from other parts of the world, and the prices must be fixed accordingly. It was the price brought by this exportable surplus that set the price for the entire domestic output.[15]

But the farmers and their defenders refused to place much stock in the overproduction theory. Admitting that the output from the farm had increased perhaps even more rapidly than population, they could still argue that this in itself was not sufficient to account for the low prices and the consequent agricultural depression. They pointed out that, with the general improvement of conditions among the masses, consumption had greatly increased. Possibly the demand attendant upon this fact alone would be nearly, if not quite, sufficient to offset the greater yearly output. There would be, moreover, even heavier consumption were it possible for those who needed and wanted more of the products of the farm to buy to the full extent of their ability to consume. In spite of all the advances of the nineteenth century the world was not yet free from want. "The makers of clothes were underfed; the makers of food were underclad."[16] Farmers used corn for fuel in the West because the prices they were offered for it were so low, while at the same moment thousands of people elsewhere faced hunger and even starvation because the price of flour was so high. Why should the Kansas farmer have to sell his corn for eight or ten cents a bushel when the New York broker could and did demand upwards of a dollar for it? Were there not certain "artificial barriers to consumption?" Were there not "certain influ-

---

[15] Nixon, in the *Iowa Journal of History and Politics*, 21:387; Shelby M. Cullom, "Protection and the Farmer," *Forum*, 8:146; Dodge, in the *Century*, 21:447; Gladden, in the *Forum*, 10:317; Hammond, *Cotton Industry*, 342. A shortage outside the United States, of course, would be reflected in higher prices. For example, crop failures in Europe in the later seventies resulted in an unusually high price for wheat. See the table on page 56.

[16] Arnett, *Populist Movement in Georgia*, 90.

ences at work, like thieves in the night," to rob the farmers
of the fruits of their toil? [17]

Many of the farmers thought that there were; and they
were not always uncertain as to the identity of those who
stood in the way of agricultural prosperity. Western farmers
blamed many of their troubles upon the railroads, by means
of which all western crops must be sent to market. There
was no choice but to use these roads, and as the frontier
advanced farther and farther into the West, the length of the
haul to market increased correspondingly. Sometimes western wheat or corn was carried a thousand, perhaps even two
thousand, miles before it could reach a suitable place for
export or consumption. For these long hauls the railroads
naturally exacted high rates, admittedly charging "all the
traffic would bear." The farmers of Kansas and Nebraska
and Iowa complained that it cost a bushel of corn to send
another bushel of corn to market, and it was commonly believed that the net profit of the carrier was greater than the
net profit of the grower. The farmers of Minnesota and
Dakota were accustomed to pay half the value of their wheat
to get it as far towards its final destination as Chicago. Small
wonder that the farmer held the railroads at least partly
responsible for his distress! He believed that if he could only
get his fair share of the price for which his produce eventually sold he would be prosperous enough.[18] "How long," a
Minnesota editor queried, "even with these cheap and wonderfully productive lands, can . . . any agricultural community pay such enormous tribute to corporate organization
in times like these, without final exhaustion?" [19]

[17] W. A. Coutts, "Agricultural Depression in the United States," *Publications
of the Michigan Political Science Association*, 2:224–227, 237–244; *Public
Opinion*, 9:111 (May 10, 1890), quoting an editorial, "Why Is Agriculture
Depressed?" from *Farm, Stock and Home* (Minneapolis).

[18] *Tenth Census of the United States*, 1880, Vol. 3, p. 533; Henry C. Adams,
"The Farmer and Railway Legislation," *Century*, 21:780–781; Chamberlain, in
the *Chautauquan*, 13:338; Thomas L. Greene, "Railroad Stock-Watering,"
Political Science Quarterly, 6:475, 489.

[19] *Pioneer Press*, January 8, 27, 1885. See also Nixon, in the *Iowa Journal*

Local freight rates were particularly high. The railroads figured, not without reason, that large shipments cost them less per bushel to haul than small shipments. The greater the volume of traffic the less the cost of carrying any portion of that traffic. Accordingly, on through routes and long hauls where there was a large and dependable flow of freight the rates were comparatively low — the lower because for such runs there was usually ample competition. Rates from Chicago to New York, for example, were low in comparison with rates for similar distances from western points to Chicago, while between local points west of Chicago the rates were even more disproportionate. Sometimes the western local rate would be four times as great as that charged for the same distance and the same commodity in the East.[20] The rates on wheat from Fargo to Duluth were nearly double those from Minneapolis to Chicago — a distance twice as great. It cost as much as twenty-five cents a bushel to transport grain from many Minnesota towns to St. Paul or Minneapolis, while for less than as much more it could be transported all the way to the seaboard. Indeed, evidence was at hand to show that wheat could actually be sent from Chicago to Liverpool for less than from certain points in Dakota to the Twin Cities. Iowa farmers complained that it cost them about as much to ship in corn from an adjoining county for feeding purposes as it would have cost to ship the same corn to Chicago; and yet the Iowa rates seemed low to the farmers of Nebraska, who claimed that they paid an average of fifty per cent more for the same service than their neighbors across the Missouri River.[21]

of History and Politics, 21:381, and Hallie Farmer, "The Railroads and Frontier Populism," Mississippi Valley Historical Review, 13:387–390.

[20] Tenth Census of the United States, 1880, Vol. 3, p. 155; Adams, in the Century, 21:780; Frank B. Tracy, "Menacing Socialism in the Western States," Forum, 15:338, and "Rise and Doom of the Populist Party," Forum, 16:242.

[21] Nixon, in the Iowa Journal of History and Politics, 21:383; Pioneer Press, January 27, 1886; Farmers' Alliance, December 30, 1890; Nebraska State Journal, February 26, 1891; Nebraska, Biennial Report of the Attorney-General to the Governor, 1887–1888, 12–20.

Undoubtedly it cost the railroads more to haul the sparse freight of the West than it cost them to haul the plentiful freight o' the East. Railway officials pointed out that western traffic was nearly all in one direction. During one season of the year for every car of wheat hauled out an empty car had to be hauled in, while the rest of the time about ninety per cent of the traffic went from Chicago westward. They asserted that the new roads were often in thinly settled regions and were operated at a loss even with the highest rates. James J. Hill maintained that the roads were reducing rates as fast as they could, and to prove it he even declared himself "willing that the state make any rates it see fit," provided the state would "guarantee the roads six per cent on their actual cost and a fund for maintenance, renewal and other necessary expenditures." [22] President Dillon of the Union Pacific deplored the ingratitude of the farmers who grumbled about high rates. "What would it cost," he asked, "for a man to carry a ton of wheat one mile? What would it cost for a horse to do the same? The railway does it at a cost of less than a cent." Moreover, he thought that unreasonable rates could never long survive, for if a railroad should attempt anything of the sort competition would come immediately to the farmers' aid, and a parallel and competing line would be built to drive the charges down. [23]

But critics of the railroads saw little that was convincing in these arguments. As for the regulation of rates by competition, it might apply on through routes, providing the roads had no agreement among themselves to prevent it, but competition could scarcely affect the charges for local hauls for the simple reason that the average western community depended exclusively upon a single road. Only rarely did the shipper have a choice of two or more railway companies with which to deal, and even when he had this choice there was

[22] *Pioneer Press*, January 3, 27, 29; February 7, 1885.
[23] Dillon, in the *North American Review*, 152:444, 448.

not invariably competition. The roads reached agreements among themselves; more than that, they consolidated. "The number of separate railroad companies operating distinct roads in Minnesota was as high as twenty, three years ago," wrote the railway commissioner of that state in 1881. "Now the number is reduced to substantially one-third that number." [24] Nor did Minnesota differ particularly in this respect from any other frontier state. Throughout the eighties as the number of miles of railroad increased, the number of railroad companies tended to decrease. Communities that prided themselves upon a new "parallel and competing line" were apt to discover "some fine morning that enough of its stock had been purchased by the older lines to give them control." [25] Thus fortified by monopoly, the railroads, as the farmer saw it, could collect whatever rates they chose.

How monopoly might operate to increase rates is well shown by a practice common during the early eighties among certain railroads of the Northwest. Selfishly determined to get every dollar of revenue they could from their customers, these roads made use of the device of charging "transit," or through rates, on all the wheat they carried. They demanded that the wheat shipper pay in advance the full rate to Chicago or Milwaukee, or whatever city happened to be the easternmost terminal. They refused entirely to quote local rates to Minneapolis or St. Paul, fearing that if they did so the grain might be transferred at these points to some other road. It was well understood, however, that the shipper who paid transit rates might unload his grain at any milling center or at any mill on the route, have it ground, and then ship it out again as flour on the same rate contract. And if he disposed of his grain finally at such a point, he might sell the balance of his unused freight for what it would bring. [26]

[24] *Annual Report of the Railway Commissioner of Minnesota, 1881,* 4–5.
[25] *Pioneer Press,* January 31, 1884.
[26] The subject of transit rates is fully discussed in the *Annual Report of the Railway Commissioner of Minnesota, 1884,* 23.

Suppose, for example, that a grain dealer at Milbank, Dakota Territory, wished to sell his wheat in Minneapolis. According to the rules of the Hastings and Dakota road, over which he must ship, he was not allowed to pay local freight to Minneapolis but was compelled to pay full transit rates to Milwaukee, forty cents a hundred pounds instead of the twenty cents a hundred that should normally have been charged. After disposing of his wheat in Minneapolis, the dealer still had on his hands a quantity of unexpended freight from Minneapolis to Milwaukee, for which he had paid approximately twenty cents a hundred. This he offered for sale on the open market, but because "transit," as the unused freight was called, was too plentiful and sold only at a discount of from two and a half to five cents per hundred pounds, the Milbank shipper was fortunate to get fifteen cents — or a little more if the market was good — for transportation that had cost him twenty cents. This loss he learned to look upon as inevitable, and as inevitably he protected himself against it liberally in advance by lowering the price paid the farmer for his grain. "Transit wheat," that is, wheat which had to be shipped over roads quoting transit rates only, always brought from three to five cents a bushel less than "free wheat." Small wonder that farmers living in "transit" regions felt themselves to be defrauded of their rightful profits! There is "probably in no other portion of this country," commented the conservative *Pioneer Press*, "any class of people subjected to such miserable oppression, exercised by a power to which there is no resistance to be offered, and from whose dominion there is no escape. . . . Those who would submit quietly to such outrage must be either more or less than men." [27]

And yet the railroads in much of the frontier area were so greatly overbuilt that, try as they might, they could not always find business enough to enable them to make a rea-

[27] *Pioneer Press*, January 26, 1884.

sonable profit on their investment. Some of the new lines
had been built by boomers and promoters without thought
as to future earning power. Others were projected into
sparsely settled regions by established corporations in order
to preempt the field for future expansion. In both cases the
new roads could in all probability be operated only at a loss,
however high they might place their rates. Certainly investments of this sort should never have been made except by
corporations fully capable of sustaining a present loss in the
hope of future profits. But in any event the farmer was
likely to complain. It made little difference to him whether
the high rates were charged by the local companies whose
lines lay entirely in too thinly settled regions or by the larger
companies who demanded from the farmers living in civilization rates high enough to cover the losses attendant upon
overbuilding at the fringe of settlement. High rates were the
net result in each case.[28]

It was commonly believed also that the practice of stock-
watering had much to do with the making of high rates.
The exact extent to which the railroads watered their stock,
or to which a particular railroad watered its stock, would be a
difficult matter to determine, but that the practice did exist
in varying degrees seems not to be open to question. A writer
in Poor's *Manual* for 1884 stated that the entire four billion
dollars at which the railways of the United States were capi-
talized represented nothing but so much "water."[29] So
sweeping a statement seems rather questionable, but the
belief was general that railroad companies got their actual
funds for investment from bond issues and passed out stocks
to the shareholders for nothing. The roads, indeed, did not
deny the existence of a certain amount of stock-watering.
They argued that their property was quite as likely to increase in value as any other property — farm lands, for

[28] Frank H. Dixon, " Railroad Control in Nebraska," *Political Science Quarterly*, 13:631–633; *Pioneer Press*, January 27, February 7, 1885.
[29] Poor, *Manual of the Railroads, 1884*, i–vi.

example — and that they were justified in increasing their
capital stock to the full extent that any increase in value had
taken place. Some of their apologists held also that the value
of the road was determined by its earning power rather than
by the amount actually invested in the enterprise. It followed, therefore, that new capital stock should be issued as
fast as the earnings of the road showed that the old valuation had been outgrown.[30]

But to those who suffered from the high rates all these
arguments seemed like so many confessions of robbery. The
governor of Colorado, considering especially the sins of the
Denver and Rio Grande, declared it "incredible that the
legitimate course of business can be healthfully promoted by
any such inflated capitalization. There must be humbug, if
not downright rascality, behind such a pretentious array of
figures."[31] The *Kansas Alliance* saw in the prevalent custom
of stock-watering an evil "almost beyond comprehension."
It placed the total amount of railway overcapitalization at a
sum far in excess of the national debt and described these
inflated securities as "an ever present incubus upon the labor
and land of the nation."[32] Jerry Simpson of Kansas figured
that the 8,000 miles of road in his state cost only about
$100,000,000, whereas they were actually capitalized at
$300,000,000 and bonded for $300,000,000 more. "We who
use the roads," he argued, "are really paying interest on
$600,000,000 instead of on $100,000,000 as we ought to."[33]
Such statements could be multiplied indefinitely. The un-
prosperous farmers of the frontier saw nothing to condone
in the practice of stock-watering. Honest capitalization of
railroad property would, they felt, make possible a material
reduction in rates. And, in spite of the assertion of one who

[30] Nebraska, *Biennial Report of the Attorney-General, 1889–1890*, 8–13;
Greene, in the *Political Science Quarterly*, 6:474–476, 478–480; Dillon, in the
North American Review, 152:446; John Moody, *The Masters of Capital*, 23–24.
[31] *Appletons' Annual Cyclopaedia, 1887*, 142.
[32] Barr, in *Kansas and Kansans*, 2:1141.
[33] Greene, in the *Political Science Quarterly*, 6:476–477.

defended the practice of stock-watering that a citizen who
questioned " the right of a corporation to capitalize its prop-
erties at any sum whatever committed an 'impertinence,' " [34]
the farmers had no notion that the matter was none of their
business.

High rates due to overcapitalization and other causes
were not, however, the sole cause of dissatisfaction with the
railways. It was commonly asserted that the transportation
companies discriminated definitely against the small shipper
and in favor of his larger competitors. The local grain merchant without elevator facilities or the farmer desirous of
shipping his own grain invariably had greater and graver
difficulties with the roads than did the large elevator companies. These latter, the farmers contended, were favored by
" inside rates," by rebates, and by preferential treatment
with regard to cars.

Secret rate understandings between the railroads and the
elevator companies were hard to prove, but discrimination
with respect to cars was open and notorious. A regulation in
force in the Hill system was interpreted to mean that " parties desiring to ship grain, whether producers or purchasers,
where there is an elevator, must ship through it, or construct
an elevator of at least 30,000 bushels, or cars will not be
furnished." [35] A farmer complained that although a certain
railroad had ruined his farm by taking a right of way through
it, there was no means by which he could force the road to
" give him a car to ship his wheat." [36] The small town with
its small shipments was at a disadvantage in competition
with the larger town and its larger shipments. No doubt,
as a governor of Minnesota pointed out, the railroads really
found it a matter of " economy, profit, and convenience " to
receive " large, frequent and easily regulated shipments
under contracts with a small number of shippers " rather

[34] Dillon, in the *North American Review*, 152:445.
[35] *Annual Report of the Railway Commissioner of Minnesota, 1883*, 15–16.
[36] *Pioneer Press*, February 5, 1885.

than to bother with small and irregular shipments from many different sources.[37] But this could not be regarded as a sufficient apology for the failure of supposedly common carriers to give equal treatment to all those who desired to use their lines. "The railways cannot have a choice of customers," declared an impotent state railway commissioner. "Railways, like inn-keepers, must take all that come until the quarters are full. If it is an inconvenience to furnish cars to flat-houses or merchants, the answer is, that is just what the railways are paid for; to serve the public generally is their proper function."[38] "Equality before the law," wrote a prominent publicist, "is a canon of political liberty; equality before the railways should become a canon of industrial liberty."[39]

There were cases in which the railroads were guilty of fairly transparent frauds. The scandal of the *Crédit Mobilier* had not passed so far into history but that critics of the Union Pacific could still find occasion to refer to it. Nor were the thefts that this particular construction company had perpetrated without close parallels in later western road building.[40] There were scandals, too, in connection with the reorganization of roads. Minor stockholders were frozen out in order to benefit a secret clique on the inside; roads were robbed by their operators and forced into a dishonest and avoidable bankruptcy. There were scandals without number in the manipulation of railroad securities. The operations along this line of Jay Gould alone almost baffle description. It is true also that many a western community smarted under heavy taxation levied to pay off subsidies to roads that were unnecessary or in some cases never even built.

[37] *Ibid.*, January 8, 1885; *Biennial Message of Governor Hubbard, 1885*, 40.
[38] *Annual Report of the Railway Commissioner of Minnesota, 1883*, 17–19.
[39] Adams, in the *Century*, 21:781. See also Nixon, in the *Iowa Journal of History and Politics*, 21:383; Dixon, in the *Political Science Quarterly*, 13:634; and Farmer, in the *Mississippi Valley Historical Review*, 13:390–391.
[40] Poor, *Manual of the Railroads, 1881*, lxxi; *Farmers' Alliance*, December 13, 1890; Dodge, in the *Century*, 21:452; Tracy, in the *Forum*, 16:241–242.

The burden of public indebtedness thus incurred to help the railways was indeed staggering. In the days when railroads were not so plentiful the constitutions and laws of the western states were so devised as to permit almost unlimited assistance to the railroads from state and local governmental units, and skillful railway promoters seldom failed to secure maximum sums. In Kansas alone during a single period of sixteen months, from July, 1885, to November, 1886, the total amount of municipal contributions to the railroads reached ten million dollars, while by 1890 the grand total of financial assistance from state and local sources was only a little less than seventy-five millions. It is estimated that in this state eighty per cent of the municipal debt was incurred to help finance the railroads.[41] These colossal debts — for other states than Kansas were similarly afflicted — the average taxpayer refused to regard as a fair charge upon him. He might have voted for some of them, but he nevertheless believed that his vote had been obtained under false pretenses. Resentment in the West against the high rates that the railroads charged was thus tinged with something a little more bitter. There was a firm conviction that the railroads were "crooked," that they existed not to serve the West but to plunder it.[42]

The indictment against the railroads was the stronger in view of their political activities. It is not unfair to say that normally the railroads — sometimes a single road — dominated the political situation in every western state. In Kansas the Santa Fe was all-powerful; in Nebraska the Burlington and the Union Pacific shared the control of the state; everywhere the political power of one or more of the roads was a recognized fact. Railway influence was exerted

[41] Farmer, in the *Mississippi Valley Historical Review*, 10:413, 416; Miller, *ibid.*, 11:471.
[42] Frank L. McVey. *The Populist Movement*, 172; Robert E. Riegel, *The Story of the Western Railroads*, ch. 11; Anderson, "Influence of Railway Advertising," 17–18.

in practically every important nominating convention to insure that no one hostile to the railways should be named for office. Railway lobbyists were on hand whenever a legislature met to see that measures unfavorable to the roads were quietly eliminated. Railway taxation, a particularly tender question, was always watched with the greatest solicitude and, from the standpoint of the prevention of high taxes, usually with the greatest of success. How much bribery and corruption and intrigue the railroads used to secure the ends they desired will never be known. For a long time, however, by fair means or foul, their wishes in most localities were closely akin to law. Beyond a doubt whole legislatures were sometimes bought and sold.[43]

In the purchase of men of influence railway passes were ever of the greatest potency. Members of the legislatures pocketed the mileage they were allowed by the state and rode back and forth to the capital on passes furnished by the railroads. Governors, judges, railway commissioners, and all other public officials were given passes and were encouraged to use them freely. Prominent attorneys were similarly privileged and in addition were generally retained by the railroads. The makers of public opinion — editors, ministers, and local politicians — were not neglected; when they were too insignificant to merit the regulation annual pass, they were given occasional free trips or half-fare permits. This "railroad invention for corrupting state officers" was not confined to the frontier but was general throughout the country.[44] "Do they [the railroads] not own the newspapers? Are not all the politicians their dependents? Has not every Judge in the State a free pass in his pocket? Do they

[43] *Farmers' Alliance*, July 17, 1889; April 26, July 26, 1890; Barr, in *Kansas and Kansans*, 2:1152; Tracy, in the *Forum*, 16:242; Farmer, in the *Mississippi Valley Historical Review*, 10:424–425; James W. Witham, *Fifty Years on the Firing Line*, 33; Virginia Bowen Jones, "The Influence of the Railroads on Nebraska State Politics," unpublished master's thesis, University of Nebraska, 1927.

[44] *Progressive Farmer*, January 29, 1889; Tracy, in the *Forum*, 16:242; Jones, "Influence of the Railroads," 25–28, 79.

not control all the best legal talent of the State?" Thus the *Progressive Farmer* of Raleigh, North Carolina, complained.[45] "To stand in with the railroads in order to get free transportation," wrote a more tolerant observer, "seemed to be the main object in life with about one-half of the population."[46]

It is true enough that there was for a long time a certain blindness on the part of many who accepted the passes to the implications of such gifts. Anti-railroad agitators such as Ignatius Donnelly of Minnesota and "Calamity" Weller of Iowa in their earlier careers not only accepted passes but actually begged for them.[47] Among those who received no passes, however, and paid the bill for those who did, a feeling of hostility to the practice gradually evolved. "A railroad pass," ran one convincing argument, "is not properly a 'courtesy.' It is money. . . . What grocer, plaintiff or defendant in a suit, would venture to give the judge a free pass for his yearly sugar and tea?" "The man who will accept railroad transportation," ran another argument, "which may be worth hundreds of dollars every year, and feel under no sort of obligation for it is a very contemptible sort of a man, and as rare as he is contemptible."[48] By 1886 Donnelly was returning his passes and, along with others of similar views, was denouncing the pass evil as a system of covert bribery.[49]

But from the standpoint of the western pioneer the crowning infamy of the railroads was their theft, as it appeared to him, of his lands. Free lands, or at least cheap

[45] August 14, 1888.
[46] Witham, *Fifty Years*, 61. In 1885 a railroad expert estimated that free transportation over the Union Pacific alone cost that road two thousand dollars a day. Nebraska, *Biennial Report of the Attorney-General to the Governor, 1887–1888*, 16.
[47] Chamberlain to Donnelly, December 22, 1879; Donnelly to Chamberlain, January 28, 1881; Curtin to Donnelly, April 14, 1880, all in the Donnelly Papers, in the possession of the Minnesota Historical Society. See also correspondence dated August 7, 1883, in the Weller Papers, in the possession of the State Historical Society of Wisconsin.
[48] *Farmers' Alliance*, March 15, November 9, 1890.
[49] Donnelly to the C. and N. W. RR, December 18, 1886, and to the Winona and St. Peter RR, December 23, 1886, both in the Donnelly Papers.

lands, had been his ever since America was. Now this " price-
less heritage" was gone, disposed of in no small part to the
railroads. To them the national government had donated an
area "larger than the territory occupied by the great German empire," land which, it was easy enough to see, should
have been preserved for the future needs of the people.[50] For
this land the railroads charged the hapless emigrant from
"three to ten prices" and by a pernicious credit system
forced him into a condition of well-nigh perpetual "bondage."[51] "Only a little while ago," ran one complaint, "the
people owned this princely domain. Now they are *starving
for land* — starving for an opportunity to labor — starving
for the right to create from the soil a subsistence for their
wives and little children."[52] To the western farmers of this
generation the importance of the disappearance of free lands
was not a hidden secret to be unlocked only by the researches
of some future historian. It was an acutely oppressive reality.
The significance of the mad rush to Oklahoma in 1889 was by
no means lost upon those who observed the phenomenon.
"These men want *free land,*" wrote one discerning editor.
"They want *free land* — the land that Congress squandered
. . . the land that should have formed the sacred patrimony
of unborn generations."[53] Senator Peffer of Kansas understood the situation perfectly. "Formerly the man who lost
his farm could go west," he said, "now there is no longer any
west to go to. Now they have to fight for their homes instead
of making new." And in no small measure, he might have
added, the fight was to be directed against the railroads.[54]

Complaints against the railways, while most violent in the
West, were by no means confined to that section. Practi-

[50] Fred E. Haynes, *James Baird Weaver,* 164.
[51] *Farmers' Alliance,* September 27, 1890.
[52] *Alliance* (Lincoln), June 26, 1889.
[53] *Ibid.*
[54] Barr, in *Kansas and Kansans,* 2:1160. See also J. W. Gleed, in the *Review
of Reviews,* 10:45, and Erastus Wiman, "The Farmer on Top," *North American
Review,* 153:14.

cally every charge made by the western farmers had its counterpart elsewhere. In the South particularly the sins that the roads were held to have committed differed in degree, perhaps, but not much in kind, from the sins of the western roads. Southern railroads, like western railroads, were accused of levying "freight and fares at their pleasure to the oppression of the citizens" and of making their rates according to the principle, "take as much out of the pockets of the farmers as we can without actually taking it all." [55] Southerners believed, in fact, that the general decline in freight rates that had accompanied the development of the railroads throughout the country was less in the South than anywhere else and that their section was for this reason worse plagued by high rates than any other. [56]

Local rates were certainly as unreasonably high, when compared with through rates, as anywhere in the country. North Carolina farmers claimed that such articles as turnips, potatoes, cabbages, and apples could be brought to the city of Raleigh from northern markets and sold for less than the similar products of near-by farms. Competition was stifled as relentlessly in the South as in the West. A group of south-eastern railroads, for example, "abdicated" their rate-making functions in favor of one supreme rate umpire, whose decisions were not to be questioned. There were discriminations as to persons and as to places. The individual or the company that had capital and could do business on a large scale received favors denied to individuals or companies of less importance. Cities of consequence were given rates that unimportant towns and villages similarly situated could not obtain. There was the usual abundance of stock-watering. There were frauds — frauds of construction companies like

[55] *Progressive Farmer*, July 31, 1888; "Governor's Message," *Public Documents of the State of North Carolina*, legislative session of 1889, No. 1, pp. 17–18.

[56] Hamilton, *North Carolina since 1860*, 222; Hammond, *The Cotton Industry*, 172.

the *Crédit Mobilier*, frauds of stock manipulation, frauds of enforced bankruptcies. Nor were the operators of southern railroads novices at the political game. They gave away passes freely and with satisfactory results. Politicians pointed with pride to the railroads and urged that "no embarrassing restraint upon their development or prosperity should be imposed by a legislative body." State railway commissions were nonexistent or relatively innocuous, and the taxation of railways was light.[57]

These common grievances of South and West against the railroads promised to supply a binding tie of no small consequence between the sections. Whether they were westerners or southerners, the orators of revolt who touched upon the railway question spoke a common language. Moreover, the common vocabulary was not used merely when the malpractices of the railroads were being enumerated. Any eastern agitator might indeed have listed many of the same oppressions as typical of his part of the country. But the aggrieved easterner at least suffered from the persecutions of other easterners, whereas the southerner or the westerner was convinced that he suffered from a grievance caused by outsiders. In both sections the description of railway oppression was incomplete without a vivid characterization of the wicked eastern capitalist who cared nothing for the region through which he ran his roads and whose chief aim was plunder. This deep-seated antagonism for a common absentee enemy was a matter of the utmost importance when the time came for bringing on joint political action by West and South.

In the northwestern grain-growing states the problem of the railroads was closely related to the problem of the ele-

[57] *Progressive Farmer*, October 13, 1887; March 6, August 14, 1888; January 29, 1889; *News and Observer* (Raleigh, North Carolina), February 3, 4, 8, 1887; "Inaugural Address of Governor G. Fowle," *Public Documents of the State of North Carolina*, legislative session of 1889, No. 18, p. 9; Arnett, *The Populist Movement in Georgia*, 68–71. One North Carolina editor declared that "the roads taxed the state more and the state taxed the roads less" than any other in the Union. *News and Observer*, February 10, 1887.

vators. Some grain houses were owned by individuals or local companies who ran one or more elevators in neighboring towns. A few were owned by the railway companies themselves. Still others were the property of large corporations that operated a whole string of elevators up and down the entire length of a railway line. These larger companies naturally built better and more commodious houses than the smaller ones; they were more efficient in their manner of doing business; and they were easily the favorites of the railroads. All the companies, large or small, must obtain on whatever terms the railway companies saw fit to impose such special privileges as the right to build upon railroad land and the right to proper sidetrack facilities. By refusing these favors the roads could prevent, and did prevent, the erection of new elevators where they deemed the old ones adequate, but once an elevator was authorized it could usually count on railway support.[58]

Thanks mainly to their satisfactory relations with the railroads, the first elevator companies to cover a territory enjoyed in their respective localities almost a complete monopoly of the grain business, both buying and selling. Wherever elevators existed the roads virtually required that shipments of grain be made through them, for in practice if not in theory cars were seldom furnished to those who wished to avoid the elevator and to load their grain from wagons or from flat warehouses. On the face of it this rule seemed harmless enough, for the elevator companies were supposedly under obligations to serve the general public and to ship grain for all comers on equal terms. This they might have done with fair impartiality had they not been engaged themselves in the buying and selling of grain. But since it

[58] *Annual Report of the Railway Commissioner of Minnesota, 1884,* 17–20; *Annual Report of the Railroad and Warehouse Commission of Minnesota to the Governor, for the Year Ending November 30, 1889,* 13–14; Farmer, in the *Mississippi Valley Historical Review,* 13:390–391; Fossum, *The Agrarian Movement in North Dakota,* 11–17, 33.

was the chief concern of the elevator operator to purchase
and ship all the grain he could get, he could hardly be expected to take much interest in providing facilities for the
farmer who wished to ship directly or for the competitive
grain merchant who lacked an elevator of his own. The
result was that the independent buyer was speedily "frozen
out," and the farmer found that if he was to get rid of his
grain at all he must sell to the local elevator for whatever
price he was offered. He claimed rightly that under such a
system he was denied a free market for his grain. To all
except the privileged elevator companies the market was
closed.[59]

This absence of a free market was the chief reason assigned by many farmers for the low prices they were paid for
their grain. Since the elevator men had a monopoly of the
grain-buying business, what was to prevent them from paying only such a figure as their pleasure and interests might
dictate? If there was only one elevator at a station it was
clear that the operator was a law unto himself and might pay
what he chose. Even if there were several elevators there
was only rarely competition as to price. Pooling was sometimes resorted to, but usually agreements with regard to
prices could be reached without this device, each elevator
taking its share of grain without attempting to capture the
business of its neighbor.[60]

When the elevator companies paid lower prices than were
justifiable, they naturally made an effort to conceal the fact.
They knew that the price of wheat in Minneapolis or Chicago
at any given time was public information. The railway rates
to such a terminal were also known to all. If, therefore, they
openly exacted more than a reasonable profit for handling
grain their practice would be subjected to an unpleasant and

[59] *Annual Report of the Railway Commissioner of Minnesota, 1883*, 19;
1884, 19–20; *Pioneer Press*, January 19, 1884; Farmer, in the *Mississippi Valley
Historical Review*, 13:392–395.
[60] *Pioneer Press*, January 12, 19, 1884.

"pitiless" publicity. So they generally quoted as good prices
as could be reasonably expected, considering the high freight
they had to pay and the current market values of wheat at
the terminals to which they shipped. For their long profits
they relied upon more skillful means. Wheat, as presented
for sale, was of course of uneven quality and must be graded
before a price could be assigned. Custom had established
certain standards of grading; for example, wheat ranked as
number one hard must weigh fifty-eight pounds or more to
the bushel, must be clean, of a good color, and at least ninety
per cent pure hard wheat. Grain that was too light in weight,
contained foul seed, or was dirty, off color, or frosted was
graded accordingly. It was in determining these grades that
the elevator men had the best opportunity to reduce the
price paid to the farmer, for the buyer fixed the grades at
will, after making such an examination of the wheat as he
saw fit. The farmer had nothing to say about it. If he objected to the grade and price he was offered, he had no
recourse but to take his grain to another elevator, probably
only to find that there the same condition prevailed. Moreover, another elevator was not always available.[61]

Undoubtedly there was great irregularity and unfairness
in the grading of grain. Farmers at one station claimed that
they had never received a grade of number one hard and very
little, if any, of number one regular. They therefore appointed a committee of three to follow a shipment of thirty
cars from their town to Duluth, where they found that four
of the cars had been graded number one hard, ten number
one regular, and nearly all above the grades the farmers
received. The local elevator had profited accordingly. It was
alleged in another community that of a shipment of wheat
from the same field, grown from the same seed, and harvested
at the same time, some had been graded number one, some

[61] *Annual Report of the Railway Commisioner of Minnesota, 1883,* 12–15;
Pioneer Press, January 26, 1884; January 27, 1885; Fossum, *Agrarian Movement in North Dakota,* 16–17.

number two, and some rejected. At best the elevator operators were anxious to grade low enough to protect themselves against losses. Charles A. Pillsbury, one of the most prominent elevator men in the Northwest, himself admitted that at the beginning of the crop season grades at Minneapolis and Duluth were much more liberal than later, when the supply was greater and the demand had diminished. Often enough elevator operators graded according to their honest judgment in the early part of the season, but later on, when inspection at the terminal points had tightened up, they became frightened and refused "for days at a time to grade anything above number two hard, no matter what the quality offered." The state railway commissioner of Minnesota estimated, conservatively he believed, that the farmer lost on an average about five cents a bushel through unfair grading. Many said it was twice that amount.[62]

If the farmer had little part in fixing the price at which his produce sold, he had no part at all in fixing the price of the commodities for which his earnings were spent. Neither did competition among manufacturers and dealers do much in the way of price-fixing, for the age of "big business," of trusts, combines, pools, and monopolies, had come. These trusts, as the farmers saw it, joined with the railroads, and if necessary with the politicians, "to hold the people's hands and pick their pockets."[63] They "bought raw material at their own price, sold the finished product at any figure they wished to ask, and rewarded labor as they saw fit."[64] Through their machinations "the farmer and the workingman generally" were "overtaxed right and left."[65]

One western editor professed to understand how all this

[62] Annual Report of the Railway Commissioner of Minnesota, 1883, 15; 1884, 9–11; Appletons' Annual Cyclopaedia, 1882, 560; Pioneer Press, January 19, 1884.
[63] Farmers' Alliance, September 20, 1890.
[64] Barr, in Kansas and Kansans, 2:1123.
[65] George E. Waring, Jr., "Secretary Rusk and the Farmers," North American Review, 152:753.

had come about. The price-fixing plutocracy, he argued, was but the "logical result of the individual freedom which we have always considered the pride of our system." The American ideal of the "very greatest degree of liberty" and the "very least legal restraint" had been of inestimable benefit to the makers of the trusts. Acting on the theory that individual enterprise should be permitted unlimited scope, they had gone their way without let or hindrance, putting weaker competitors out of business and acquiring monopolistic privileges for themselves. At length the corporation "had absorbed the liberties of the community and usurped the power of the agency that created it." Through its operation "individualism" had congealed into "privilege."[66]

The number of "these unnatural and unnecessary financial monsters"[67] was assumed to be legion. An agitated Iowan denounced the beef trust as "the most menacing" as well as the most gigantic of "about 400 trusts in existence."[68] A Missouri editor took for his example the "plow trust. As soon as it was perfected the price of plows went up 100 per cent . . . who suffers? . . . Who, indeed, but the farmer?"[69] Senator Plumb of Kansas held that the people of his state were being robbed annually of $40,000,000 by the produce trust.[70] Southern farmers complained of a fertilizer trust, a jute-bagging trust, a cottonseed oil trust. Trusts indeed there were: trusts that furnished the farmer with the clothing he had to wear; trusts that furnished him with the machines he had to use; trusts that furnished him with the fuel he had to burn; trusts that furnished him with the materials of which he built his house, his barns, his fences. To all these he paid a substantial tribute. Some of them, like the

---

[66] *Farmers' Alliance*, February 28, 1891. See also the *Progressive Farmer*, August 21, 1888; Farmer, in the *Mississippi Valley Historical Review*, 10:423; and Witham, *Fifty Years*, 52.

[67] *Alliance*, June 26, 1889.

[68] Nixon, in the *Iowa Journal of History and Politics*, 21:386.

[69] *Kansas City Times*, quoted in the *Omaha Herald*, April 20, 1888.

[70] Barr, in *Kansas and Kansans*, 2:1142.

80 THE POPULIST REVOLT

manufacturers of farm machinery, had learned the trick of installment selling, and to such the average farmer owed a perpetual debt.[71]

The protective tariff, while not universally deplored in farm circles, met with frequent criticism as a means of "protecting one class at the expense of another — the manufacturer against the farmer, the rich against the poor."[72] Because of the tariff the American market was reserved for the exclusive exploitation of the American manufacturer, whose prices were fixed not in accordance with the cost of production but in accordance with the amount of protection he was able to secure. The genial system of logrolling, which on occasion made Democrats as good protectionists as Republicans, insured a high degree of protection all around. And so the tariff became a veritable "hot-bed for the breeding of trusts and combines among all classes of men thus sheltered by the law."[73] This situation was the more intolerable from the farmer's point of view for the reason that he must sell at prices fixed by foreign competition. The protective tariff gave the eastern manufacturer "the home market at protective prices," but the prices of western and southern farm produce were "fixed by the surplus sold in foreign free-trade markets."[74] Thus the farmers were as a class "a priori 'unprotected,' the victims of a system of free-trade selling and 'protected' purchasing — in their economic relation as consumers paying heavy prices for high tariff goods, and as producers most of them selling against the competition of the world's markets."[75] Southern farmers were probably more keenly aware of the oppressive nature of

[71] *Progressive Farmer*, January 29, April 16, 1889; Hammond, *Cotton Industry*, 172; Arnett, *Populist Movement in Georgia*, 71–72, 104.
[72] "Governor's Message," *Public Documents of the State of North Carolina*, legislative session of 1889, No. 1, p. 32.
[73] John T. Morgan, "The Danger of the Farmers' Alliance," *Forum*, 12:402.
[74] William E. Smythe, in "A Bundle of Western Letters," *Review of Reviews*, 10:44.
[75] Editorial, "The Vote of the Farm," *Nation*, 50:329.

the tariff than the westerners, and they denounced it roundly
as an important contribution to their distress. They could
see plainly enough that because of the tariff they were effectively debarred from exchanging their cheap farm products
for the cheap manufactured goods of European nations.[76]
Westerners, while by no means silent on the subject, were
inclined to regard the tariff as "subordinate to . . . finance,
land, and transportation."[77]

It was the grinding burden of debt, however, that aroused
the farmers, both southern and western, to action. The widespread dependence upon crop liens in the South and farm
mortgages in the West has already been described. In the
South as long as the price of cotton continued high and in the
West as long as the flow of eastern capital remained uninter-
rupted, the grievances against the railroads, the middlemen,
and the tariff-protected trusts merely smouldered. But when
the bottom dropped out of the cotton market and the western boom collapsed, then the weight of debt was keenly felt
and frenzied agitation began. The eastern capitalists were
somehow to blame. They had conspired together to defraud
the farmers—"to levy tribute upon the productive energies of West and South." They had made of the one-time
American freeman "but a tenant at will, or a dependent
upon the tender mercies of soulless corporations and of absentee landlords."[78]

> There are ninety and nine who live and die
>   In want, and hunger, and cold,
> That one may live in luxury,
>   And be wrapped in a silken fold.
> The ninety and nine in hovels bare,
> The one in a palace with riches rare.

[76] North Carolina, *Annual Report of the Bureau of Labor Statistics, 1887*,
95; J. G. Carlisle, "The Tariff and the Farmer," *Forum*, 8:476.
[77] *Farmers' Alliance*, August 23, 1890; William V. Allen, "Western Feeling
towards the East," *North American Review*, 162:591; Farmer, in the *Mississippi
Valley Historical Review*, 10:426–427.
[78] Smythe, in the *Review of Reviews*, 10:44; Goodloe, in the *Forum*, 10:355.

. . . .
And the one owns cities, and houses and lands,
And the ninety and nine have empty hands.[79]

As one hard season succeeded another the empty-handed
farmer found his back debts and unpaid interest becoming an
intolerable burden. In the West after the crisis of 1887 interest rates, already high, rose still higher. Farmers who needed
money to renew their loans, to meet partial payments on
their land, or to tide them over to another season were told,
truly enough, that money was very scarce. The flow of eastern capital to the West had virtually ceased. The various
mortgage companies that had been doing such a thriving
business a few months before had now either gone bankrupt
or had made drastic retrenchments. Rates of seven or eight
per cent on real estate were now regarded as extremely low;
and on chattels ten or twelve per cent was considered very
liberal, from eighteen to twenty-four per cent was not uncom-
mon, and forty per cent or above was not unknown. Naturally the number of real estate mortgages placed dropped off
precipitately. Instead of the six thousand, worth nearly
$5,500,000, that had been placed in Nebraska during the
years 1884 to 1887, there were in the three years following

[79] *Alliance*, July 31, 1889. The stages by which the pioneer farmer arrived
at this state of mind were graphically described by a writer for the *Pioneer
Press*, January 19, 1884: " A farmer with only a few dollars in his pocket comes
out here and takes a claim. It only costs $15 for the preliminary fees, and he
has six months to make his improvements. These improvements usually consist
of a sod shanty, a well four feet deep, and from five to twenty acres of breaking. When he has done this much he can mortgage his farm for sufficient money
to prove up and buy a horse or two. When he is known to be in possession of
this amount of property, his credit is good for a plow and he obtains his seed by
giving a mortgage on his crop in advance. Then he goes in debt for the necessary machinery to harvest his crop, and by the time his grain is ready to sell
he is pretty well buried under a pile of debts. He takes his wheat to market, of
course firmly believing that it is nothing less than No. 1 hard, but the elevator
man's eagle eye promptly discovers that it has been " frosted " or is " damp "
and instead of getting 80 cents a bushel, as he expected, he is forced, from
his necessities, to sell for what he can get. It is then that he begins to kick. The
sun which shone so brightly upon him in the spring is now obscured by two or
three blanket mortgages; and he sits down in his lonely cabin on the bleak
prairies and imagines that he is being ground down by the despotic heel of
monopoly. Therefore he kicks, and keeps on kicking, and never ceases till he's
dead or out of debt."

1887 only five hundred such mortgages, worth only about
$650,000, while only one out of four of the farm mortgages
held on South Dakota land in 1892 had been contracted prior
to 1887.[80] When the farmer could no longer obtain money
on his real estate, he usually mortgaged his chattels, with the
result that in many localities nearly everything that could
carry a mortgage was required to do so. In Nebraska during
the early nineties the number of these badges of "dependence and slavery" recorded by the state auditor averaged
over half a million annually. In Dakota many families were
kept from leaving for the East only by the fact that their
horses and wagons were mortgaged and could therefore not
be taken beyond the state boundaries.[81]

Whether at the old rates, which were bad, or at the new,
which were worse, altogether too often the western farmer
was mortgaged literally for all he was worth, and too often
the entire fruits of his labor, meager enough after hard times
set in, were required to meet impending obligations. Profits
that the farmer felt should have been his passed at once to
someone else. The conviction grew on him that there was
something essentially wicked and vicious about the system
that made this possible. Too late he observed that the money
he had borrowed was not worth to him what he had contracted to pay for it. As one embittered farmer-editor wrote,

There are three great crops raised in Nebraska. One is a crop of
corn, one a crop of freight rates, and one a crop of interest. One
is produced by farmers who by sweat and toil farm the land.
The other two are produced by men who sit in their offices and
behind their bank counters and farm the farmers. The corn is
less than half a crop. The freight rates will produce a full average. The interest crop, however, is the one that fully illustrates

[80] *Farmers' Alliance*, August 16, 1890; *First Biennial Report of the Bureau of
Labor and Industrial Statistics of Nebraska*, 1887–1888, pp. 210–213; Nixon, in
the *Iowa Journal of History and Politics*, 21:378; J. W. Gleed, in the *Forum*,
11:470; Tracy, in the *Forum*, 16:243; Farmer, in the *Mississippi Valley Historical Review*, 10:419.
[81] Barr, in *Kansas and Kansans*, 2:1138; *Farmers' Alliance*, August 30, 1890;
Farmer, in the *Mississippi Valley Historical Review*, 10:419.

the boundless resources and prosperity of Nebraska. When corn
fails the interest yield is largely increased.[82]

What was the fair thing under such circumstances? Should
the farmer bear the entire load of adversity, or should the
mortgage-holder help? Opinions varied, but certain extremists claimed that at the very least the interest should be
scaled down. If railroads were permitted to reorganize,
reduce their interest rates, and save their property when
they got into financial straits, why should the farmer be
denied a similar right? [83]

The only reorganization to which the farmer had recourse, as a rule, was through foreclosure proceedings, by
which ordinarily he could expect nothing less than the loss of
all his property. Usually the mortgagor was highly protected
by the terms of the mortgage and could foreclose whenever
an interest payment was defaulted, whether the principal
was due or not. In the late eighties and the early nineties
foreclosures came thick and fast. Kansas doubtless suffered
most on this account, for from 1889 to 1893 over eleven
thousand farm mortgages were foreclosed in this state, and
in some counties as much as ninety per cent of the farm
lands passed into the ownership of the loan companies. It was
estimated by one alarmist that "land equal to a tract thirty
miles wide and ninety miles long had been foreclosed and
bought in by the loan companies of Kansas in a year." [84]
Available statistics would seem to bear out this assertion,
but the unreliability of such figures is notorious. Many
farmers and speculators, some of them perfectly solvent,
deliberately invited foreclosure because they found after the
slump that their land was mortgaged for more than it was
worth. On the other hand, many cases of genuine bankruptcy were settled out of court and without record. But
whatever the unreliability of statistics the fact remains that

[82] *Farmers' Alliance*, August 23, 1890.
[83] Barr, in *Kansas and Kansans*, 2:1155, 1157, 1161.
[84] *Ibid.*, 2:1151; Farmer, in the *Mississippi Valley Historical Review*, 10:420.

in Kansas and neighboring states the number of farmers who
lost their lands because of the hard times and crop failures
was very large.[85]

In the South the crop-lien system constituted the chief
mortgage evil and the chief grievance, but a considerable
amount of real and personal property was also pledged for
debt. Census statistics, here also somewhat unreliable because of the numerous informal and unrecorded agreements,
show that in Georgia about one-fifth of the taxable acres
were under mortgage, and a special investigation for the
same state seemed to prove that a high proportion of the
mortgage debt was incurred to meet current expenditures
rather than to acquire more land or to make permanent improvements. Similar conditions existed throughout the cotton South. Chattel mortgages were also freely given, especially by tenants, but frequently also by small proprietors.
Interest rates were as impossibly high as in the West, and
foreclosures almost as inevitable. Evidence of foreclosures on
chattels could be found in the " pitiful heaps of . . . rub-
bish " that " commonly disfigured the court house squares." [86]
Foreclosures on land, or their equivalent, were numerous,
serving alike to accelerate the process of breaking down the
old plantations and of building up the new " merchant-owned
'bonanzas.'" Many small farmers lapsed into tenantry; indeed, during the eighties the trend was unmistakably in the
direction of " concentration of agricultural land in the hands
of merchants, loan agents, and a few of the financially strongest farmers." [87]

Taxation added a heavy burden to the load of the farmer.
Others might conceal their property. The merchant might
underestimate the value of his stock, the householder might
neglect to list a substantial part of his personal property, the
holder of taxable securities might keep his ownership a secret,

[85] Barr, in *Kansas and Kansans*, 2:1139; J. W. Gleed, in the *Forum*, 9:96;
Miller, in the *Mississippi Valley Historical Review*, 11:484.
[86] Arnett, *Populist Movement in Georgia*, 58.
[87] *Ibid.*, 61; Hamilton, *North Carolina since 1860*, 222.

but the farmer could not hide his land. If it was perhaps an
exaggeration to declare that the farmers "represent but one-
fourth of the nation's wealth and they pay three-fourths of
the taxes," [88] it was probably true enough that land bore the
chief brunt of taxation, both in the South and in the West.
Tax-dodging, especially on the part of the railroads and
other large corporations, was notorious. Some North Carolina railroads had been granted special exemptions from
taxation as far back as the 1830's, and they still found them
useful. In Georgia the railroads paid a state tax but not a
county tax. Nearly everywhere they received special treatment at the hands of assessors, state boards of equalization,
or even by the law itself.[89] Western land-grant railroads
avoided paying taxes on their huge holdings by delaying to
patent them until they could be sold. Then the farmer-
purchaser paid the taxes. Meantime the cost of state and
local government had risen everywhere, although most dis-
proportionately in the West, where the boom was on. In the
boom territory public building and improvement projects out
of all proportion to the capacity of the people to pay had
been undertaken, and railways, street-car companies, and
other such enterprises had been subsidized by the issuing of
state or local bonds, the interest and principal of which had
to be met by taxation. For all this unwise spending the
farmers had to pay the greater part. The declaration of one
Kansas farmer that his taxes were doubled in order "to pay
the interest on boodler bonds and jobs voted by non-
taxpayers to railroad schemes and frauds and follies which
are of no benefit to the farmer" was not without a large
element of truth.[90] The farmer was convinced that he was
the helpless victim of unfair, unreasonable, and discrimina-

[88] Nixon, in the *Iowa Journal of History and Politics*, 21:380.
[89] Arnett, *Populist Movement in Georgia*, 72–73; *News and Observer*, February 3, 4, 8, 1887.
[90] Farmer, in the *Mississippi Valley Historical Review*, 10:413, 423; Miller,
ibid., 11:477–488; Sheldon, in *Nebraska History and Record of Pioneer Days*,
7:17.

tory taxation. Here was another reason why he was "gradually but steadily becoming poorer and poorer every year."[91]

Beset on every hand by demands for funds — funds with which to meet his obligations to the bankers, the loan companies, or the tax collectors and funds with which to maintain his credit with the merchants so that he might not lack the all-essential seed to plant another crop or the few necessities of life that he and his family could not contrive either to produce for themselves or to go without — the farmer naturally enough raised the battle cry of "more money." He came to believe that, after all, his chief grievance was against the system of money and banking, which now virtually denied him credit and which in the past had only plunged him deeper and deeper into debt. There must be something more fundamentally wrong than the misdeeds of railroads and trusts and tax assessors. Why should dollars grow dearer and dearer and scarcer and scarcer? Why, indeed, unless because of the manipulations of those to whom such a condition would bring profit?[92]

Much agitation by Greenbackers and by free-silverites and much experience in the marketing of crops had made clear even to the most obtuse, at least of the debtors, that the value of a dollar was greater than it once had been. It would buy two bushels of grain where formerly it would buy only one. It would buy twelve pounds of cotton where formerly it would buy but six. The orthodox retort of the creditor to such a statement was that too much grain and cotton were being produced — the overproduction theory. But, replied the debtor, was this the whole truth? Did not the amount of money in circulation have something to do with the situation? Currency reformers were wont to point out that at the close of the Civil War the United States had

[91] "Memorial of the Farmers Alliance of Chatham County to the General Assembly of North Carolina," *Public Documents of the State of North Carolina,* legislative session of 1889, No. 25, p. 2.
[92] Miller, in the *Mississippi Valley Historical Review,* 11:488; Hamilton, *North Carolina since 1860,* 222–223.

nearly two billions of dollars in circulation. Now the population had doubled and the volume of business had probably trebled, but the number of dollars in circulation had actually declined! Was not each dollar overworked? Had it not attained on this account a fictitious value? [93]

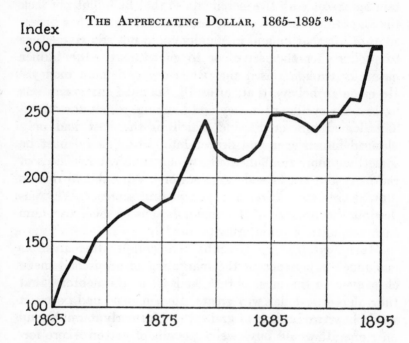

Index   THE APPRECIATING DOLLAR, 1865–1895 [94]

Whatever the explanation, it was clear enough that the dollar, expressed in any other terms than itself, had appreciated steadily in value ever since the Civil War. The depreciated greenback currency, in which all ordinary business was transacted until 1879, reached by that year a full parity with

[88] *Alliance,* June 6, 12, 1889; *Progressive Farmer,* April 20, 1889; Peffer, in the *Forum,* 8:467.

[94] Reproduced from Arnett, *Populist Movement in Georgia,* 69, with the permission of the author. Mr. Arnett derives the index numbers to 1891 "from those for prices in the Aldrich Report, column for weighted averages of commodities said to comprise 68.6 per cent of the total expenditures. Thereafter they are based on the figures of the Bureau of Labor."

gold. But the purchasing power of the dollar still continued its upward course. For this phenomenon the quantity theory may be — probably is — an insufficient explanation, but in the face of the figures from which the accompanying chart has been drawn, the fact of continuous appreciation can hardly be denied.

For those farmers who were free from debt and were neither investors nor borrowers such a condition might have had little meaning. The greater purchasing power of the dollar meant fewer dollars for their crops, but it meant also fewer dollars spent for labor and supplies. Conceivably, the same degree of prosperity could be maintained on the smaller income. But in the West and in the South the number of debt-free farmers was small indeed, and for the debtor the rising value of the dollar was a serious matter. The man who gave a long-term mortgage on his real estate was in the best position to appreciate how serious it was. Did he borrow a thousand dollars on his land for a five-year term, then he must pay back at the end of the allotted time a thousand dollars. But it might well be that, whereas at the time he had contracted the loan a thousand dollars meant a thousand bushels of wheat or ten thousand pounds of cotton, at the time he must pay it the thousand dollars meant fifteen hundred bushels of wheat or fifteen thousand pounds of cotton. Interest, expressed likewise in terms of produce, had mounted similarly year by year so that the loss to the borrower was even greater than the increase in the value of the principal. What it cost the debtor to borrow under such circumstances has been well expressed by Arnett in the table on the following page, which is based on statistics taken from the census of 1890.

Add to this the unreasonably high interest rates usually exacted and the commissions and deductions that were rarely omitted, and the plight of the debtor farmer becomes painfully clear. He was paying what would have amounted to about a twenty or twenty-five per cent rate of interest on a non-appreciating dollar.

It was, moreover, far from comforting to reflect that in such a transaction what was one man's loss was another's gain. Nor was it surprising that the harassed debtor imputed to the creditor, to whose advantage the system worked, a

DEBT APPRECIATION, 1865–1890 [95]

AVERAGE FIVE-YEAR DEBT CONTRACTED IN	APPRECIATION (IN TERMS OF DOLLAR'S PURCHASING POWER)
1865–1869	35.2
1870–1874	19.7
1875–1879	4.5
1880–1884	11.7
1885–1890	11.6

deliberate attempt to cause the dollar to soar to ever greater and greater heights. Had not the creditor class ranged itself solidly behind the Resumption Act of 1875, by which the greenback dollar had been brought to a parity with gold? Was not the same class responsible for the "crime of 1873," which had demonetized silver and by just so much had detracted from the quantity of the circulating medium? Was there not, indeed, a nefarious conspiracy of creditors — eastern creditors, perhaps with English allies — to increase their profits at the expense of the debtors — western and southern — by a studied manipulation of the value of the dollar? "We feel," said Senator Allen of Nebraska, "that, through the operation of a shrinking volume of money, which has been caused by Eastern votes and influences for purely selfish purposes, the East has placed its hands on the throat of the West and refused to afford us that measure

[95] Arnett, *Populist Movement in Georgia,* 68.

of justice which we, as citizens of a common country, are entitled to receive."[96] And the grievance of the West against the East was also the grievance of the South.[97]

Nor was this grievance confined to resentment against the steadily mounting value of the dollar. There was in addition an undeniable and apparently unreasonable fluctuation in its purchasing power during any given year. At the time of crop movements, when the farmers wished to sell — indeed, had to sell, in most cases — the dollar was dear and prices were correspondingly depressed. When, on the other hand, the crop had been marketed and the farmers' produce had passed to other hands, the dollar fell in value and prices mounted rapidly. Wall Street speculators and others bought heavily when prices were low and sold later when prices were high at handsome profits — profits which, the farmers firmly believed, should have gone to the original producer. Why should these things be?

Southerners and westerners who studied the question found an answer in the inelasticity of the currency. In agricultural sections, they reasoned, money was needed most when crops were harvested, for at such times the fruit of an entire year's productive effort was thrown upon the market. The farmer could not hold his crop, for he must have money with which to meet pressing obligations — store debts, interest charges, mortgages. Sell he must, and sell he did. But this tremendous demand upon the currency increased the value of the dollar, since there were no more dollars in circulation at crop-moving time than at any other time! When at last the crop was moved and the abnormal demand for money had subsided, then the value of the dollar declined and prices rose correspondingly. These fluctuations within a

[96] Allen, in the *North American Review*, 162:590; Edward B. Howell, in "A Bundle of Western Letters," *Review of Reviews*, 10:43.
[97] *Proceedings of the Fourth Annual Session of the North Carolina Farmers' State Alliance*, 1890, p. 9.

given year were often large enough to make the difference between prosperity and adversity for the farmer. Could he have sold at the maximum price, or even at the average price, instead of at the minimum, he would have had far less of which to complain. As it was, he could see no good reason why the supply of money should remain constant when the need for it was variable. Unless, perchance, this were the deliberate intent of those who stood to profit from the situation. Was this not another device by which it was ordained that those who had might profit from those who had not? [98]

This inelasticity of the currency, as well as its inadequacy, was blamed in large part upon the national banking system, against which every critic of the monetary situation railed. According to these critics the laws establishing national banks were " of the same character of vicious legislation that demonetized silver." They were "conceived in infamy and . . . for no other purpose but to rob the many for the benefit of the few." [99] It was at least susceptible of demonstration that the national bank-note circulation had steadily dwindled. The law of 1864 provided that national bankers must buy bonds of the United States to the extent of not less than one-third of their capital stock. These bonds were then deposited with the government and upon them as security the banks might issue notes up to ninety per cent of the par value of the bonds. This right to issue paper money, however, was exercised more freely in the years directly following the Civil War than later on, and in 1873 the total bank-note circulation had mounted to $339,000,000. From that time on it tended to decline. In 1876 it was $291,000,000; in 1891 it was only $168,000,000. By just so

[98] *Progressive Farmer*, May 13, 1890; Walker, in the *Annals of the American Academy of Political and Social Science*, 4:791; Chamberlain, in the *Chautauquan*, 13:338; Arnett, *Populist Movement in Georgia*, 93, 169.
[99] Frank Drew, "The Present Farmers' Movement," *Political Science Quarterly*, 6:295.

much was the volume of the currency diminished, and, if the quantity theory held, the appreciation of the dollar was correspondingly accelerated.[100]

The explanation of this cut in the note issues of national banks is not hard to find. The national debt was being paid off and some of the bonds upon which note issues had been based were retired on this account. Indeed, with its yearly revenues far in excess of its necessary expenditures, the government relaxed the rule with regard to the minimum bond deposit required of the banks and even entered the market to buy back national bonds before they were due. Naturally the bonds rose to a heavy premium, and national bankers found that they could make a greater profit by selling their bonds and retiring their notes than by keeping their notes in circulation. But in the whole proceeding the western and southern debtors saw something sinister. Was not the government in league with the money sharks to increase the value of the dollar? Why should the government bonds be paid off in gold, or its equivalent, rather than in new issues of greenbacks, the kind of money that had been used ordinarily in the purchase of these bonds? Why should there be any national banks at all? Why should "the business of issuing money and controlling its volume" be turned over "to a few persons who used their power to their own interest?" Why should not the government itself issue money direct to the people and at reasonable rates?[101]

The argument for government issues of greenbacks had at least one dependable leg to stand upon. Such a policy would promote flexibility in the currency. The volume of note issues might then be regulated in accordance with the

[100] Horace White, *Money and Banking*, 351; E. L. Bogart, *An Economic History of the United States*, 392–393.
[101] *Progressive Farmer*, April 17, 1888; *Farmers' Alliance*, July 19, 1890; Ernest D. Stewart, "The Populist Party in Indiana," *Indiana Magazine of History*, 14:334; McVey, *Populist Movement*, 147–148; Arnett, *Populist Movement in Georgia*, 95; Bogart, *Economic History of the United States*, 393.

demand, whereas under the national banking system the volume was regulated chiefly by the possible profit to the bankers concerned. If it paid the banks to buy bonds and turn them into currency, then the banks could be expected to follow that course. If it paid better, however, to keep their bond holdings static or even to dispose of bonds and retire notes, then they followed that course. Not the needs of the country but the probable profits of the bankers determined the amount of the currency. The charge was made that at certain critical times, especially during crop-moving seasons, scarcity of money was deliberately connived at, since higher rates of interest could then be charged, presumably with greater profits. Probably this charge was baseless in the main, but that the currency was virtually inflexible under the national banking system is a matter of common knowledge.[102]

Hostility to the national banks was increased by the argument, plausible enough but fallacious, that these institutions were able to exact double interest, once on the bonds they purchased from the government and once on the notes secured by these bonds and lent to their customers. As a western editor put it:

If a man is worth a million and wants to start a national bank, he buys one million in bonds which he deposits with the government who pays him interest for every dollar in gold twice a year. These bonds are not taxed one cent to help defray the expenses of the government. Now then, the government carrying out its policy to make the rich richer issues him $900,000 in greenbacks to bank on and get all the interest he can and charge [sic] him 1 per cent on the greenbacks to cover expense of engraving, etc. The man was worth $1,000,000, the government makes him worth $1,900,000, and untaxed. Is this right farmer? You bear the burden and your land cannot escape taxation.

[102] *Progressive Farmer*, August 21, 1888; *Farmers' Alliance*, December 6, 1890; Barr, in *Kansas and Kansans*, 2:1123; White, *Money and Banking*, 401–403.

Moreover, the attitude of the farmer towards the bankers was not improved by the reflection that the men who bought Civil War bonds in the first instance had bought at a low price and with depreciated paper currency.[103]

Such were the grievances of which the farmers complained. They suffered, or at least they thought they suffered, from the railroads, from the trusts and the middlemen, from the money-lenders and the bankers, and from the muddled currency. These problems were not particularly new. Always the farmer had had to struggle with the problem of transportation. He had never known a time when the price of the things he had to buy was not as much too high as the price of the things he had to sell was too low. He had had his troubles with banks and bankers. But those earlier days were the days of cheap lands, and when things went wrong the disgruntled could seek solace in a move to the West. There was a chance to make a new start. Broader acres, more fertile fields, would surely bring the desired results. And with the restless ever moving to the West, the more stable elements of society left behind made pleasing progress. Now with the lands all taken and the frontier gone, this safety valve was closed. The frontier was turned back upon itself. The restless and discontented voiced their sentiments more and fled from them less. Hence arose the veritable chorus of denunciation directed against those individuals and those corporations who considered only their own advantage without regard to the effect their actions might have upon the farmer and his interests.[104]

[103] *Alliance,* July 31, 1889; J. T. Holdsworth, *Money and Banking,* 171.
[104] Turner, *Frontier in American History,* 275–281.

CHAPTER FOUR

THE FARMERS' ALLIANCE[1]

FROM the early seventies on, the "embattled farmers" showed an unmistakable tendency to organize for the redress of their grievances. For a time the Patrons of Husbandry was the favorite farm order, and the main attack was directed against the malpractices of the railroads. Out of this agitation grew the Granger movement, which for various reasons was short-lived and failed of its chief objectives, but did at least succeed in establishing the right of the states to regulate in some degree the business of common carriers.[2] Local political parties of the same period, such as the People's Antimonopoly party of Minnesota, were also thoroughly imbued with the spirit of farmer organization, and even the National Greenback party, with its emphasis upon currency reform, was designed primarily as a vehicle through which the debt-ridden farmers of the country could express their views. During the late seventies and the eighties a whole new crop of farm orders appeared. In the old Granger states of the Northwest the National Farmers' Alliance began to grow; in the South the National Farmers' Alliance and Industrial Union, shadowed presently by the

[1] This chapter follows in the main an article by John D. Hicks and John D. Barnhart, "The Farmers' Alliance," in the *North Carolina Historical Review,* 6:254–280, and is reprinted by permission.
[2] Solon J. Buck, *The Granger Movement: A Study of Agricultural Organization and Its Political, Economic, and Social Manifestations, 1870–1880.*

Colored Farmers' Alliance, made notable headway; in Illinois and neighboring states the Farmers' Mutual Benefit Association, in Michigan and elsewhere in the old Northwest the Patrons of Industry, and even in the Northeast a Farmers' League, all — not to speak of numerous minor orders — gave promise of ever increasing vitality.[3]

Among these various farmers' organizations the National Farmers' Alliance and the National Farmers' Alliance and Industrial Union were the most important. The former was more commonly known as the Northern or the Northwestern Alliance to designate the territory in which it was strongest, the upper Mississippi Valley. The latter was usually called the Southern Alliance, as it was predominant in the South.

Ordinarily the date of the founding of the Northern Alliance is fixed at March 21, 1877, and the credit is given to a group of New Yorkers, mainly Grangers, who thought it expedient to create a "political mouthpiece" through which the Patrons of Husbandry could speak. The redress of grievances against the railroads, the reform of taxation, and the legalization of Granger insurance companies were among the specific ends sought.[4] A somewhat shadowy claim was made, however, that the New York Alliance was preceded by a similar society, the Settlers' Protective Association, which a number of Kansas squatters had organized about 1874 to defend their land titles against railroad claimants. After accomplishing this end, the society turned to cooperative enterprises and appointed an eastern purchasing agent, through whom the plan of organization became known in

[3] Solon J. Buck, "Independent Parties in the Western States, 1873–1876," *Essays in American History Dedicated to Frederick Jackson Turner*, 149–152, and *The Agrarian Crusade*, ch. 6; Drew in the *Political Science Quarterly*, 6:282–310; Emory A. Allen, *Labor and Capital*, chs. 14–21. An excellent account of the Northern Alliance is given in John D. Barnhart's "The History of the Farmers' Alliance and of the People's Party in Nebraska," a doctoral dissertation submitted to Harvard University in 1929.

[4] W. J. Fowler, "The Farmers' Alliance: Letter from President Fowler," *Western Rural* (Chicago), November 20, December 4, 1880. See also the issue of October 23, 1880.

New York. According to this tradition, the New York Alliance was copied directly from the Kansas model.[5]

However this may be, it seems clear that the first really effective Alliance organization was founded by a Chicago editor named Milton George. Prior to becoming a newspaper man George had farmed successfully in Fulton County, Illinois, and had learned at first hand the difficulties under which the rural classes labored. His advent into the newspaper world came about more or less by accident. It chanced that he had lent some money to H. N. F. Lewis, the editor of a Chicago farm journal known as the *Western Rural*. Lewis failed in business, and George, in order to recover his investment, decided to take over the paper and edit it himself. Accordingly he moved to the vicinity of Chicago, hired a good editorial writer by the name of David W. Wood, acquired a small farm, and for a number of years divided his interest between farming and editing. Railway discriminations against the rural classes goaded him especially, and his paper teemed with editorials denouncing the roads, demanding their regulation by the government, and criticizing the free pass evil by which railway legislation was checked and the makers of public opinion bribed. Declaring that "the railroads are literally starving some of our farmers to death," he urged the formation of cheap transportation clubs and farmers' alliances to aid the farmer in the defense of his rights. The alliance idea he seems clearly to have borrowed from the New York experiment.[6]

George's first move was to organize, on April 15, 1880, a Farmers' Alliance for Cook County, Illinois. Through this body as a central agency he proposed to establish other alliances of "practical farmers" and ultimately a national order. Popularized in the columns of the *Western Rural*, the

[5] F. G. Blood, *Handbook and History of the National Farmers' Alliance and Industrial Union*, 35; Barr, in *Kansas and Kansans*, 2:1140.

[6] Witham, *Fifty Years*, 25-26, 61; *Western Rural*, January 24, March 6, 1880; March 12, 1881.

idea caught hold readily, and requests for charters began to come in. The first charter was granted to a group of farmers near Filley, Nebraska, and others were soon scattered liberally throughout the states of the Middle West. To organize the movement on a national basis a "Farmers' Transportation Convention" was called to meet at Chicago on October 14, 1880, and on the appointed date several hundred delegates from Alliance chapters, farmers' clubs, and granges put in their appearance.

Long resolutions were adopted condemning the railway system of the country as "a virtual monopoly . . . defiant of all existing law . . . oppressive alike to the producer and consumer, corrupting to our politics, a hindrance to free and impartial legislation, and a menace to the very safety of our republican institutions." To correct these evils Congress was called upon to inaugurate a program of governmental control, and the political parties were warned to nominate for office men in sympathy with the farmers' program. A constitution was adopted providing for local, state, and national alliances, the national organization to be a delegate body of little authority, serving only to link together the various state and local units, which were left free each to work out its own problems in its own way. Dues and salaries apparently were not contemplated, and the early expenses of the national body, including the cost of holding the initial convention, seem to have been paid for by Milton George himself, whose generous financial assistance was extended also to some of the state alliances.[7]

The success of the organization that thus came into existence was probably due more to hard times than to the efforts of the farmer-editor. The late seventies were for the most part years of large crops and low prices, years in which the price per bushel, the total value of the crop, and the return

---

[7] *Chicago Tribune*, October 15, 1880; *Western Rural*, October 23, 1880; Witham, *Fifty Years*, 26–28, 51; Peffer, in the *Forum*, 8:472.

per cultivated acre steadily declined.[8] This period of declining prices naturally gave stimulus to agricultural discontent and was an important reason for the rise of the Alliance. An increased foreign demand brought higher prices for the large crops of 1879 and 1880 and temporary prosperity,[9] but drought in much of the upper Mississippi Valley in 1881 reduced the yield of wheat twenty-two per cent and of corn thirty-two per cent and again plunged the northwestern farmers into gloom.[10] In its earliest phase the Alliance was most popular where the drought hit hardest — in Kansas, Nebraska, Iowa, and Minnesota — and somewhat less popular in Illinois, Wisconsin, Michigan, and Missouri.

Aided by this unfortunate agricultural situation, Milton George succeeded well with the movement he had begun, in spite of the opposition of rival agricultural journals, the editors of which seemed to think that the Alliance was but a shrewd scheme to increase the circulation of his paper, the *Western Rural*. Within a month some two or three hundred locals had been chartered and within a year, optimists claimed, not less than a thousand.[11] It was harder to start state organizations than the locals, and in the early years of Alliance history the few that came into existence were by no means flourishing. In Nebraska an advance in freight rates helped along the cause, and on January 5 and 6, 1881, delegates from about twenty-five counties met at Lincoln and organized the first state Alliance, adopting a constitution rather more elaborate than that of the national order.[12]

[8] *Report of the Commissioner of Agriculture for the Years 1881 and 1882*, 580–586; Wholesale Prices, Wages, and Transportation, Report by Mr. Aldrich from the Committee on Finance, March 3, 1893, *Senate Report* No. 1394, Parts 1–4, 52 Congress, Session 2, pp. 9-10, 26–27, 29, 34–35, 61.
[9] In addition to the references cited in note 8 see also the *Report of the Commissioner of Agriculture for the Year 1879*, 128, and the *Annual Report of the Commissioner of Agriculture for the Year 1880*, 194.
[10] *Report of the Commissioner of Agriculture for the Years 1881 and 1882*, 577–593.
[11] *Western Rural*, November 13, 1880; October 15, 1881; Witham, *Fifty Years*, 51, 62.
[12] *Official Proceedings of the Nebraska State Farmers' Alliance, at Its First Annual Meeting at Lincoln, Nebraska, January 5 and 6, 1881; Omaha Daily Bee*, January 7, 8, 1881.

Before eighteen months had elapsed similar action had been taken in Kansas, Iowa, Wisconsin, Illinois, Minnesota, and Michigan. In New York the state organization had preceded the formation of the national order, and an independent Alliance in Texas, from which the Southern Alliance later grew, was for a time regarded as a part of the National Farmers' Alliance. When the second annual convention was held at Chicago, on October 5 and 6, 1881, the secretary reported that the Alliance had acquired a total membership of some 24,500 farmers, Nebraska, Kansas, and Iowa taking the lead in the order named. At the third annual meeting in St. Louis, on October 4, 1882, it was claimed that two thousand alliances with a total membership of 100,000 farmers were represented.[13]

Once this early enthusiasm had spent itself, however, the Alliance entered into a period of decline. For a few years, beginning about 1883, times were not inordinately hard in the Northwest — crops were generally good, and prices were fair — and there was little call for the Alliance type of protest. The fourth annual meeting, held in Chicago during the month of October, 1883, was so poorly attended that the officers chosen were commissioned to hold over until their successors could be elected — a wise provision, for it turned out that there was no annual meeting in 1884. The secretary, Milton George, was given the task of acting in place of the national officers, but even the founder and financier of the movement seems for a time to have lost heart. His journal ceased almost altogether to publish Alliance news. With a few exceptions the state alliances were as lifeless as the national order, and the marvel of it is that the organization did not die out entirely.[14]

Poor wheat prices in the fall and winter of 1884-85 probably had much to do with revival of interest in the Alliance. Certain it is that in the northwestern wheat-growing area the

[13] *Western Rural*, October 15, 1881; May 27, 1882; November 4, 1882.
[14] *Ibid.*, September 8, October 13, 1883; November 20, 1886.

alliances suddenly developed an unwonted vitality. Letters
in the familiar strain began to pour in once more upon the
editor of the *Western Rural*, who took courage and began to
work anew. "The Farmers' Alliance has never shown so
much life as it is now showing," he rejoiced. "More alliances
are being organized than ever before, especially in the West."
Early in 1885 the Dakota Territorial Alliance was organized
and in the same year the order was introduced into Colorado.[15] In 1886 a statement of Alliance principles, which was
mailed out to all subscribers to the *Western Rural* and to
such others as were thought to be interested, elicited considerable response. Farmers who indicated their belief in the
principles enunciated were made honorary members and
encouraged to join together into active locals.[16] Many new
alliances were formed, old ones were resuscitated, and in November, 1886, another national convention, the first since
1883, was successfully held. By counting in the activities of
the Southern Alliance, which by this time was also getting
under way, the secretary gave a most exaggerated idea of the
newly attained prosperity, but after all due allowances are
made, it is obvious that the order was growing with considerable rapidity.[17]

As the hard times of the late eighties set in, the strength
of the Alliance movement increased correspondingly. The
national meeting of 1887, held in Minneapolis during the
month of October, was able to break new ground. Assured
that the Alliance had passed the period of infancy, the convention adopted a more adequate constitution, made Milton
George a life member, and proposed to become self-sustaining
by introducing a system of dues and fees. The state leaders,
hitherto little interested in the affairs of the national order,
now took over its management, and the resolutions adopted
exhibited a more pronounced tinge of radicalism. It was sug-

[15] *Ibid.*, December 27, February 9, 1884; January 3, 17, March 7, 1885.
[16] *Ibid.*, May 22, 1886. The declaration was printed in many other issues also.
[17] *Ibid.*, November 20, 1886.

gested that government control of the railways might well be supplemented by the actual ownership of one or more of the transcontinental lines, and the financial problem was represented by a demand for the free and unlimited coinage of silver at the customary ratio. Only seven states of perhaps twice that number entitled to representation had delegates at the Minneapolis convention, but the interest apparently was great and hope for the future was high. Friendly exchanges were made with the Knights of Labor, who were in session at the same time and place. Jay Burrows of Nebraska and August Post of Iowa were made president and secretary, respectively.[18]

Within the next few years the Farmers' Alliance increased enormously in membership and became a power to be reckoned with in the whole Northwest. "The people are aroused at last," wrote the jubilant editor of the *Western Rural.* "Never in our history has there been such a union of action among farmers as now."[19] By 1890 the secretary's office reported new members coming in at the rate of a thousand a week, and with pardonable exaggeration predicted a total of two million members in the immediate future. Kansas alone claimed a hundred and thirty thousand, and the other frontier states where times were hard, especially Nebraska, the Dakotas, and Minnesota, did not lag far behind.[20] In November, 1890, the official Farmers' Alliance paper of Nebraska announced that the order had ten fully organized state alliances, five others in process of organization, and numerous locals in other states.[21] But the strength of the movement was not to be reckoned in the constantly shifting number of states in which the Alliance was fully organized or to which its locals had spread. Its strength lay rather in the

[18] *Ibid.,* October 8, 15, 22, 1887.
[19] *Ibid.,* August 20, 1890.
[20] *Ibid.,* May 10, August 20, 1890; *National Economist* (Washington, D. C.), 2: 72 (October 19, 1889); Barr, in *Kansas and Kansans,* 2: 1141.
[21] *Farmers' Alliance* (Lincoln), November 20, 1890.

complete and thoroughgoing awakening of class consciousness among the farmers of the Middle Border.

The Southern Alliance, meanwhile, had gone through a
somewhat similar course of development. Out in Lampasas
County, Texas, a group of frontier farmers had organized an
Alliance as early as 1874 or 1875, the main purpose of which
was to promote cooperation in such matters as catching
horse thieves, rounding up estrays, and purchasing supplies.
The Alliance was also calculated to furnish effective opposition to the activities of land sharks and cattle kings, whose
disregard of the rights of the small farmer was notorious.
To help along the good work a secret ritual was devised, the
lodge was introduced into a number of neighboring counties
where comparable problems existed, and in 1878 a Grand
State Alliance was established. But when the Greenback
agitation was at its height the order was drawn into politics
and was virtually killed by the resulting dissensions.[22]

Seed was saved from this early beginning, however, and in
1879 some of it was planted in Parker County, Texas, by a
man named Baggett, who had once been a member of the old
Lampasas County society. On changing his residence to
Parker County, Baggett had brought with him a copy of the
original constitution and a knowledge of the original Alliance
methods. He promptly founded a new lodge, which, by
maintaining a strictly nonpartisan attitude, escaped the pit-
falls that had wrecked its predecessor. The progress of the
revived order was rapid, and, with a dozen newly organized
locals, another Grand State Alliance was set up. In 1880
the order was incorporated by the state of Texas as a " secret
and benevolent association." [23] Officially known as the
"Farmers' State Alliance," it soon achieved considerable
prominence throughout central and northern Texas and even

[22] W. Scott Morgan, History of the Wheel and Alliance, and the Impending
Revolution, 281; J. E. Bryan, The Farmers' Alliance; Its Origin, Progress, and
Purposes, 3–5; Buck, Agrarian Crusade, 112.
[23] William L. Garvin, History of the Grand State Farmers' Alliance of Texas,
6; Appletons' Annual Cyclopaedia, 1890, 299–300.

expanded across the border into Indian Territory. By December, 1885, the claim was made that the Alliance had about fifty thousand members scattered among not less than twelve hundred locals. The next year eighty-four counties and twenty-seven hundred suballiances were represented in the state meeting held at Cleburne.[24]

The unrest among Texas farmers, which was clearly indicated by this sudden banding together, arose from somewhat the same general conditions that operated in the frontier states farther to the north.[25] Not only in Texas, however, but all through the South the time was ripe for a movement of protest against the low prices, the pernicious crop-lien system, and the tyranny of the country merchants. Conditions must be changed, and farmer cooperation to that end seemed the logical means.[26]

For a considerable time, however, the Texas Alliance posed primarily as a social organization. Its true nature was supposedly set forth in resolutions adopted at Cleburne, which committed the order to labor for "the education of the agricultural classes, in the science of economical government, in a strictly non-partisan spirit," and "a better state mentally, morally, socially and financially," but declared that "the brightest jewels which it garners are the tears of widows and orphans, and its imperative demands are to visit the homes where lacerated hearts are bleeding to assuage the suffering of a brother or sister; bury the dead; care for the widows and educate the orphans," and so on. A change in Alliance policy was foreshadowed, however, by the adoption, in addition to these harmless, if rhetorical, resolutions, of a

[24] Bryan, *Farmers' Alliance*, 9; Garvin, *Farmers' Alliance of Texas*, 84; *National Economist*, 2:196 (December 14, 1889); *Handbook of Facts and Alliance Information*, 113. The notion set forth in F. G. Blood, *Handbook and History of the National Farmers' Alliance and Industrial Union*, 35, that a man from Kansas, versed in the lore of the Settlers' Protective Association, had something to do with the origin of the Texas Alliance, is not to be taken too seriously.

[25] See *ante*, ch. 1.

[26] Hammond, *Cotton Industry*, chs. 5–6; Otken, *Ills of the South*, chs. 2–3.

long set of demands definitely designed to throw the weight of the Alliance into the political scales. These demands put the Alliance on record as favoring the higher taxation of lands held for speculative purposes, the prohibition of alien landownership, the prevention of dealing in futures, so far as agricultural products were concerned, more adequate taxation of the railways, new issues of paper money, an interstate commerce law, and sundry other political novelties. A committee of three was to present this ambitious program to the state legislature and to Congress.[27]

The motto of the lodge, "In things essential, Unity; and in all things, Charity," did not avail to prevent serious factional strife as a result of this attempt to draw the Alliance into politics. Texas had a lively political contest in view for the fall of 1886, and some members professed to fear that the Cleburne proposals worked towards complicating matters further by launching the Alliance as an independent political party. So acute was the dissatisfaction that after the convention a minority met, organized themselves into an opposition Alliance, secured a state charter, and elected officers. The dissolution of the order now seemed imminent, but at this juncture there appeared on the scene one C. W. Macune, a man of varied talents who came to play an important rôle in Alliance history.

Macune was born in Kenosha, Wisconsin, on May 20, 1851. His father, a Canadian of Scotch-Irish descent, was by trade a blacksmith, by profession a Methodist preacher, and by instinct perhaps a rover, for he had just moved from Canada to Wisconsin when his son was born, and a year later he was headed for California. The California venture, however, proved ill-advised, for the elder Macune was stricken with cholera on the way and died at Fort Laramie. Thereafter the year-old boy was taken by his mother to Fremont, Illinois, where he received some elementary schooling; but at ten years of age he quit school because of his mother's straitened financial circumstances and went to work on a farm. Inheriting, apparently, his father's tendency to seek new environments, he tried out successively in his early manhood California, Kansas, and Texas. Texas seemed to suit him best, for there he married and made his home for many years. His versatility was surprising. He read law, practiced medicine, and finally made his influence so keenly felt in agricultural matters that in 1886 he was elected chairman of the executive committee of the Texas State Alliance. He was presently to figure as business manager of the Texas Alliance Exchange and still later as editor of the *National Economist,* the most conspicuous of all the Alliance papers. Macune had a ready wit, he could speak and write almost without effort, and he possessed much personal magnetism; but time proved him to be a man of erratic judgment, poor business ability, and uncertain ethical standards.

In the critical situation in which the Texas Alliance found itself in 1886 Macune, with his tact and power of persuasion, proved to be the man of the hour. As chairman of the executive committee of the Alliance he was able to secure a meeting of the old officers and the newly chosen leaders of the seceding faction, after which the president and the vice president of the old Alliance resigned and the seceders agreed to hold their charter in abeyance, pending the next meeting of the state Alliance. Macune, now acting president by virtue of the resignation of his superior officers, promptly called a meeting of the state Alliance to be held at Waco in January, 1887, and laid plans to avert the impending split.[28]

By the time of the Waco meeting Macune's plans were fairly well matured. To begin with, he would dazzle the imaginations of Texas Alliancemen and draw attention from

[28] William L. Garvin and S. O. Daws, *History of the National Farmers' Alliance and Co-operative Union of America,* 65, 146–147; Bryan, *Farmers' Alliance,* 11; *National Economist,* 2:196 (December 14, 1889); Morgan, *Wheel and Alliance,* 298–300.

petty squabbles at home by proposing a program of active expansion. For a time he toyed with the idea of joining forces with the Northern Alliance to secure immediate organization on a national scale. This would have been an easy and natural development, for the aims of the two orders were by no means irreconcilable, and in the early eighties the Texas Alliance had been considered a part of the National Farmers' Alliance — for statistical purposes was still so considered. That the return of the prodigal with such accretions as it could obtain in the southern states would have been gladly received, Macune knew well. He even wrote about the matter to Streeter of Illinois and Burrows of Nebraska, past president and president, respectively, of the northern order; but ultimately he decided that union between the two Alliances was not the part of wisdom.

His expressed objections to joining forces with the Northern Alliance were three-fold. First, the Northern Alliance was a loose, nonsecret organization, having at the time no system of fees or dues and being still dependent upon the good graces and charity of its founder, Milton George. Second, colored persons were eligible to membership — a condition of affairs unthinkable in the South. Third, by a ruling that any person raised on a farm was to be considered a farmer, the way was left open for members to be recruited from the nonagricultural classes. What Macune had in mind was a strongly centralized order composed of farmers only, bound together by ties of secrecy and unified in purpose and procedure. Nothing of the sort could be obtained, he felt, under the existing constitution of the Northern Alliance, nor could such changes in that document as he desired be counted upon with any degree of certainty. In any event Macune decided to proceed at once with a new national order to have its main strength in the South. It would thus be in no sense a rival of the Northern Alliance, for it would not enter the same territory. Its purpose should be " to organize the cotton belt of America so that the whole world of cotton raisers might be united for self-protection." [29] At the Waco meeting Macune scored a complete success. Not only was the threatened split in the Texas Alliance averted, but union with a similar order in Louisiana, as a first step towards active expansion, was successfully consummated.

The beginnings of the Farmers' Union, as the Louisiana order was called, dated back to 1880, when a farmers' club had been founded in Lincoln Parish of that state. About 1885 secret work, patterned upon the Granger ritual, was introduced; a constitution, borrowed in the main from the Texas Alliance, was adopted; and a number of new lodges were formed. The Louisiana Farmers' Union was by no means so strong as the Texas Alliance, having in 1887 perhaps not more than ten thousand members, but Macune believed that a fusion of forces would greatly hearten both orders and would pave the way for further expansion. Accordingly, after some correspondence, he dispatched a trusted lieutenant to confer with the Louisiana leaders, and the negotiations thus begun were so successful that delegates from the Farmers' Union appeared at Waco, ready to effect the desired consolidation. The two orders were therefore declared one under the name of the National Farmers' Alliance and Co-operative Union of America, Macune was elected president in his own right, and his program of expansion throughout the cotton belt was made the duty of the hour.[30]

In order further to divert attention from the political dissensions that had rent the Texas Alliance, Macune now proclaimed the doctrine that the Alliance was primarily

[29] Bryan, *Farmers' Alliance*, 11–13, 60; Garvin and Daws, *Farmers' Alliance*, 66; Drew, in the *Political Science Quarterly*, 6:283; *National Economist*, 2:196 (December 14, 1889).

[30] *National Economist*, 1:56; 2:196 (April 13, December 14, 1889); *Appletons' Annual Cyclopaedia, 1890*, 299; Peffer, in the *Forum*, 8:471; Morgan, *Wheel and Alliance*, 293; Garvin and Daws, *Farmers' Alliance*, 46–47; Bryan, *Farmers' Alliance*, 12; Buck, *Agrarian Crusade*, 115 note.

what its new name would indicate, a business organization. Its first purpose must be the protection of the cotton raisers of America. All other occupations were "either organized or were rapidly organizing" in their own defense. Why should the farmer not do likewise? By means of business cooperation something could and should be done to rescue the southern farmer from his present economic distress. When the combined orders came together at Shreveport, Louisiana, for their next meeting, Macune's mind was fully made up. "Let the Alliance be a business organization for business purposes," he said, "and as such necessarily secret, and as secret necessarily non-political." This declaration did not, however, prevent the Alliance from speedily adopting demands not unlike those that only a short time before had so nearly destroyed the Alliance in Texas.[31]

With evangelistic fervor Macune threw himself into the work of spreading the Alliance gospel throughout the South. After the union of the Farmers' Alliance of Texas and the Farmers' Union of Louisiana, articles of incorporation for the new order were filed in the District of Columbia, and a national charter was procured. This helped to fix attention upon the national scope of the work. A board, charged with the task of extending the Alliance into the various southern states, was then appointed, and in the spring of 1887 numerous lecturers and organizers were sent forth. Introducing the Alliance as "a strictly white man's non-political secret business association,"[32] they found little to oppose their efforts. "The farmers seem like unto ripe fruit," they reported, "you can garner them by a gentle shake of the bush."[33] Indeed, the way was often made ready for the Alliance organizer by a revival of interest in the formerly well-nigh extinct Grange or by the appearance, almost spon-

[31] Garvin and Daws, *Farmers' Alliance*, 65–68, 74, 80; Bryan, *Farmers' Alliance*, 17–20; *National Economist*, 1:8 (March 14, 1889); Buck, *Agrarian Crusade*, 116.
[32] *Progressive Farmer*, September 9, 1887.
[33] Garvin and Daws, *Farmers' Alliance*, 50.

taneously, of such new societies as the Farmers' Clubs of North Carolina, which were already under way when the Alliance entered that state. Throughout the South the Alliance workers got results far in excess of their fondest expectations. Macune had read the minds of the southern cotton growers aright. With prices steadily declining, their plight was desperate, and they knew it. They were ready to follow the lead of anyone who could point the way of escape from the ills that beset them, and the Alliance as a cooperative business order seemed to offer a ray of hope. Practically every southern state was fully organized by the end of the year.[34]

In some of the states of the Southwest, particularly in Arkansas, Alliance organizers found the field almost preempted by another farm order, nearly as old as their own, having similar aims and a similar nature. The Agricultural Wheel, as it was called, traced its origin to a meeting at McBee's schoolhouse, ten miles west of Des Arc in Prairie County, Arkansas. There, in February, 1882, seven farmers banded together to form a sort of neighborhood debating club. The arguments that their meetings produced took a political and economic turn, being directed mainly against corruption in politics and monopoly in business. Similar clubs in the same county and others farther away soon came into existence, which talked of the evils of the country store and the oppressive " anaconda " mortgage system by which the Arkansas merchant was protected. In 1883 state organization was achieved, and the name Wheel was adopted. When, in 1885, another farm order known as the Brothers of Freedom, was absorbed, the joint membership was placed at forty thousand. By 1886 this had grown to fifty thousand, and by the end of 1887, aided by the same conditions that had favored the expansion of the Southern Alliance, the

[34] *National Economist*, 1: 8, 297 (March 14, 1889; July 27, 1889); *Progressive Farmer*, April 14, June 2, September 9, 1887; Garvin and Daws, *Farmers' Alliance*, 48–50; Bryan, *Farmers' Alliance*, 14.

Wheel had extended into eight states and could claim half a million members. Obviously a union of forces between the Wheel and the Alliance was the next proper procedure.[35]

It was not easy, however, to unite two orders of such considerable pretensions, and much time elapsed before the desired end could be fully attained. Each order appointed committees to confer. Finally, in December, 1888, both national gatherings were held at Meridian, Mississippi, and consolidation was officially agreed upon. The details of union were to be worked out by a joint committee, however, and the first meeting of the combined order was to be held at St. Louis in December, 1889. By September, 1889, the work of consolidation was declared complete, and the Wheel and the Alliance gave way to the new Farmers' and Laborers' Union of America. In sending out new rituals and secret work much care was exercised to restrict membership to farmers and their natural allies — country ministers, country teachers, and the editors of farm journals. Lawyers, merchants, merchants' clerks, and individuals owning an interest in any banking or mercantile establishment, except a farmers' cooperative store, were specifically debarred. The authority of the new order was in general fully recognized, although complete consolidation was not effected in Arkansas until 1891.[36]

With the chance for friction thus reduced to a minimum, the growth of the Alliance went on apace. Convinced by the arguments of the organizers that the Alliance could help them get out of debt and make money — that it was " the substance of things hoped for, the evidence of things not seen " — southern farmers flocked into it by the hundreds of thousands, if not, indeed, by the millions. By 1890 the most conservative estimate gave the Southern Alliance well over a

[35] Morgan, *Wheel and Alliance*, 60–69; Bryan, *Farmers' Alliance*, 22–25; *Handbook of Facts*, 114; *Appletons' Annual Cyclopaedia, 1886*, 42.

[36] *National Economist*, 1:3; 2:24–25, 99; 4:364 (March 14, September 28, November 2, 1889; February 21, 1891); Drew, in the *Political Science Quarterly*, 6:284; Blood, *Handbook*, 39, 45.

million members, and the number was more frequently fixed
at three millions. Undoubtedly one of the Alliance leaders
claimed too much when he said that " civilized history furnishes nothing to parallel the general and rapid spread of the
Alliance," but there was truly plenty of food for thought in
the fact that the various orders of which the Alliance was
composed had " sprung into existence in widely separated
parts of our country at about the same time, and that they
have grown until they now number in their ranks nearly half
of our entire farming population." [37]

With two separate Alliances at work, one in the South and
the other in the North, consolidation into one gigantic national order seemed the next logical development. Indeed,
for some time the idea of union had been taking root. In
accordance with a previously extended invitation a delegate
from the Southern Alliance, Evan Jones of Texas, attended
an executive committee meeting of the Northern Alliance at
Des Moines in February, 1888, to explain the nature of the
southern order, with a view to effecting a consolidation. The
northern executive committee went on record as favorable to
union, and a delegation was selected to attend the Meridian
meeting of December, 1888. This delegation failed to put in
its appearance, but the proceedings that culminated in the
union of the Wheel and the Alliance as the Farmers' and
Laborers' Union of America clearly had in mind the possibility of drawing the Northern Alliance into the new organization. Southern delegates appeared before the annual
meeting of the National Alliance, held in Des Moines on
January, 1889, to explain the action taken at Meridian and
to work for union. Little objection was offered to the proposition, and the plan presented was referred to the various
state Alliances for further consideration. It was agreed that

[37] *National Economist*, 2:134. 197; 4:185, 198 (November 16. December 14,
1889; December 6. 13, 1890); Polk, in the *North American Review*, 153:11;
Morgan, in the *Forum*, 12:407.

both Alliances should hold their next annual meeting at St. Louis in December, 1889, and the assumption was that at that meeting consolidation could be fully effected.[38]

The plan for the St. Louis meeting was broader, however, than any mere attempt to unite the Northern and the Southern Alliances. For the ambitious leaders of the latter organization nothing short of a confederation of all farmer orders and an agreement with the forces of labor would suffice.

Foremost among the organizations which, it was hoped, would join forces with the Alliances was the Farmers' Mutual Benefit Association. This organization had sprung from a venture in cooperative marketing attempted by some farmers in Johnson County, Illinois, as early as 1883. These farmers found it to their advantage to pool their grain and ship it without the assistance of middlemen. Since meetings were necessary to make arrangements for shipments and to effect settlements, evolution in the direction of a secret farm order was not unnatural. The Farmers' Mutual Benefit Association always featured its cooperative business activities, but it became a lodge somewhat after the Odd Fellow pattern, and it sometimes made its influence felt in politics. By 1887 it claimed, and probably had, about a hundred and fifty thousand members — some estimates were much higher. Representatives from this group were present at the St. Louis meeting.[39]

The Colored Farmers' National Alliance and Cooperative Union was also strong. The Southern Alliance did not admit colored members, but the wisdom of having them lined up in a parallel organization early appealed to some of the white leaders. A southern white man, R. M. Humphrey, who had

[38] *National Economist*, 1:8; 2:72–73 (March 14, October 19, 1889); *Western Rural*, October 5, 1889; February 16, 1890.

[39] *National Economist*, 2:134–135 (November 16, 1889); Blood, *Handbook*, 59; Bryan, *Farmers' Alliance*, 67–68; Chamberlain, in the *Chautauquan*, 13:339; Fred E. Haynes, *Third Party Movements since the Civil War, with Special Reference to Iowa*, 231–232.

been a Baptist missionary among the negroes, is entitled to the chief credit for organizing and extending the work among the colored people. Attracted by the ritual as well as by the possible economic benefits of the order, members flocked to the colored Alliance in prodigious numbers. The first organized group was formed at Houston, Texas, in December, 1886, and by January, 1891, a million and a quarter members were claimed, with a dozen complete state organizations, and locals wherever negro farmers were sufficiently numerous. In 1888 a national organization was established, and the second annual meeting was called for St. Louis at the time designated by the white Alliances for their meeting. Consolidation with the white orders was not anticipated, but since the Colored Alliance was little more than an appendage to the Southern Alliance, close cooperation was inevitable.[40]

The Knights of Labor had had a long, and of late years a none too successful, career. It was invited to send delegates to the St. Louis convention, and it responded by authorizing the attendance of three well-known labor leaders, Powderly, Beaumont, and Wright. Shortly before the St. Louis convention the Knights of Labor had met in general assembly at Atlanta, Georgia, and had listened to addresses by officers of the Georgia State Alliance, which voiced "a desire for affiliation, possibly amalgamation, between the organized tillers and the organized toilers of the country."[41] It was largely in response to these overtures that the Knights were represented at St. Louis. The American Federation of Labor, whose cooperation would also have been gladly received, refused to have anything to do with the Alliances, because they were "composed of employing farmers."[42]

The genuine need for farm relief, which was accentuated with each succeeding year, coupled with the hope of attain-

[40] National Economist, 1:5, 409, 200, 201 (March 14, September 14, December 14, 1889); Drew, in the Political Science Quarterly, 6:287–288.
[41] St. Louis Globe-Democrat, December 5, 1889.
[42] Drew, in the Political Science Quarterly, 6:290; Bryan, Farmers' Alliance, 27.

ing "a more perfect union" of the agricultural orders, led to a great convergence of farmers — delegates and nondelegates — at St. Louis in the early days of December, 1889. The railroads, not much concerned as yet about what the farmers did, offered round-trip tickets for one and one-third fares. Representatives from Dakota to Florida, from Texas to New York, reflecting "every shade of political opinion," put in their appearance. There were high tariff men and low tariff men. There were Democrats, Republicans, Greenbackers, Union Laborites, and Prohibitionists. There were old soldiers who had worn the blue and old soldiers who had worn the gray. The meeting was widely heralded as the "first . . . in the history of the country when the plain farmers from so many States and Territories have been called together for consultation and united action." [43]

But the "plain farmers" were by no means destitute of effective leadership. C. W. Macune, now editor of the *National Economist* of Washington, D. C., official journal of the Southern Alliance, was present as the retiring president of the old Farmers' Alliance and Co-operative Union, with well-laid plans. "His parlor at Hurst's hotel," wrote one reporter, "is the gathering place for all the most influential officers and delegates and he is undeniably dictating the policy of the convention." [44]

Colonel Leonidas L. Polk of North Carolina was also present and willing to lead. Polk was born in Anson County, North Carolina, on April 24, 1837. His biographers were able to boast that he was "one of the Mecklenburg family of Polks, to which belonged Colonel Thomas and James K. Polk," but his parents were inconspicuous farm people who died while Polk was only a boy. Polk's public career began in 1860 when as a young man of twenty-three he was elected to the lower house of the state legislature. He seems, in

[43] *National Economist*, 2:73, 123 (October 19, November 9, 1889); Bryan, *Farmers' Alliance*, 31.
[44] *St. Louis Republic*, December 4, 1889.

common with many other southerners, to have doubted the wisdom of secession, and when war actually came he refused at first to accept a commission, preferring to enlist as a private. He saw military service in the 25th and 43d North Carolina regiments from the outbreak of the war until 1864, when he was again chosen to the legislature, this time as an "army candidate." The next year he sat in the first of the many constitutional conventions held in North Carolina during Reconstruction times.

Always a farmer at heart, Polk advocated at an early date the establishment of a state department of agriculture, and when in 1877 this step was finally taken he became the first commissioner, holding office until 1880. It was while serving in this capacity that he gained the intimate knowledge of southern agriculture that served him so well in his later career. In 1886 he founded a weekly newspaper, the *Progressive Farmer*, in which, from its first issue, he began to urge a union of the farmers of the state in support of such political measures as he thought would best serve their interests. With this project he had already made good progress when organizers sent out by the Southern Alliance entered North Carolina. He accepted the outside aid, however, without hesitation, merged his organization with that of the national order, and became himself the secretary of the North Carolina State Alliance. Polk's pungent editorials in the *Progressive Farmer* and his easy facility as an orator had made him a marked man by the time the St. Louis convention met. Apparently he was in perfect accord with Macune, but he was not the type of individual to be dominated wholly by the ideas of another, and he was quite as ambitious as Macune himself.

Probably the most prominent of the Northern Alliance leaders in attendance was its president, Jay Burrows, who, like Polk, was the editor of a farm journal. Burrows was a reformer of long standing; he had participated in the Gran-

ger movement and had probably done more than any other
man to make the Nebraska State Alliance a success. He was
born at Mayville, Chautauqua County, New York, on March
4, 1833; and as a resident of New York he grew to manhood,
engaged in newspaper work, married, and enlisted for service
in the Civil War. He remained in the army during the full
four years and saw active fighting with Sheridan and Custer
in the Shenandoah Valley. His wife, meantime, removed to
Iowa, and there Burrows went when the war was over to
make his way in the world as a farmer. In 1880 he followed
the frontier still farther into the West and settled near Filley,
Gage County, Nebraska. He became secretary of the state
Alliance, then president of the Northern Alliance, and finally
in 1889 moved to Lincoln, Nebraska, to become editor of the
Alliance, a weekly newspaper that came to have great influence throughout the entire Northwest.

Burrows was at his best in the scathing editorials he wrote
in denunciation of those whom he deemed responsible for the
farmers' woes. He read extensively, thought deeply, and
held opinions of his own on nearly every subject. His tenacious adherence to his convictions, however, made him a
hard man to work with, and he repeatedly fell out with other
reformers who failed to see eye to eye with him. Unlike
Macune he was a man of such high standards of conduct that
his honesty and sincerity of purpose were never seriously
questioned. At the St. Louis convention he forgot neither
the battles and issues of the Civil War nor the differences in
organization and program that eventually were to hold the
Northern and the Southern Alliance apart. Probably Burrows hoped to head the consolidated farm order, in case such
an organization grew out of the St. Louis meeting; but he
was determined at any rate that the southern leaders should
not have things entirely their own way.[45]

[45] Morgan, *Wheel and Alliance,* 285–288; *Appletons' Annual Cyclopaedia,
1890,* 299; *St. Louis Republic,* December 4–7, 1889; Bryan, *Farmers' Alliance,*
31; *Nebraska State Journal,* January 17, 1900.

The official delegates from the Southern Alliance at the St. Louis meeting probably numbered two hundred; the Northern Alliance sent only about seventy-five; and these two numbers reflected not unfairly the relative strength of the orders. The southerners held their meeting in the Entertainment Hall of the Exposition Building, while the northerners met at the Planters' House; but it was confidently expected that, following a few preliminaries, the latter would adjourn to sit with the former. Two months before the convention met, the old Farmers' Alliance and Co-operative Union had ceased to exist, being succeeded by the Farmers' and Laborers' Union, having a newly drawn constitution and a newly chosen set of officers. Into this order the Northern Alliance, or more properly the National Farmers' Alliance, was now called upon to submerge itself, and, superficially at least, there seemed little occasion to resist the call.[46]

Actual consolidation proved, however, to be no easy matter. Soon after the two conventions met they appointed committees on conference to work out the details of organic union, and trouble at once began. Representatives of the Northern Alliance found in the existing constitution of the Farmers' and Laborers' Union at least three insurmountable barriers to consolidation. First, they could not accept the new name, from which, possibly to satisfy the Wheel or to offer a bait to the forces of labor, had been dropped the word "Alliance." The northerners suggested that, instead of the Farmers' and Laborers' Union, the name should be the National Farmers' Alliance and Industrial Union. Secondly, they objected to the exclusion of negroes, and thirdly, they felt that secrecy, which had rarely had a part in the work of the Northern Alliance, should be left optional with each state.[47]

[46] St. Louis Republic, December 3, 4, 1889; National Economist, 2:196, 210 (December 14, 21, 1889).
[47] Nebraska had used a secret ritual for a time. See J. M. Thompson, "The Farmers' Alliance in Nebraska: Something of Its Origin, Growth, and Influence," Proceedings and Collections of the Nebraska State Historical Society, 10:200.

To satisfy these objections the southerners were willing to go a considerable distance. The new name they accepted without objection and wrote into their constitution. They were willing also to strike out the word " white " from their qualifications for membership, leaving to each state the right to prescribe the eligibility of colored persons within its jurisdiction, although it was stipulated that only white men might be elected as delegates to the national legislative body, the Supreme Council. As a last concession the southern order was said to be willing to let negroes into the Supreme Council, and probably if this had been the only obstacle to union, it could easily have been overcome. On the matter of secrecy the South found it harder to give in. The success of the Southern Alliance had always been regarded as due in no small part to the hold that secret work gave the national order over its members. Largely by virtue of this character-istic the Southern Alliance could act with a unity that the Northern Alliance had never been able to achieve. It was suggested that secrecy might properly be introduced gradu-ally. Let the states that were not ready to receive secret work at once have a year's time in which to prepare for it.[48]

These concessions failed to satisfy at least a majority of the northern delegates, and they were rejected. As the chief excuse the northern conferees pled that the representatives of two northern states, Iowa and Minnesota, were without instructions on the matter of union and would find it neces-sary to refer the subject to their state Alliances. This situa-tion, however, should have furnished no insurmountable barrier to union, for the accession of the Iowa and Minnesota Alliances could have been delayed indefinitely.

Determined not to bear the entire odium of failure in the negotiations, the northerners proposed as an alternative to consolidation, temporary confederation. While keeping the

[48] *St. Louis Globe-Democrat*, December 7, 1889; *National Economist*, 2:213, 215 (December 21, 1889); Drew, in the *Political Science Quarterly*, 6:284-285.

two orders intact and separate for the present, they suggested that a third united body of some kind be formed under a constitution to be adopted in joint session. This common organization, it was hoped, might be so perfected in the future as to form the basis for a complete amalgamation of forces. But the southern delegates turned the new plan down " for lack of time," and, out of patience at the rejection of what seemed to them the most liberal terms, they voted " to stand firmly on the propositions made " and to invite the northern delegates " to appear before this body for secret work." [49]

Needless to say, this ended immediate prospect of union, although the somewhat petulant invitation of the southern convention was actually accepted by the Kansas and Dakota representatives, who claimed full authority to act on behalf of their constituencies. Their action, subsequently indorsed by local state conventions, resulted in the virtual secession of the Kansas, North Dakota, and South Dakota Alliances from the northern order and their absorption into the newly formed Farmers' Alliance and Industrial Union. Some hope was entertained that, when the various state conventions were held elsewhere in the North, the example of the seceders would be generally followed, and thus by steady accretions union would ultimately be obtained. But these hopes were not justified by the events. Efforts to join the two orders continued for some time, but they never accomplished anything.[50]

The reasons for the failure of the Northern and the Southern Alliance to combine are not fully revealed by the unsuc-

[49] St. Louis Globe-Democrat, December 7, 1889; National Economist, 2:195, 212, 215, 229 (December 14, 21, 28, 1889); Alliance, December 14, 1889; January 11, 1890. The National Economist, 2:210–218 (December 21, 1889), printed the official proceedings of the southern order of the Alliance in full.

[50] National Economist, 2:215–217, 229, 242; 3:232 (December 21, 28, 1889; January 4, June 28, 1890); Alliance, January 11, 1890; Drew, in the Political Science Quarterly, 6:285; Blood, Handbook, 1. In Kansas a strong southern order had existed side by side with the northern, and a union of forces was highly desirable.

cessful negotiations looking towards union. Continuing sectional prejudice undoubtedly played a far greater part than was admitted. The southern order, being greatly superior in numbers, could afford to be generous in its overtures, and it was. The northern order, drawing its membership in no small part from the veterans of the Civil War, looked askance at anything that meant submission to southern supremacy. One hostile Kansas editor shrewdly branded the whole Alliance movement as a " rebel yell." With taunts like this in prospect how could the Northern Alliance be expected to forego a separate existence? Confederation it might survive, but consolidation might well be its death. Moreover, there was a real question as to the identity of interest between northern and southern farmers. Some northern questions were of comparatively small consequence to the South, and some problems that seemed vital to the South meant far less to the North. Would the two sections ever be able to work together in harmony within one organization? Would not the southern element, greater in size and firmly intrenched in its control of the national offices, shape the policy for the whole? This reflection rather than mere desire for place may have inspired the determination of some of the northern leaders to seek prominent offices in the national order should consolidation take place.

One further consideration undoubtedly influenced many Alliancemen against union. Already the idea of a third party had appeared, and in the North it was receiving serious thought. But apparently the South was wedded irretrievably to the one-party system. Perhaps there were southerners who could look with complaisance upon the formation of a new party, but even to these the matter was one of the greatest delicacy, to be mentioned only with bated breath. Here was a problem that, in its relation to amalgamation, southerner as well as northerner must well consider. Would union prove embarrassing to those who wanted to work

towards independent political action? Would union prove
embarrassing to those who hoped to prevent independent
political action? Possibly each question might be answered
in different ways by different individuals, but the mere suggestion of a third party caused a certain amount of hesitation
on all sides. When finally the third party did materialize, it
so supplanted the Alliances in interest and importance that
no one cared particularly whether they were united or not.[51]
From the standpoint of diminishing the number of farmer
and labor orders, the St. Louis meeting was a complete failure. The Northern and the Southern Alliance each went its
respective way when the meeting was over, just as before.
The Colored Alliance was no more and no less completely
under the domination of the Southern Alliance than it had
always been. The Farmers' Mutual Benefit Association resisted all overtures towards organic union and agreed merely
upon mutual cooperation. The Knights of Labor scarcely
considered union with the farmers on any terms. "The
Knights and the farmers ask for the same things," said Powderly, "only the Knights ask for more, such as the environment of the farmer does not call for. . . . Hence, complete
consolidation is not feasible."[52]
This gathering did reveal, however, a notable willingness
on the part of both northern and southern farmers to work
together for the attainment of common ends and to accept
whatever assistance they could get from organized labor.
They seemed to agree with one accord that the time was
ripe for a series of political demands that, if carried into
effect, might accomplish a sort of "bloodless revolution."
The Northern and Southern Alliances drew up separate platforms, but the two documents exhibited such a surprising

[51] St. Louis Globe-Democrat, December 7, 1889; Alliance, January 11, 1890;
December 20, 1890; Blood, Handbook, 39; McVey, Populist Movement, 137;
Herman C. Nixon, "The Cleavage within the Farmers' Alliance Movement,"
Mississippi Valley Historical Review, 15:22-23.
[52] St. Louis Republic, December 5, 1889; National Economist, 2:242 (Janu-
ary 4, 1890); Bryan, Farmers' Alliance, 75.

degree of similarity that they might almost have been inter-
changed. There was, to be sure, a difference in emphasis.
The Southern Alliance, for example, was interested more in
financial problems than in anything else, whereas the Northern Alliance was terribly in earnest about the railroads
and the disappearance of the public lands. But both organizations called for financial reforms of an inflationist
nature, both protested against the excessive landholdings of
the railroads and against any landholding by aliens, and
both asked that the government take over and operate all
means of transportation and communication. Barring the
fact that the Northern Alliance seemed to have overlooked
the free-silver issue, there was on the three fundamental
issues of land, transportation, and finance virtually no North
and no South. Each order, indeed, was consciously attempting to state a creed that would satisfy the greatest possible
number, whether within or without its organization. Southern Alliancemen were far from unanimous on the desirability
of government ownership of the railroads, and yet they put
such a clause into their platform, no doubt merely to satisfy
the wishes of the northern farmers and the laborites. Both
sets of demands trimmed a good deal on the tariff, although
difference of opinion on this subject was mainly confined to
the North.[53]

The Southern Alliance scored the sensation of the meeting by announcing publicly and officially a plan of cooperation between "the millions who till the soil" and "the millions who consume the product of their labor." This was
embodied in a wordy agreement that preceded the formal
demands: [54]

The undersigned committee representing the Knights of
Labor having read the demands of the National Farmers' Alli-

[53] Both platforms are given in full, with comments by Drew, in the *Political
Science Quarterly*, 6:291–293. See also Appendix A of this work.
[54] *St. Louis Republic*, December 3, 7, 1889; *National Economist*, 2:214–215
(December 21, 1889); Bryan, *Farmers' Alliance*, 33–34.

ance and Industrial Union which are embodied in this agreement hereby endorse the same on behalf of the Knights of Labor, and for the purpose of giving practical effect to the demands herein set forth, the legislative committee of both organizations will act in concert before Congress for the purpose of securing the enactment of laws in harmony with the demands mutually agreed.

And it is further agreed, in order to carry out these objects, we will support for office only such men as can be depended upon to enact these principles in statute law uninfluenced by party caucus.

To emphasize further the newly cemented bond with labor, the constitution of the Southern Alliance was amended so as to permit mechanics to join the order. One might even think that the absorption of the Knights was the ultimate goal of the farmers, whatever the laborers may have thought. While the Northern Alliance could point to no formal agreement with a labor organization, it stated in no uncertain terms its sympathy " with the just demands of labor of every grade " and it recognized " that many of the evils from which the farming community suffers oppress universal labor, and that therefore producers should unite in a demand for the reform of unjust systems and the repeal of laws that bear unequally upon the people." [55]

The Alliance demands at St. Louis were not consciously designed to be the platform of a new political party. Political action was contemplated, but obviously it was still the plan to work within existing parties. Men should be chosen to office who would do their utmost to enact into law the policies outlined, if the Alliance-Knights' declaration were to prove effective, wholly regardless of the will of a party majority. Some of the more regular party men were aghast at

[55] Drew, in the *Political Science Quarterly*, 6:285, 292. The clause in the Southern Alliance constitution adopted at St. Louis " opened membership to farmers, preachers, teachers and doctors residing in rural districts as well as to mechanics and editors of agricultural journals." In 1890, however, this was again amended, and mechanics to be eligible must thereafter be " country " mechanics, and the editors of " farm " journals must profess Alliance views. See also the *St. Louis Globe-Democrat*, December 7, 1889.

the idea of ignoring party discipline as expressed in the will of a caucus, but to others this seemed merely to be good sense. Let the farmers of America, and the laborers, too, if they would, make use of the power they possessed. Let them force the government, whether in state or nation, to redress just grievances. It was in this expression of a common class program that the St. Louis action was significant. Union of the various orders was not accomplished; indeed, union was deemed by many to be wholly undesirable. But a unity of purpose, never so well expressed before, was definitely asserted.[56]

Following the St. Louis gathering something closely akin to a unity of leadership was also accomplished. With some justification the southern leaders now assumed the right to direct the whole farmers' movement. The northern leaders furnished them no real rivalry, for the loosely drawn ties of the Northern Alliance made only for impotence in matters of national importance. On the other hand, the recently revised constitution of the Southern Alliance retained the highly centralized organization that had proved so effective in promoting the rapid growth of the order and that, presumably, would prove equally effective in pushing forward the various items of the Alliance program. Delegates — two from each state and one additional for each ten thousand members — constituted the Supreme Council, which met each year to make laws and determine policies for the whole order. Between sessions a president, assisted by a powerful executive committee, carried on the work. The national officers were given power, were well paid, and were expected to accomplish something.[57] For the presidency the St. Louis meeting chose Colonel L. L. Polk of North Carolina.

The Southern Alliance had not only a constitution better

[56] St. Louis Republic, December 6, 7, 1889; National Economist, 2:305 (February 1, 1890).
[57] National Economist, 2:213 (December 21, 1889); Drew, in the Political Science Quarterly, 6:285.

adapted to the needs of the moment, but it had also the advantage of greater numbers. At the time of the St. Louis gathering it was already two or three times as large as the Northern Alliance, and following that meeting it drew the Dakotas and Kansas into its fold, leaving the northern order still weaker in comparison. The southern leaders determined, moreover, to give up their former policy of keeping out of northern territory and to "make a grand break across the Mason and Dixon line." As President Polk expressed it, "We took the farmers and laborers of the North and of the East and of the West by the hands, and to-day we are trampling sectionalism under our feet." In spite of some opposition from the "Open" Alliance, as the National Farmers' Alliance was now often called to distinguish it from the secret Farmers' Alliance and Industrial Union, the southern workers pushed successfully into states as far west as Washington, Oregon, and California, and as far east as Ohio, Pennsylvania, and New York. The rather ably edited *National Economist* poured Alliance propaganda over the whole country and published without discrimination such news of the activities of all the farmer organizations as it could gather. Perhaps the expressed desire of the southern leaders "to break down sectional lines" was not wholly gratified, but they did succeed in focusing public attention, in the North as well as in the South, upon themselves. What they said and did was generally construed to represent the will of most Alliancemen.[58]

[58] *New York Times*, December 1, 1890; Blood, *Handbook*, 47–58.

CHAPTER FIVE

ALLIANCE ACTIVITIES

THE work of the farm orders, according to a prominent Allianceman, was fourfold: "first, social; second, educational; third, financial; fourth, political." [1]

Certainly the Alliances and their imitators gave abundant opportunity for neighborhood gatherings of the type common in Granger times — lodge meetings, picnics, conventions. Meetings of the suballiances, or "lodges," were held as often as once or twice a month. In the South a fairly elaborate secret ritual, and elsewhere some definite order of procedure, gave foundation to the program; but special features such as music, speeches, and debates were usually provided to enhance the interest of the occasion. Holidays and anniversaries were made the pretexts for all-day picnics or rallies; and the chronic need of funds suggested box suppers and entertainments. Committees were appointed to carry aid and comfort to the sick or afflicted members of the order and to relieve the destitute. County, district, and state meetings — "conventions" — were attended not only by the authorized delegates but also by hundreds of other farmers and their families, who thus satisfied a natural craving to visit the local metropolis. [2]

---

[1] H. L. Loucks, "Alliance Business Effort in Dakota," *National Economist*, 1:21 (March 14, 1889).

[2] Morgan, *Wheel and Alliance*, 256; Barnhart, "Farmers' Alliance and People's Party," 131-133. Further light is thrown on the subjects discussed in this chapter by Raymond C. Miller in "The Populist Party in Kansas," an unpublished doctoral dissertation submitted to the University of Chicago in 1928.

These various activities met a real and obvious social need, but they lent themselves also to more strictly educational purposes. Even in the South, where the ritualistic side of the Alliance was most strongly emphasized, the Alliance leaders made it a point to disseminate as widely as possible scientific agricultural information. There, as elsewhere, visiting officers and lecturers urged upon the farmers the necessity of crop rotation; they pointed out the wisdom of careful seed selection; they set forth the merits of newly devised agricultural machines; they deplored the lack of crop diversification; and they scolded the farmers roundly for their generally bad business methods.[3]

Ordinarily every local organization had an officer known as the "lecturer," whose duty it was to suggest subjects for discussion and to take the lead in expounding them. Spurred into action by the efforts of the lecturer and the attractiveness of the subjects, Alliance members learned to express themselves in public. They learned also to seek ammunition for their speeches and debates in books and papers that they might otherwise never have read. Their horizons widened; the variety of the topics they discussed increased. Local study groups were organized which, by "taking up the questions that were agitating the minds of the people and discussing them in an earnest manner," so aroused the interest of the participants that the meetings often lasted far into the night.

Jay Burrows advocated that the Northern Alliance, of which he was president, establish a " general system of home culture, somewhat on the plan of the Chautauqua," with organized classes and a supervised course of study. The lifetime of the Alliance was too short to permit of putting so ambitious a project into full effect, but circulating libraries were formed in the various locals, which introduced to the readers such exciting literature as Bellamy's *Looking Backward*, Donnelly's *Caesar's Column*, Powderly's *Thirty Years*

[3] Arnett, *Populist Movement in Georgia*, 78.

of Labor, Peffer's *The Farmer's Side,* and Weaver's *A Call to Action;* while questions and answers based upon these and similar books were distributed through the medium of the Alliance press. Some magazines, notably the *Arena* and the *Forum,* which gave considerable space to the farmer's point of view, were widely read. Like the Grange before it, the Alliance became a great "national university." [4]

The educational work of the Alliance led to similar activities under other auspices. In North Carolina farmers' institutes were organized by state authority "to set farmers studying their vocations"; [5] and especially in the older states of the South a strong demand was made that the funds available for the use of agricultural colleges under the Morrill Act of 1862 should not be diverted to serve other ends. [6] A National Farmers' Congress, composed of one farmer from each congressional district appointed by the various state executives, met yearly, beginning in 1881; and in the later eighties, when the Alliance was strong and agricultural distress acute, the Congress repeatedly spoke out its mind on the causes of the farmers' ills and the remedies that should be forthcoming. The creation in Washington of a national Department of Agriculture, the business of which in part was to help educate the farmer, may be attributed very largely to the activities of these Farmers' Congresses. [7]

During the Alliance period farm papers increased amazingly in numbers and in circulation; and while many of them came to devote a disproportionate amount of space to politics, practically all of them gave serious attention also to the business of educating the farmer along other lines. The most notable of the Alliance papers was the *National Economist* of Washington, D. C., official journal of the Southern Alliance.

[4] Walker, in the *Annals of the American Academy of Political and Social Science,* 4:793; Barnhart, "Farmers' Alliance and People's Party," 134–136.

[5] *Greensboro Daily Record,* November 22, 1890.

[6] *Progressive Farmer* (Raleigh), February 17, 1886; June 9, 1887; Arnett, *Populist Movement in Georgia,* 121; Simkins, *Tillman Movement,* 57.

[7] Drew, in the *Political Science Quarterly,* 6:290; Walker, in the *Annals of the American Academy of Political and Social Science,* 4:792; Haynes, *Third Party Movements,* 227.

When this order began its program of rapid expansion throughout the South, after the Meridian meetings, ten prominent members subscribed two hundred dollars each towards the formation of a stock company that should undertake to provide an Alliance newspaper. The money subscribed was not all paid in, but the paper was nevertheless started, in March, 1889, with C. W. Macune as its editor.[8] It appeared weekly, well printed and well edited, and for several years it exerted an enormous influence in farmer circles.[9]

Other reform papers of consequence were the *Non-Conformist*, first edited at Winfield, Kansas, by two brothers named Vincent, but later moved to Indianapolis;[10] the *Progressive Farmer* of Raleigh, North Carolina, edited by Colonel L. L. Polk, for several years president of the Southern Alliance; and the *Farmers' Alliance* (originally the *Alliance*), published at Lincoln, Nebraska, by Jay Burrows, for a time president of the Northern Alliance. The number of minor Alliance sheets was legion; there were said to be a hundred and fifty in Kansas alone.[11] Organized finally into a National Reform Press Association, these papers, perhaps a thousand strong, must have had a total circulation reaching into the hundreds of thousands. Sometimes their methods of securing subscribers helped along the sale of books that the editors thought the farmers should read. For example, in 1889 the *Alliance* offered a copy of Bellamy's *Looking Backward* and a year's subscription to the paper for a dollar and a quarter. The book alone in paper covers could be obtained for fifty cents.[12]

[8] *National Economist*, 1:7 (March 14, 1889).
[9] *St. Louis Globe-Democrat*, November 4, 1892; *People's Party Paper* (Atlanta, Georgia), April 28, 1893.
[10] Barr, in *Kansas and Kansans*, 2:1136; Stewart, in the *Indiana Magazine of History*, 14:353.
[11] Barr, in *Kansas and Kansans*, 2:1141. According to the *National Economist*, 7:328 (August 6, 1892), there were nearly nine hundred such papers in the United States in 1892.
[12] The *Alliance* (Lincoln), December 14, 1889; Walker, in the *Annals of the American Academy of Political and Social Science*, 4:794.

It is difficult to estimate the importance of all this educational work and its accompanying newspaper propaganda.
Perhaps a contemporary observer was right when she said:

People commenced to think who had never thought before,
and people talked who had seldom spoken. On mild days they
gathered on the street corners, on cold days they congregated in
shops and offices. Everyone was talking and everyone was thinking. . . . Little by little they commenced to theorize upon their
condition. Despite the poverty of the country, the books of
Henry George, Bellamy, and other economic writers were bought
as fast as the dealers could supply them. They were bought to
be read greedily; and nourished by the fascination of novelty and
the zeal of enthusiasm, thoughts and theories sprouted like
weeds after a May shower. . . . They discussed income tax and
single tax; they talked of government ownership and the abolition of private property; fiat money, and the unity of labor; . . .
and a thousand conflicting theories.[13]

Out of this "intellectual ferment" came the demands for
reform that found their way into Alliance and Populist national platforms and in a surprising number of instances
ultimately into state or national law.

Accompanying these various social and educational activities, oftentimes springing directly from them, were certain
other activities of a strictly business nature. Alliance leaders
were quick to point out that other occupations were rapidly
organizing to protect their business interests and that to
compete with them on equal terms the farmers, too, must
organize. "Let the Alliance be a business organization for
business purposes," was an oft-repeated slogan.[14]

Early Alliance experiments in business, however, were
along fairly simple lines. One type of cooperative effort
aimed merely to secure from some particular merchant in
each county or locality an agreement to sell to members of

[13] Elizabeth Higgins, *Out of the West*, 133, 136.
[14] *National Economist*, 1:8 (March 14, 1889); Garvin and Daws, *Farmers'
Alliance*, 65–68, 80.

the order on especially favorable terms in return for assurance that they would not trade elsewhere. Such agreements were not easily enforced and were rarely long-lived.[15] Cooperative stores, elevators, and gins were frequently undertaken by local alliances, the funds needed to start such enterprises being subscribed by interested members. Sometimes these cooperatives underbid their competitors by as much as twenty-five or fifty per cent, and when they were well managed they were able to exist over a period of years. For the most part, however, they had to confine themselves to a strictly cash business, whereas what the average farmer required was credit.[16]

Efforts at cooperative marketing were also common. In 1886 the Texas State Alliance made an earnest effort to market cotton directly. Local Alliancemen were advised to bulk their cotton and on special sale days to invite in the city buyers. It was thought that buyers would compete against one another and thus raise the price; but buyers from a distance failed to show up, and the effect of the scheme on prices was negligible, although it did stimulate the establishment of Alliance warehouses.[17] More successful for a time were the attempts of the Farmers' Mutual Benefit Association in Illinois to pool grain for shipment directly to the central markets.[18] Cooperative fire, hail, and life insurance was tried out by the Dakota Alliance and by a few others with good results temporarily; and in 1890 a National Alliance Aid Association, operating on the Dakota plan, was opened in Washington, D. C.[19]

[15] N. B. Ashby, *The Riddle of the Sphinx*, 375; Edward Wiest, *Agricultural Organization in the United States*, 464.

[16] Ashby, *Riddle of the Sphinx*, 375; Drew, in the *Political Science Quarterly*, 6:306; Garvin and Daws, *History of the National Farmers' Alliance*, 153–158; *Farmers' Alliance*, March 29, 1890; Haynes, *Third Party Movements*, 313; Arnett, *Populist Movement in Georgia*, 80.

[17] Wiest, *Agricultural Organization in the United States*, 464–465.

[18] Blood, *Handbook*, 59.

[19] The *Alliance*, June 26, August 21, 1889; *National Economist*, 1:21; 2:65; 3:28 (March 14, October 19, 1889; March 29, 1890).

More important than these sporadic efforts at cooperation were the so-called business agencies. The "agency" in
some cases consisted only of an Alliance purchasing agent
who accepted cash orders for goods and bought for cash
direct from wholesalers and manufacturers, merely adding a
commission to cover the cost of operation and making no
attempt to accumulate capital. This system was especially
successful in Iowa, Minnesota, and Nebraska, although some
trouble was experienced in obtaining certain lines of goods
and in communicating to distant purchasers the proper instructions as to the operation of complicated machines.[20]
Sometimes, as in North Carolina and in Dakota Territory,
business agencies were started as joint stock companies,
shares being sold in small sums to Alliance members to secure a modest amount of capital with which to begin operations. Usually county agents were appointed to assist the
state agents in the ordering and the distributing of goods.
These state agencies did a brisk business among those farmers who had the cash to patronize them.[21]

Still more ambitious were the efforts of the Southern
Alliance under the leadership of Macune "to organize the
agriculturalists of the cotton belt for business purposes."
According to Macune, experience proved that it was not
sufficient for the farmers of localities to act together or even
for the farmers of each state to unite. All the cotton growers
must come into a "strong, solid, secret, and binding organization," welded together for the express purpose of
"breaking the power of monopoly."[22] Apparently Macune
had no scruples about creating a farmers' monopoly if he
could, especially a monopoly of the world's supply of cotton.

[20] *Farmers' Alliance* (Lincoln), November 8, December 27, 1890; Ashby,
Riddle of the Sphinx, 374.
    [21] *National Economist*, 1:14, 20 (March 14, 1889); Blood, *Handbook*, 46, 52,
53; Drew, in the *Political Science Quarterly*, 6:286; *Appletons' Annual Cyclopae-
dia, 1890*, 299; *Proceedings of the Fourth Annual Session of the North Carolina
State Alliance*, 1890, pp. 13–15.
    [22] Garvin and Daws, *Farmers' Alliance*, 67, 74; *National Economist*, 1:8
(March 14, 1889).

By this means he would force purchasers to pay the farmers a fair price. Similarly, the farmers, acting together,
might refuse to buy the commodities they needed, except at a
fair price. The tables would be turned. Instead of being
forced to take whatever prices were offered for his goods, the
farmer would himself, through the Alliance organization,
be able to dictate his prices to the world. Instead of being
forced to pay whatever prices were demanded of him for
supplies, the farmer would himself, through the Alliance,
state what prices he would pay.

As a first step towards the realization of this desirable
state of affairs a Farmers' Alliance Exchange was organized
in Texas under Macune's direct supervision. The plan was to
sell farm produce and to buy farm supplies through the Exchange headquarters at Dallas. The capital needed to inaugurate this business was to be obtained by levying an
assessment of two dollars upon each member of the state
Alliance; and a board of elected trustees was constituted to
direct the work of organization. The scheme raised high
hopes that the pernicious crop-lien system, which left so
many Texas farmers to the tender mercies of the country
storekeepers, might now be brought to an end. Dreamers
even conjured up the vision of Alliance cotton and woolen
mills, of Alliance implement and wagon factories, of an Alliance printing press, and of an Alliance university.[23]

In September, 1887, provided with temporary quarters by
the city of Dallas, the Texas Exchange opened its doors. The
city also promised a building site and ten thousand dollars
cash subsidy. The Exchange at first confined its business to
marketing cotton and grain on a commission basis and to
purchasing farm implements at a heavy discount, which was
passed on to the farmers. Gradually dry goods, groceries,
and other farm necessities were added to the list of articles
bought and sold. It was speedily discovered, however, that

[23] Ousley, in the *Popular Science Monthly*, 36: 821–822.

as long as the business of the Exchange was kept on a strictly
cash basis, it was of no particular use to the ordinary farmer,
for the farmer with cash was a notable exception. To meet
this situation a plan for credit sales was hit upon. For this
purpose each suballiance was to act as a unit, the members
to make a schedule of their joint needs and to execute a joint
note, bankable supposedly because of the signatures of responsible farmers, who protected themselves in turn by crop
liens.

This scheme proved to be the undoing of the Texas Exchange. Notes enough were executed, and upon them the
Exchange contracted to deliver goods; but the banks refused
money upon these notes as security except at a ruinous discount. In Alliance quarters it was claimed that "those who
controlled the moneyed institutions of the State either did
not choose to do business with us or they feared the ill will
of a certain class of business men who consider their interests antagonistic to those of our order and corporation." But
the bankers held that any other course on their part would
have been sheer madness. The Exchange, they observed, had
never been able to collect much of the money it had expected
to raise by assessments upon Alliance members; and it was
attempting to do a far larger business than its meager capital warranted. As for the suballiance notes taken in, the
bankers held that they were worth only from twenty-five
cents to seventy-five cents on the dollar — never more.[24]

The Texas Exchange managed to continue in business for
about twenty months. It erected in Dallas a four-story
building, which not only housed the Exchange itself but
served also as headquarters for the Texas State Alliance and
its official newspaper, the *Southern Mercury*. Perhaps fifty
thousand dollars worth of merchandise was carried in stock,
and the first year it was claimed that a million-dollar business was conducted.[25]

[24] Ousley, in the *Popular Science Monthly*, 36:823–825; Wiest, *Agricultural
Organization in the United States*, 466–477.
[25] *National Economist*, 1:16 (March 14, 1889).

But eventually the Exchange collapsed. When signs of its impending doom could no longer be ignored, a frantic effort was made to collect the two-dollar assessment fee from each of the two hundred and fifty thousand members of the Alliance in Texas, but from this source only fifty or sixty thousand dollars, all told, was ever realized.[26] This sum, together with the ten thousand dollars paid in by the city of Dallas, constituted the entire capital upon which the Exchange could draw. An expert accountant who went over the books after the failure pointed out that a capital of half a million dollars would have been required to conduct such a business on sound principles. Moreover, the profit charged — ten per cent — was altogether too small and barely paid the cost of handling, with no allowance for overhead. Macune, who had served as business manager of the Exchange before he became editor of the *National Economist,* came in for much criticism when the bankruptcy of the Exchange was announced. He was charged with having robbed it of over a million dollars, but he was probably guilty of nothing worse than bad business methods.[27]

While the Texas Exchange was still in operation, it was cited by Alliance organizers everywhere in the South as an example of what the Alliance in business could do; and an effort was made to introduce the " Macune Business System " into every southern state. An exchange organized in Mississippi in February, 1888, was reported within a month to be doing a business of about seven hundred dollars a day. In June, 1888, steps were taken in Georgia to organize an exchange on the Texas model, but with an interesting modification of the credit system. Whereas in Texas the state Alliance itself accepted the joint notes executed by local Alliancemen, in Georgia the exchange was to sell for cash only, and county trade agents were charged with the busi-

[26] Ousley, in the *Popular Science Monthly,* 36: 826.
[27] *National Economist,* 1: 297, 386; 3: 396; 5: 102 (July 27, 1889; August 30, 1890; May 2, 1891); Modern I. Argus (pseud.), *Minor Chronicles of the Goodly Land of Texas,* a pamphlet in the Library of Congress that purports to expose Macune. See also Ashby, *Riddle of the Sphinx,* 376–377.

ness of providing credit for needy members through the
local Alliances. Credit purchasers must give to the suballiance notes secured by the same sort of liens on their crops
or lands that a prudent merchant would require, and the
county agent, by discounting these notes at banks, could
obtain the cash needed to deal with the state exchange.[28]
The Georgia Exchange turned out to be one of the strongest
in the South and was said during its first year of existence to
have saved its patrons $200,000 in fertilizers alone. In spite
of much opposition from the merchant class, it continued in
operation until 1893.[29] In July, 1889, a Farmers' Alliance
Exchange was organized in South Carolina,[30] and in August
of the same year the *National Economist* reported that,
while there were only four perfectly organized state exchanges, a majority of the states had business agencies.[31]
So numerous were the agencies and exchanges that a State
Business Agents' Association was organized at St. Louis
during the December, 1889, meeting of the Alliances in that
city,[32] and next year at Ocala some eighteen state agencies
and exchanges were represented.[33]

The idea of organizing one great national agency to head
the state agencies and exchanges was natural enough. According to one enthusiast, such an institution "with a central head, extending its branches and striking its roots into
every nook and corner of the land, and embracing as its
members all who are interested in the success of agriculture,
became an absolute necessity."[34] The secretary of the State
Business Agents' Association at Ocala voiced a common
sentiment when he said:

[28] *National Economist*, 1:5, 74, 75 (March 14, April 20, 1889).
[29] Arnett, *Populist Movement in Georgia*, 80–81.
[30] Blood, *Handbook*, 57.
[31] *National Economist*, 1:326 (August 10, 1889).
[32] *Ibid.*, 2:217, 218 (December 28, 1889).
[33] *Proceedings of the Supreme Council of the National Farmers' Alliance and Industrial Union*, 1890, pp. 34–35. Cited hereafter as *Ocala Proceedings*.
[34] Garvin and Daws, *Farmers' Alliance*, 86.

Our enemy cannot meet us successfully if we stand united, but if every agent attempts to work out his problem single handed and alone, each will fall an easy prey to the powers of monopoly. I am convinced that we have gone as far as we can as individual agents.[35]

But Macune himself, sobered by the disaster that had overtaken his Texas Exchange, now held that there was no point in establishing a national exchange — that nothing more than cooperation between the various state exchanges should be attempted. Indeed, his attention was now riveted upon the working out of a political remedy for the farmers' ills.[36] At its Ocala meeting the Southern Alliance recommended the plan of the Georgia Exchange as " the best and most successful," but no further steps were taken by the Southern Alliance itself or by any other organization towards the establishment of the once much-heralded national agency.[37]

It is difficult to estimate the importance of the Alliance business ventures. The state agents in session at Ocala claimed that the annual business done by the various members of their association amounted in 1890 to about ten million dollars,[38] and if this was the case they doubtless had saved the farmers immense sums. Stores and other less pretentious cooperative ventures likewise helped as long as they lasted, not only directly but also indirectly, by causing competitors to reduce their prices and to give better service.

But eventually practically all the cooperatives disappeared. They met with much discrimination on the part of wholesalers and manufacturers; they had difficulties with the railroads that private dealers escaped; they often suffered from the bad judgment of managers who bought unwisely and sold too cheaply; they were never able to solve the problem of selling on credit; and they could rarely depend upon

---

[35] *Ocala Proceedings*, 47.
[36] *National Economist*, 2:198 (December 14, 1889).
[37] *Ocala Proceedings*, 49.
[38] *Ibid.*, 35.

the loyalty of their customers when, as frequently happened, cutthroat competition set in.[39] The Southern Alliance conducted one successful large-scale boycott. When the price of jute bagging soared beyond all reason, supposedly because of the formation of a trust, the farmers of the cotton belt stood together under Alliance leadership, substituted cotton bagging for jute, and brought the monopolists to terms.[40]

Almost from the first Alliance leaders, in the North and in the South, confessed that their respective organizations hoped to bring about certain political reforms. They talked much of their educational propaganda, their social activities, and their business ventures; but for results they came more and more to depend upon favorable legislation. The farmer program of reform differed somewhat from state to state and from section to section, but in general it expressed a debtor protest against what was esteemed to be unreasonable oppression by the creditor classes. Railways should be compelled to lower their rates and to abandon various other evil practices, either by laws directly devised for the purpose or by railway commissions endowed with adequate authority. Land monopolies should be broken down. Trusts of every sort and kind should be destroyed. Interest rates should be lowered, laws protecting mortgage holders made less stringent, and national banks abolished outright. More money — paper or silver or both — should be supplied to help carry the increasing volume of business that the growth of the country made inevitable.[41]

In the earlier years responsible Alliancemen were careful to point out that the Alliance was strictly nonpartisan, both in the sense that it did not necessarily prefer one of the

[39] Ashby, *Riddle of the Sphinx*, 375–378.
[40] *Progressive Farmer*, August 7, 1888; *Alliance*, October 5, 1889; *National Economist*, 1:296, 297 (July 27, 1889); Blood, *Handbook*, 49; Arnett, *Populist Movement in Georgia*, 104.
[41] Drew, in the *Political Science Quarterly*, 6:293, gives a summary of Alliance demands. See also Morgan, *Wheel and Alliance*, 141–144, and Ashby, *Riddle of the Sphinx*, 408–411.

political parties to another and in the sense that it had no intention of becoming itself a political party. Candidates for membership in the Alliance were assured that nothing would ever be required of them that would in any way conflict with their political or religious views. That all political parties could be represented in Alliance ranks was considered a matter for self-congratulation. In theory at least Alliance demands were never to be directed at a particular party but at all parties.[42] Individual Alliancemen were to exert their utmost influence in the party to which they happened to belong to have candidates nominated and platforms adopted that were in accord with Alliance principles.[43] The assumption was that if the farmers as individuals could only be induced to stand up for their rights, they could insure that both parties would respond in a satisfactory manner.[44]

This rather specious reasoning, by which the Alliance as an organization was held to be entirely free from political activity although it encouraged Alliancemen as individuals to work hard for the realization of Alliance political demands, gradually gave way; and as time went on the Alliances, Northern and Southern, came to admit with increasing frankness their direct interest in politics. In the South, however, out-and-out political action could be taken only with the greatest caution, for there anything that might threaten the political solidarity of the whites was regarded with grave suspicion. In at least two southern states, Texas and Arkansas, the risks attendant upon such a course were soon demonstrated.

As previously stated, the Texas Alliance in its early days had dabbled somewhat in politics. The earliest organization had gone to pieces in the seventies because of its affinity for

[42] Chamberlain, in the *Chautauquan*, 13:339.
[43] *Farmers' Alliance*, November 29, 1890. See also *Appletons' Annual Cyclopaedia, 1890*, 299, and the *Progressive Farmer*, June 16, September 22, 1887.
[44] Characteristic utterances on this subject may be found in the *National Economist*, 1:5, 51 (March 14, April 13, 1889); the *Great West* (St. Paul), October 18, 1889; and the *Alliance*, November 29, 30, 1889.

the doctrines of the Greenbackers, and again in 1886, when a number of the leaders of the revised order sought to make political use of the Alliance machinery, the order was brought to the verge of disintegration. The day was saved only by a prompt disavowal of politics and a determination to feature the business side of the Alliance.[45] When, two years later, the Union Laborites sought to set up a third-party organization in the state and attempted to draw into it farmers as well as laborers, the Alliance turned a cold shoulder. Two conventions were held, the second of which nominated a full state ticket headed by Evan Jones, president of the state Alliance. But Jones promptly decided that he could not accept the proffered nomination, and the revised ticket of Prohibitionists, Union Laborites, and Republicans that was finally constituted made little headway with the voters. Apparently the experience of 1886 had cured the Texas Alliance of all desire for aggressive political action.[46]

It was in Arkansas that the Agricultural Wheel originated, and this order from its inception showed a marked tendency towards participation in politics. When in 1886 a national Wheel was projected and a national constitution was under consideration, it was specifically stated that the order, while in no sense a political party, did contemplate obtaining what it wanted through legislation. That same year the Arkansas State Wheel put an independent ticket in the field, and the " Wheeler " candidate for governor polled upwards of twenty thousand votes.[47] The legislature chosen at this election, while overwhelmingly Democratic, hastened to put a drastic antirailroad law on the statute books, even going so far as to abolish free passes to public officials.[48] But the revolters were not bought off. In 1888 the Union Labor party adopted

[45] *National Economist*, 2:196 (December 14, 1889); Bryan, *Farmers' Alliance*, 11; Morgan, *Wheel and Alliance*, 100–102; Buck, *Agrarian Crusade*, 112.
[46] *Appletons' Annual Cyclopaedia, 1888*, 766–777; Morgan, *Wheel and Alliance*, 304–305.
[47] *Appletons' Annual Cyclopaedia, 1886*, 41–42.
[48] *Ibid., 1887*, 30.

a platform favoring "such legislation as will secure the reforms demanded by the Agricultural Wheel, the National Farmers' Alliance, and the Knights of Labor "; it nominated a full state ticket that the Republicans agreed to support; and it polled with this assistance only fifteen thousand fewer votes for its candidates than the victorious Democrats received. The danger of an acute party division growing out of this independent movement was made fully manifest. Furthermore, the election was characterized by an unusual amount of fraud and violence. Clearly the Arkansas example was not such as the average southern state would wish to follow.[49]

Nevertheless the time was ripe for a political revolution in the South, and the methods by which it was to be accomplished were soon discovered. In South Carolina, well before Alliance organizers entered the state, " Pitchfork " Ben Tillman had laid his plans for the overturn of the reigning Bourbon dynasty and for the control of the Democratic party by the hitherto inarticulate rural whites. Tillman was himself a product of the back country in which the bulk of the lower class white population of his state lived. He was born on August 11, 1847, at Chester, Edgefield County, South Carolina, not far from the Georgia boundary. His parents were not poor; neither were they of the socially elect. His father, who died when Benjamin was only two years old, was an owner of land and slaves and an innkeeper. His mother, a woman of great physical vitality and considerable business ability, materially improved the financial rating of the family after her husband's death and gave her children what educational advantages the times afforded. There must have been a wild streak in the Tillman nature, however, for of Benjamin's five brothers two were murdered as a result of unseemly escapades and a third, George, himself killed a man during a drunken dispute over a faro game. This same

[49] Ibid., 1888, 40; 1889, 35–36.

George, who prided himself on the intemperance of his
speech and the violence of his manner, became a successful
politician and ultimately landed in Congress. His influence
upon his younger brother Benjamin was marked.

Another of the Tillman brothers died of wounds received
while fighting as an officer in the Confederate army, and
Benjamin, too, narrowly escaped death as a soldier. In 1864,
when only seventeen years of age, he enlisted, but before he
saw service he was totally incapacitated by an abscess in one
eye, which soon lost him the sight of the member and
caused him the most acute suffering for two years. Shortly
after his recovery he married and settled down on a four-
hundred-acre farm given him by his mother. Negro and
carpetbag rule roused him to wrathful action, and he participated in more than one affray that was calculated to strike
terror to the heart of the freedmen. Then, with the negroes
and carpetbaggers out of the way, he turned his energies
against the Bourbon rulers of the state, whose misdeeds, he
convinced himself, had much to do with the economic distress of the back-country farmers. In 1886 he and his supporters sought to secure a governor and a legislature that
would respond to the wishes of the farmers, and while the
result of the election showed that neither goal had been
reached, a program of reform, including such items as the
abolition of the lien law and the renovation of the state agricultural department, had been formulated, and a start in the
direction of farmer organization had been made.[50]

In North Carolina, Colonel L. L. Polk, editor of the
Progressive Farmer, had inaugurated a somewhat similar
movement. Largely through his activities two mass conventions of farmers were held in Raleigh in January, 1887, to
consider the agricultural conditions of the state and, if possible, to influence the actions of the legislature, then in
session. A state-wide organization of farmers, to be known

[50] Simkins, *Tillman Movement,* chs. 2–4.

as the North Carolina Farmers' Association, was planned, and Polk himself undertook through the columns of his paper to encourage the formation of farmers' clubs in every community, from which the all-essential local backing for the state association should come.[51] Alliance organizers who stumbled upon these movements in the Carolinas could not fail to be impressed by them. Thereafter they were apt to preach not only the gospel of cooperative buying and selling that they had learned in Texas but also the gospel of political cooperation with a view to obtaining legislative reforms. Not a few of these organizers must have been well aware of the fact that should the Alliance, as such, enter the political arena it might mean offices and political preferment for themselves.[52]

The plan of action that gradually unfolded itself in the Carolinas was simplicity itself; there was to be no threat of a new party such as had brought discredit upon farmer organizations in Texas and in Arkansas; instead, the Alliance was merely to capture the existing Democratic machinery. Loyalty to the one-party system as a guarantee of continued white supremacy would insure that the Alliance, once it had won control, would have no formidable opposition party to contend with, and the Alliance program could be put through. In South Carolina the Tillmanites accepted the Alliance — merged with it, one might say — and used it to accomplish the desired end. There in the election of 1888 the revolting farmers again lost the governorship, but they won a majority in the lower house of the legislature, and they elected almost half the members of the state Senate. In every instance the farmer candidates had been regularly nominated and elected as Democrats. After the election at least one proposition satisfactory to the farmers, a law authorizing the existing railroad commission to fix maximum rates, was

[51] *News and Observer* (Raleigh), January 19, 27, 28, 1887; *Progressive Farmer*, April 14, May 26, June 2, 1887.
[52] Simkins, *Tillman Movement*, 111–113.

adopted. The chances seemed good that with another election the South Carolina farmers would have the party and the state government wholly in their control.[53]

In North Carolina the Alliance leaders had advised against bringing politics into the farmer organizations, but they had not hesitated to "take agricultural questions into politics."[54] At the outset of the campaign of 1888 they proposed a farmers' candidate for the Democratic gubernatorial nomination, and they got enough support for him to be able to deadlock the convention for thirty-two consecutive ballots before they had to acknowledge defeat. The Alliance was mainly responsible, also, for the fact that the legislature chosen that year in North Carolina contained a larger proportion of farmers than any other since the Civil War, and a farmer majority in the House dictated the choice of an Alliance leader as speaker.[55]

In North Carolina as in South Carolina all that the Farmers' Alliance had done had been accomplished apparently without risk of reviving the negro menace. The Alliance had merely gone as far as it could towards taking possession of the regular Democratic machinery. Once nominations had been made, the farmers had considered the matter closed, whether the nominees were satisfactory to them or not. To enhance the future power of the farmers' organization there was a demand, originated probably by Tillman in South Carolina but voiced in both states, that direct primaries be instituted.[56] Not only was it believed that these primaries would eliminate the jockeying, so disadvantageous to the inexperienced farmer, that characterized conventions, but it was also believed that the primaries would insure a greater voting strength to the rural whites. Under the convention system the number of votes assigned

[53] *Ibid.*, 100–102.
[54] *Progressive Farmer*, June 16, September 22, 1887.
[55] Hamilton, *North Carolina since 1860*, 228; *Progressive Farmer*, April 24, June 5, 1888; January 15, 1889.
[56] *Progressive Farmer*, February 23, 1888; Simkins, *Tillman Movement*, 86–88.

to each county was based upon the number of its inhabitants and not upon the number of its voters. Consequently a given number of voters in the white upcountry counties had less political strength in the conventions than a much smaller number of voters in the black counties of the low country, where the city classes and the great planters were in full control.[57]

In the northwestern states ever since Granger times the farmers had been accustomed to the idea of concerted political action; hence in this region it is not surprising to find that from the very beginning of Alliance agitation Alliance leaders had occasionally attempted to deliver the vote of their followers where it would do the most good. Party loyalties were strong, and in theory they were not to be disturbed, but if one party agreed to the Alliance demands and the other did not, or if a candidate of one party was deemed satisfactory and his opponent was not, Alliancemen were encouraged to vote for their own interests rather than for their party. It made little difference whether the concessions came from Democrats or from Republicans. Ignatius Donnelly of Minnesota expressed the idea correctly when he said, " Remember that the Alliance is not a political organization, in the sense of organizing a party, or setting up a state ticket " but that " it cannot stop corruption . . . without sending the right kind of men to the State Legislature."[58] Obviously " the right kind of men " were farmers or men who sympathized with the farmers.

Grievances against the railroads furnished the chief pretext for the political activity of the Alliance in the Northwest. Here all Alliancemen agreed " that the consolidation and combination of railroad capital and its influence in the United States in maintaining an oppressive and tyrannical transportation system " demanded " instant, vigorous and

[57] *Ibid.*, 239–240.
[58] *Mankato Review* (Minnesota), June 17, 1884; the *Great West*, October 18, 1889.

increasing action on the part of the producers," and that it would be wise "to subordinate other political questions to the emancipation of the people from this terrible oppression." But other matters, such, for example, as discriminations by the elevator companies, unfair practices in the grading of grain, the mortgage evil, unequal taxation, and the need of ballot reform, all came in for consideration.[59]

As early as 1882 there was unmistakable evidence in Nebraska of an intention on the part of the farmers to stand together; and in the campaign of this year some of them went the extreme length of forming an Antimonopolist party, which nominated a state ticket and polled a considerable vote. Thereafter the Nebraska Alliance freely gave or withheld its indorsement of old-party candidates and influenced elections all it could. The legislature elected in Minnesota in 1884 was notable for the number of farmers— Alliancemen—in the House, although the Senate was safely conservative; and the next year, when the Alliance leaders decided to throw the vote of their order to the Democratic ticket, the Republicans, who normally won by tremendous majorities, barely escaped defeat.[60] In 1886 the lower house of the Iowa legislature was so largely composed of organized farmers that a prominent agriculturist was elected speaker; and in 1887 Iowa farmers, in spite of numerous Greenback defections, had much to do with the renomination and reelection of Governor William Larrabee, who had taken the farmers' side against the railroads.[61]

Alliance influence was likewise strongly exerted to obtain satisfactory state platforms. For example, when a delegation from the Minnesota Alliance met with the Republican

[59] *Alliance*, November 30, 1889; *Pioneer Press*, February 5, 1887; Witham, *Fifty Years*, 51; Bryan, *Farmers' Alliance*, 62–65.
[60] Barnhart, "Farmers' Alliance and People's Party," 155–175; John D. Hicks, "The Origin and Early History of the Farmers' Alliance in Minnesota," *Mississippi Valley Historical Review*, 9:219.
[61] H. C. Nixon, "The Populist Movement in Iowa," *Iowa Journal of History and Politics*, 24:35–37; Witham, *Fifty Years*, 41; Haynes, *Third Party Movements*, 193, 436–439.

state nominating convention of 1886, the spokesman for the farmers reported: "The committee on platform . . . gave us everything we asked for, in even stronger language than we had used, and the . . . convention unanimously adopted it." [62] But generous as were the Republicans, the Democrats seem to have been still more so, for in this particular election Alliance leaders did their best to deliver to the Democrats the entire Alliance vote. The farmers seemed to be in an enviable position: Republicans and Democrats were bidding against one another in order to obtain Alliance support.[63] On the surface the victory of the Alliance in Iowa the next year was not less sweeping than the Minnesota triumphs of 1886. Not only did the farmers get a platform to their liking, but, according to one of their leaders, they actually "captured the whole machinery of the Republican party." [64]

Farmer influence extended even into the far Northwest, where six territories, from the Dakotas on the east to Washington on the west, were preparing for statehood. In the six constitutional conventions held during the summer and fall of 1889 members of the Farmers' Alliance very generally took the lead, writing into fundamental law their antipathy for the railroads and other corporations, their distrust of state officials, and their sundry demands for reform.[65] Throughout the West, Alliancemen told the parties, Democratic as well as Republican, what to put in their platforms; while farmers appeared in ever increasing numbers as members of state legislatures and holders of state offices to see that platform promises were not forgotten.

As a result of these successes considerable legislation of the sort that the Alliance favored — particularly with respect to railway regulation — was actually enacted. A law of

[62] *Pioneer Press*, September 24, 1886; E. V. Smalley, *A History of the Republican Party . . . of Minnesota*, 223.
[63] *Pioneer Press*, September 15, 26, 1886.
[64] Witham, *Fifty Years*, 41; Nixon, in the *Iowa Journal of History and Politics*, 21:382–383.
[65] John D. Hicks, *The Constitutions of the Northwest States*.

1887 in Nebraska created a state Board of Transportation, charged with the duty of preventing discriminations and maintaining rates that were reasonable and just, but not specifically authorized to set maximum rates.[66] The Minnesota legislature of 1887 passed a law that greatly enlarged the powers of the existing Railroad and Warehouse Commission and came very close indeed to giving the Commission full rate-making authority.[67] In 1888 the Iowa legislature under the leadership of Governor Larrabee actually gave to an elective commission the full right to dictate railway schedules of "reasonable maximum rates."[68] The North Dakota constitution of 1889 proposed to control public utilities through a board of railway commissioners, composed of three elected members. And so the multiplication of railway commissions and the elaboration of their duties went on. By 1892 there were thirty such commissions throughout the country having varying degrees of authority,[69] besides the Interstate Commerce Commission.

But somehow these laws, as well as sundry others of a minor nature that were enacted because of Alliance insistence, proved singularly abortive. The Nebraska Board of Transportation was described by a local Alliance newspaper as "a railroad commission whose chief end is to draw their own salary, freights going up and down at the will of the railroad companies. . . ."[70] The railroads in Minnesota successfully resisted as unconstitutional the rate-making power of the state commissioners.[71] The Iowa law proved effective for a time, but ultimately the railroads succeeded in electing too many of their own friends to membership on

[66] Dixon, in the *Political Science Quarterly*, 13:628.
[67] Hicks, in the *Mississippi Valley Historical Review*, 9:221.
[68] Nixon, in the *Iowa Journal of History and Politics*, 21:383; Haynes, *Third Party Movements*, 442.
[69] Fossum, *Agrarian Movement in North Dakota*, 32; Adams, in the *Century*, 21:783.
[70] *Alliance*, August 7, 1889; Dixon, in the *Political Science Quarterly*, 13:629–630.
[71] *134 United States Reports*, 418; *Appletons' Annual Cyclopaedia, 1890*, 555.

the commission;[72] and the North Dakota farmers were brazenly tricked out of legislation they deemed essential to make their constitutional commission effective.[73] Small wonder that Alliance leaders felt baffled! One writer, with sympathy and penetration, thus recounted the history of their political efforts:

They elected lawyers and other professional men to represent them and their interests. . . . They appealed to their party leaders. They sought to interest the press in their behalf. They brought their case before the courts. They presented themselves before the bar of public opinion. But they were disappointed. [Then they tried electing their own men — dirt farmers — to office. But it] made little difference whether they sent a farmer or a politician to the Legislature. If the farmer went to the capital fresh from the plow, among a crowd of lobbyists, he was as clay in the hands of the potter. If his constituents kept him there year after year, until he learned the ways of legislation, then he ceased to be a farmer and became a member of some other class, perhaps a stockholder in a great railroad, or manufacturing corporation, with interests in common with the opponents of agricultural classes.[74]

It was inevitable under these circumstances that there should be a strong demand for the transformation of the Alliance into an independent political party. In the years 1889 and 1890 new members flocked into the order as never before, and with these notable accessions the plausibility of third-party action was correspondingly increased. The Alliance had the strength now to enter the political field directly, and since its nonpartisan efforts had failed, what else was

[72] Haynes, *Third Party Movements*, 444.
[73] Fossum, *Agrarian Movement in North Dakota*, 34–35. For example, a law designed to compel railroad companies to cease discrimination in the granting of elevator sites carried no penalty in case it was disobeyed. Another law was intended to give the railroad commissioners authority to require on sixty days' notice the building of loading platforms from which the farmers could load their own grain, but it was so worded as to enable the attorney-general to rule that the notice had to be given within sixty days after the law was passed.
[74] Walker, in the *Annals of the American Academy of Political and Social Science*, 4:795–796. Compare Waring, in the *North American Review*, 152:753.

there left for it to do? Such a course was ably urged by professional third-party politicians — ex-Grangers, ex-Green-
backers, ex-Union Laborites — who were on hand in numbers to offer themselves as leaders of the new movement.
Some of them had long foreseen that the Alliance ultimately
would drift into politics and had urged that on this account
the Farmers' Alliance be not neglected. As early as 1882 a
correspondent wrote in this vein to Lemuel H. (" Calamity ") Weller of Iowa:

It [the Alliance] has been a great educator of the people, and
has really been the foundation of much political success. It
brings the farmers together for conference. Not being partizan,
and not even political in the common sense of the term, there is
no reason why any farmer should not join it. Once in, the inevitable result is that a partizan begins to unbend a little, and to
see that his interests are not being served in the halls of legislation; and in the majority of cases he will become an independent
voter.[75]

This was precisely what had happened. For most northwestern Alliancemen loyalty to the old party — the party that
once they had loved, but by which they had now been
betrayed — was dead. If through a new party there was
hope of reform, then let the new party come.

[75] David W. Wood to Weller, December 26. 1882, in the Weller Papers.

CHAPTER SIX

THE ELECTIONS OF 1890

BY THE year 1890, an election year, both the Southern and the Northern Alliances were earnestly at work along political lines. Both drew much inspiration from the activities, already described, of the national conventions held at St. Louis during the month of December, 1889, conventions that had failed to achieve anything like organic union but had succeeded in drawing up clear and convincing statements of their mutual political aims.[1] In the South it was generally agreed that the Alliance must make every effort to capture the existing state machinery of the Democratic party; in the Northwest this was deemed inadequate, and instead separate third-party action was demanded.

The determination that independent state tickets should be nominated by Alliancemen and their friends everywhere in the Northwest grew with the hardness of the times. By 1890 the full effects of the deflation in real estate values, following the collapse of speculation in 1887, were being felt. High taxes, necessary because of lavish civic expenditures in the days of the boom, became an intolerable burden. Because of the "mortgage evil" a large proportion of the population lived constantly in the shadow of impending bankruptcy. Crop failures came with disheartening frequency, and even a bumper corn crop, such as Nebraska had

See Appendix A.

in 1889, gave small relief, for the price declined as the yield
increased. With corn at fifty cents a bushel on the Chicago
market, the farmer in far-away Kansas or Nebraska was
lucky to get as much as fifteen or twenty cents a bushel:
the railroads and the commission men got the rest. Stories of
ten-cent corn and of corn to burn were rife. Moral suasion
was tried upon the railroads to induce them to lower their
rates so that the frontier farmer might live, but in vain; and
apparently no reforms were to be expected through the
agency of the existing political parties. Moreover, in the
summer of 1890 the drought was general all along the
frontier.[2]

Well before this time an attempt to organize the various
forces of discontent into a national party had resulted in
the formation on the twenty-second of February, 1887, at
Cincinnati of the Union Labor party.[3] The following year
the new party nominated a national ticket, and in seven
states, Kansas, Missouri, Iowa, Wisconsin, Illinois, Texas,
and Arkansas, it polled an appreciable number of votes.
As a nation-wide movement for reform, Union Laborism
was a complete failure, but in most of the states mentioned,
notably in Kansas, it furnished a convenient means for the
expression of local dissatisfaction with old-party rule. In the
West, however, the Union Labor label was a misfit,[4] even a
liability, for the suggestion was easily conveyed that the
new party condoned such acts of violence as were commonly
imputed to organized labor during and after the strikes of
1885 and 1886.[5] The party, however, did have some signifi-

[2] *Farmers' Alliance* (Lincoln), July 19, 1890; Tracy, in the *Forum*, 15:338.
Barnhart, in the *American Political Science Review*, 19:536; Barr, in *Kansas and
Kansans*, 2:1146.
[3] Considerable valuable material on this subject is to be found in the Weller
Papers for 1887 and 1888.
[4] In Minnesota local leaders discarded the national label for one of their own,
the Farm and Labor party. They nominated Ignatius Donnelly for governor,
but he soon withdrew, thereby damaging the cause materially. See the *St. Paul
Globe*, August 29, 1888, and *Appletons' Annual Cyclopaedia, 1888*, 559.
[5] During the campaign of 1888 an express package addressed to the Winfield
Non-Conformist exploded. Since the *Non-Conformist* was the leading Union

cance. Its brief career seemed to indicate that for the time
being, at least, local state parties based upon local issues
had a better chance to survive than a single unified national
party; moreover, Union Labor leaders promptly made their
services available to the Alliance in the work of organizing
that body along party lines.[6]

Particularly in Kansas was this the case. There in the
local county elections of 1889 Union Laborites had worked
for the nomination of People's tickets and had secured the
earnest cooperation of the Alliance. In Cowley County, for
example, the Republican "ring" was thus completely put
to rout. While the Union Laborites may have furnished the
leaders, it was from the Alliance that the votes came, for,
as a Republican paper, the *Winfield Courier*, observed,

In every locality where the Alliance was strong, the People's
ticket had large majorities, and in every township where there
was no Alliance, the usual vote was cast. The Alliance for this
year at least has been handled as a very compact and orderly
political machine. Whether this is to continue as its policy is not
known.[7]

That this policy was to continue was clearly indicated in
March, 1890, when a meeting of county Alliance presidents
"resolved that we will no longer divide on party lines, and
will only cast our votes for candidates of the people, by the
people and for the people."[8] It seemed best, however, not
to convert the Alliance, as such, into a political party but
rather to create a separate but more inclusive organization,
to which every good Allianceman would give his support as a
matter of course. Consequently a call was issued for a convention to be held in Topeka on June 12, 1890, to which the

Labor paper in Kansas and since its editors were members of a secret inner circle
of Union Laborites, the Videttes, whose ritual had lately been exposed, it was
easy to argue that the whole Union Labor fraternity was anarchistic and ready
to use bombs. See Miller, "Populist Party in Kansas," 60.

[6] Henry D. Wilcox to Weller, January 4, 188[9], in the Weller Papers; Barr,
in *Kansas and Kansans*, 2:1137, 1142; Haynes, *Third Party Movements*, ch. 14.

[7] Quoted by Barr, in *Kansas and Kansans*, 2:1143.

[8] *Ibid.*, 2:1146.

task of organizing a new party was to be entrusted. This convention, composed of forty-one Alliancemen, twenty-eight Knights of Labor, ten members of the Farmers' Mutual Benefit Association, seven Patrons of Husbandry, and four single-taxers, determined that full state and congressional tickets, pledged to the principles of the St. Louis platform, should be nominated, that the name of the new organization should be the People's party, and that a state nominating convention should be called to meet two months later. On the thirteenth of August the nominating convention met and named a People's party ticket, headed by John F. Willits for governor.[9]

The progress of events in Nebraska paralleled closely what was happening in Kansas. In Nebraska, however, Union Laborites figured less conspicuously, and Alliance leaders undertook third-party action only after considerable hesitation. "The organization of a small minority party will not secure the passage of any law," declared the *Alliance* in December, 1889, "unless public opinion demands it. If public opinion demands it, it will be enacted without the minority party."[10] Some argued that it would be better to capture the old-party primaries and in that way bring about the desired reforms;[11] others, that it would be sufficient to support the better candidates presented by the old parties. Yet everyone knew that where these methods had been tried before, they had usually availed nothing. Already in 1889 several Nebraska counties, following the Kansas lead, had nominated local tickets and in some instances had succeeded in electing them; and when a convention of county Alliance presidents and organizers was held in Lincoln on April 22, 1890, it decided that if the strength of the Alliance increased sufficiently, state-wide independent action would be justified. The strength of the Alliance did indeed increase; more signifi-

[9] Barr, in *Kansas and Kansans*, 2:1147.
[10] *Alliance* (Lincoln), December 28, 1889.
[11] *Farmers' Alliance*, March 1, 1890.

cant still, it was apparent that many people joined the Alliance precisely because they expected independent political action to be taken. Following the annual state meeting of the Alliance, held in May, petitions calling for a People's Independent convention were circulated with such success that by the twenty-first of June ten thousand signatures had been obtained and it could be announced that "the Convention will be called." [12] The convention was called; it met in Lincoln on July 29, 1890; and, with the support of such Grangers, Knights of Labor, and Union Laborites as cared to cooperate, it nominated a People's Independent ticket, headed by John H. Powers, president of the state Alliance, as candidate for governor.[13]

What happened in Kansas and Nebraska happened elsewhere all along the "Middle Border." South Dakota "Independents" held their first convention in Huron on the seventh of June, several days prior to the comparable convention in Kansas — hence the claim, sometimes made, that the true birthplace of Populism was South Dakota, not Kansas. A later convention nominated a full ticket, with another state Alliance president, H. L. Loucks, as the candidate for governor.[14] In North Dakota the procedure was much the same, except that the Prohibitionists were given greater prominence than was ordinarily the case. The convention call urged "farmers, prohibitionists, and friends of reform generally" to revolt against the "political bosses and ring masters" of the state. The ticket nominated gave place to four Prohibitionists and four Democrats, but it was headed by another state Alliance president, Walter Muir.[15] In Minnesota the executive committee of the state Alliance, urged

[12] *Farmers' Alliance*, May 17, June 21, July 5, December 20, 1890; Barnhart, in the *American Political Science Review*, 19:539.
[13] *Farmers' Alliance*, August 2, 1890; *National Economist* (Washington), 3:392 (August 30, 1890).
[14] Blood, *Handbook*, 1; *Nation*, 50:460; *Appletons' Annual Cyclopaedia, 1890*, 782; *National Economist*, 3:291–292 (July 19, 1890).
[15] Fossum, *Agrarian Movement in North Dakota*, 36 note; *Appletons' Annual Cyclopaedia, 1890*, 629.

on by the local chapters, called a nominating convention on
its own responsibility, and an Alliance ticket, frankly so
named, was put in the field. A feud between the president
of the Minnesota Alliance, R. M. Hall, and its most distinguished member, Ignatius Donnelly, accounts for the fact
that neither of them won the nomination for governor,
which went instead to a "dark horse," Sidney M. Owen,
editor of a Minneapolis farm journal.[16] In Colorado an
"Independent Fusion ticket" was chosen by the Alliance
and Union Labor forces; in Michigan members of the Alliance and the Patrons of Industry cooperated to organize an
Industrial party, which put up a full state ticket; in Indiana
the remnants of the old Greenback party joined with the
various farm orders to form a new People's party and to
nominate candidates. That there was no third-party organization of state-wide dimensions in Iowa is accounted for
in part by the fact that no state election was held there in
1890; this was true also for Illinois, where only minor state
officers were chosen.[17]

Wherever state tickets were nominated by the independents, local tickets, particularly candidates for the legislature, were generally chosen also. Indeed, the control of the
various state legislatures seemed to the revolters a matter
of even greater importance than the control of the state
executive departments, for only through legislative initiative
could the desired reforms be achieved. Congressional candidates were nominated in many western districts, and the
failure of the national Republican administration to produce
the better times it had promised was duly noted, but throughout the Northwest the struggle was primarily one for intra-
state reforms.[18]

[16] John D. Hicks, "The People's Party in Minnesota," *Minnesota History
Bulletin,* 5:537-538.
[17] Haynes, *Third Party Movements,* 242-243.
[18] *Farmers' Alliance,* July 19, 1890. Congressional campaigns sometimes
aroused much excitement. In the third Nebraska district Omer Kem of Broken
Bow won the Populist nomination for Congress and captivated the nominating

The campaign that followed, and sometimes accompanied, the various independent conventions of 1890 can hardly be described. Perhaps Elizabeth N. Barr, writing many years later on the uprising in Kansas, comes as near as anyone could to expressing in words the spirit of the times. According to her version,

> The upheaval that took place . . . can hardly be diagnosed as a political campaign. It was a religious revival, a crusade, a pentecost of politics in which a tongue of flame sat upon every man, and each spake as the spirit gave him utterance. For Mary E. Lease, Jerry Simpson, . . . and half a hundred others who lectured up and down the land, were not the only people who could talk on the issues of the day. The farmers, the country merchants, the cattle-herders, they of the long chin-whiskers, and they of the broad-brimmed hats and heavy boots, had also heard the word and could preach the gospel of Populism. The dragon's teeth were sprouting in every nook and corner of the State. Women with skins tanned to parchment by the hot winds, with bony hands of toil and clad in faded calico, could talk in meeting, and could talk right straight to the point.[19]

Many of the most effective of these campaigners were newcomers to the political arena. Mrs. Mary Elizabeth Lease of Kansas, "a tall, slender, good-looking woman of thirty-seven years," Irish by birth and endowed with the traditional talents of her race, was easily the most spectacular of them all. She had lived in Kansas since 1873, had witnessed at first hand all the familiar frontier tragedies, and knew whereof she spoke. Though the mother of four children, she had nevertheless undertaken the study of law and had been admitted to the bar in 1885. She tried out her

---

convention by his oratory. His speech on the evils of the times provoked one of the most famous of the quatrains ever written by "Doc" Bixby, Nebraska's pioneer columnist:

> "I cannot sing the old songs,
> My heart is full of woe;
> But I can howl calamity
> From Hell to Broken Bow."

[19] Barr, *Kansas and Kansans*, 2:1148–1149.

gift of oratory in support of the Union Labor candidates during the campaign of 1888, and thereafter she was in constant demand as an Alliance speaker. The heated atmosphere of the campaign of 1890 was exactly to her liking, and she made in that year no less than a hundred and sixty speeches for the "cause." "What you farmers need to do is to raise less corn and more *Hell*," she is reputed to have said.

Wall Street owns the country. It is no longer a government of the people, by the people and for the people, but a government of Wall Street, by Wall Street and for Wall Street. The great common people of this country are slaves, and monopoly is the master. The West and South are bound and prostrate before the manufacturing East. Money rules, and our Vice President is a London banker. Our laws are the output of a system which clothes rascals in robes and honesty in rags. The parties lie to us and the political speakers mislead us. . . . the politicians said we suffered from overproduction. Overproduction when 10,000 little children, so statistics tell us, starve to death every year in the United States, and over 100,000 shop-girls in New York are forced to sell their virtue for the bread their niggardly wages deny them. . . . Kansas suffers from two great robbers, the Santa Fe Railroad and the loan companies. The common people are robbed to enrich their masters. . . . There are thirty men in the United States whose aggregate wealth is over one and one-half billion dollars. There are half a million looking for work. . . . We want money, land and transportation. We want the abolition of the National Banks, and we want the power to make loans direct from the government. We want the accursed foreclosure system wiped out. Land equal to a tract thirty miles wide and ninety miles long has been foreclosed and bought in by loan companies of Kansas in a year. We will stand by our homes and stay by our firesides by force if necessary, and we will not pay our debts to the loan-shark companies until the Government pays its debts to us. The people are at bay, let the bloodhounds of money who have dogged us thus far beware.[20]

[20] Barr, in *Kansas and Kansans*, 2:1150–1151. See also Miller, "Populist Party in Kansas," 178.

Startlingly effective, also, was the oratory of Jerry Simpson, the People's candidate for Congress in the seventh Kansas district. Simpson was a Canadian by birth and hence was never one of the third-party candidates for a presidential nomination in spite of his outstanding personality. He was born in New Brunswick on March 31, 1842, but his parents moved to New York in 1846, and in the public schools of Oneida County he received an elementary education. For many years he was a sailor on the Great Lakes, ultimately rising to the command of large vessels there. At one time he also had an interest in a sawmill near Chicago. When the Civil War broke out he joined an Illinois regiment and served in the army until incapacitated by illness. In 1878 he emigrated to Kansas, where he became a farmer and stock-raiser near Medicine Lodge. Simpson made a deep impression upon Hamlin Garland, who met him a few months later in Washington and described him as follows:

He is about fifty years of age, of slender but powerful figure, whose apparent youthfulness is heightened by the double-breasted short sack coat he wears. His hair is very black and abundant, but his close-clipped moustache is touched with gray, and he wears old-fashioned glasses, through which his eyes gleam with ever-present humor. The wrinkles about his mouth show that he faces the world smilingly. His voice is crisp and deep and pleasant to the ear. He speaks with the Western accent mainly; and when he is making a humorous point or telling a story, he drops into dialect, and speaks in a peculiar slow fashion that makes every word tell. He is full of odd turns of thought, and quaint expressions that make me think of Whitcomb Riley. He is a clear thinker, a remarkable speaker, and has a naturally philosophical mind which carries his reasoning down to the most fundamental facts of organic law and human rights.

As a result not only of his frontier experiences but also of his reading he was driven to the espousal of political and economic heresies long before the campaign of 1890. As a Greenbacker, a single-taxer, and a Union Laborite, he was already known to third-party circles. In this campaign Simpson made it his particular business to denounce the railroads. He openly charged that the Sante Fe system dominated the politics of his state, and he demanded government ownership as the only adequate remedy. A firm believer in the doctrines of Henry George, he argued that "man must have access to the land or he is a slave. The man who owns the earth, owns the people, for they must buy the privilege of living on his earth." He called attention, also, to the profits made by the Chicago grain gamblers, who took advantage of the farmers' dire necessity to buy when the price of grain was low, and then waited for the price to go up before selling. "If the Government had protected the farmer as it protects the gamblers, this could not have happened," he claimed. Simpson's Republican opponent in the race for Congress was a well-groomed gentleman, Colonel James R. Hallowell, whom Simpson, on the defensive because of the jibes at his own rural appearance, promptly dubbed "Prince Hal" and accused of wearing silk stockings. Victor Murdock, then a young reporter, came back with the retort that Simpson wore no socks at all; hence the appellation, "Sockless Jerry," which followed Simpson the rest of his life.[21]

Ignatius Donnelly of Minnesota, perhaps the greatest orator of Populism, had broken a lance for every considerable reform cause that the United States had known, beginning with pre-Civil War Republicanism. He was born in Pennsylvania on November 3, 1831, of Irish parents and had come to Minnesota in time to suffer the full effects of the panic of 1857, had turned from real estate promotion and the law to antislavery politics, and had served three terms in Congress during and after the war. He had become a Republican when to do so branded him a born reformer,

<hr>

[21] Barr, in *Kansas and Kansans*, 2:1151–1152; *National Economist*, 5:179 (May 30, 1891); Hamlin Garland, "The Alliance Wedge in Congress," *Arena*, 5:451.

and he was once again an irregular when Liberal Republicanism won his full support. Afterwards he led the Grangers
of Minnesota in their war on the railroads; he went through
a clean-cut and complete conversion from hard-money principles to Greenbackism; he flirted with the Union Laborites
in 1888 and almost became their candidate for governor of
Minnesota that year; and he now landed fairly and squarely
in the forefront of the latest movement for reform. As the
New York Sun remarked, a reform convention in Minnesota
without Donnelly would have been " like catfish without
waffles in Philadelphia." [22]

The Minnesota " sage " — his neighbors called him the
" sage of Nininger " — was a man of varied talents: he wrote
books on popular science; delivered side-splitting lectures on
" Wit and Humor "; defended in print and on the platform
the Baconian theory of the authorship of Shakespeare; and
talked convincingly on any subject whatever that had to do
with politics or economics. No one ever denied Donnelly's
oratorical skill, although his orations showed no great pro-
fundity. He was at his best in unsparing denunciation or
encomium. His argumentative triumphs were won by reasoning that was adroit and clever but usually full of sophis-
try. No one could more easily make the worse appear the
better reason, and apparently no one delighted more in doing so. He possessed remarkable facility in the use of statistics for this purpose and could fairly breathe the breath of
life into the dullest of figures. Audiences listened to his deductions with interest and almost invariably with at least
temporary conviction.

Donnelly wished to be the Alliance candidate for governor
of Minnesota in 1890, and when the nomination was denied
him, he was almost ready to play the part of Achilles. But
he soon relented and campaigned the state with his usual
vigor, speaking to enormous crowds, for his fame as an

[22] July 22, 1890.

orator and entertainer had reached every corner of Minnesota. Already he had addressed audiences in practically every town in the state, not once or twice but repeatedly. Everyone recognized his short, plump figure, his monkish, smooth-shaven face, his tawny hair, and his genial smile. Outside of politics people liked him, and his neighbors rarely deserted him even at the ballot box. He took much credit to himself for the large vote polled by the third-party candidates in Minnesota that year, admitting in private that if only he himself had been nominated for governor, the Alliance ticket would have won.[23]

Less versatile than the eccentric Donnelly, James B. Weaver of Iowa inspired far greater confidence among those who, while thoroughly deploring existing conditions, were fearful of anything that savored of extreme radicalism. Weaver was born on June 12, 1833, at Dayton, Ohio, the son of cultivated parents who were able to give him opportunities for a good education. He graduated from the Cincinnati law school in 1854 and soon afterward emigrated to Iowa, where he practiced law until the outbreak of the Civil War. He enlisted in the Union army, attained by promotion the rank of colonel, and was brevetted brigadier general for "gallantry on the field" just as the war closed. Resuming the practice of law at Des Moines, Iowa, he achieved considerable success at the bar and in politics, where, as a Republican, he held minor offices. His successes soon turned to failures, however, perhaps more because he was strongly prohibitionist in his views on the temperance question than for any other reason. In 1877 he cut loose from the Republican party, and the year after as a Greenbacker with Democratic support he won a seat in Congress. In 1880 he was the Greenback nominee for president of the United States and toured the country "from Arkansas to Maine, and from

Lake Michigan to Mobile," speaking, as he believed, to more than a million people. From 1883 to 1887 he again served in Congress as a Greenback-Democrat, but in 1888 he was not re-elected, and in 1890 he was ready to throw in his lot with the "Union Labor Industrial Party of Iowa," as the pre-Populist organization in that state was called. His services as a distinguished and experienced campaigner, with an especial appeal for the "old soldier" vote, were available not only within his own state but in neighboring states as well.

Weaver was no such striking figure as Simpson or Donnelly. As Mrs. Diggs, herself one of the leading lights of the Alliance, somewhat reluctantly admitted, he was "a person of such symmetrical, such harmonious development, that each excellent trait is balanced by all others."

He presents a composite picture in which strength and gentleness blend. Inherent qualities have been drawn out by culture. The cannibalism of politics has snapped and bitten at him in vain. Serene while others are in tumult; clear while others are confused; secure in his orbit while others are erratic; certain while others are in doubt, — these characteristics make him a man of value second to none in a great epoch like the present. [24]

These were the most outstanding among the orators of revolt, but the number of less prominent speakers was legion. Scores and possibly hundreds of men and not a few women had been in training as Alliance lecturers for just such work as the campaign of 1890 furnished them.

Next to Mrs. Lease, Mrs. Annie L. Diggs, also of Kansas, was perhaps the best known of the Alliance women. She was a mere mite of a person, weighing less than a hundred pounds, but apparently her size never interfered with her ambition to speak and write. Her first crusade was precipitated about 1877 by her discovery that the university boys in her home town, Lawrence, were being ruined by liquor;

[24] Haynes, *James Baird Weaver*, *passim*; Annie L. Diggs, "The Farmers' Alliance and Some of Its Leaders," *Arena*, 5:601–603.

before the Alliance agitation gave her yet another field for her labors, she had gained prominence as a silkworm enthusiast and as a devotee of liberalism in religion. Both as an orator and as a writer Mrs. Diggs had a real gift of words. Her language was smoother and more dignified than that of Mrs. Lease, but hardly less vigorous.

Another Kansas woman of note was Mrs. Sarah Emery, whose book, *Seven Financial Conspiracies*, had a wide vogue. Mrs. Emery was a large, competent person who had stood the punishment of active campaigning for financial reforms ever since the Greenback campaign of 1880. Mrs. Bettie Gay of Texas was a black-eyed, calm-voiced southern woman whose husband had died in 1880 leaving her to shift for herself, her child, and the numerous brood of homeless boys and girls she insisted on mothering. She managed a farm, spoke with equal energy on woman's suffrage, prohibition, and Alliance reforms, and wrote voluminously. Mrs. Eva McDonald Valesh of Minnesota, the " jauntiest, sauciest, prettiest little woman in the whole coterie of women in the Alliance," was " quite as much at home on an improvised stove-box platform on the street corner, speaking earnestly to her toil-hardened brother Knights of Labor, as in the drawing-room, radiating sparkling wit and repartee." [25]

Ably supported by the Alliance press and by many independent papers as well, this throng of zealots

doled out an intoxicating mixture. In vain the reports of the meetings were suppressed by a partisan press. In vain the Republican and Democratic leaders sneered at and ridiculed this new gospel, while they talked tariff and war issues to small audiences. . . . The excitement and enthusiasm were contagious, and the Alliance men deserted their former parties by thousands.

Women who never dreamed of becoming public speakers, grew eloquent in their zeal and fervor. Farmers' wives and daughters rose earlier and worked later to gain time to cook picnic dinners,

[25] Barr, in *Kansas and Kansans*, 2:1152–1169; Annie L. Diggs, " The Women in the Alliance Movement," *Arena*, 6:161–179.

to paint the mottoes on the banners, to practice with the glee clubs, to march in procession. Josh Billings' saying that " wimmin is everywhere " was literally true in that wonderful picnicking, speech-making Alliance summer of 1890.[26]

For it was not only through speech-making that the crusading ardor of the western farmers made itself known. The thousands who never mounted a rostrum proved their interest by attendance at picnics, basket dinners, barbecues, and political rallies of every sort and kind. To reach these meetings it was often necessary for individual farmers to drive long distances, and in order to impress the countryside parades were arranged — seemingly endless parades of farmers and their families, mostly in cumbersome farm wagons — on the way to the appointed meeting place. As many as sixteen hundred teams were reported to have participated in a march on Hastings, Nebraska, where, according to the reports, no less than twelve thousand people gathered.[27] In Kansas, crowds of from twenty to twenty-five thousand were reported. When Colonel Polk spoke at Columbus, Kansas, "a procession four miles long entered town singing and playing, and left the same way," while at Winfield the local reform paper held that " when the head of the procession was under the equator, the tail was coming around the North pole." [28] On the march " the embattled farmers " displayed their political wares through such banners and mottoes as these:

> We are mortgaged, all but our votes.
> Vested rights will go down forever and human rights
>   will prevail.
> We vote as we pray.
> We voted with our party no matter where it went,
> We voted with our party till we haven't a cent.[29]

[26] Tracy, in the *Forum*, 16:243.
[27] Barnhart, " Farmers' Alliance and People's Party," 223.
[28] *National Economist*, 3:285 (July 19, 1890); Barr, in *Kansas and Kansans*, 2:1149. On campaign methods in Kansas see also Miller, " Populist Party in Kansas," 182–184.
[29] Barnhart, " Farmers' Alliance and People's Party," 222–223.

There was more to the rally itself than the endless speech-making. Races and games, bands, glee clubs, and group songs enlivened the occasion. Particularly did the farmers delight in song. Parodies of familiar verses that could be sung to familiar tunes served this purpose best. Songs of this description were contributed to Alliance newspapers by scores and hundreds, and the best of them, sung by the excited throngs, added immeasurably to the camp-meeting and revivalistic atmosphere of the campaign. Enterprising journalists gathered the favorite songs together into song books and sold them by the thousands. One such song, "Good-bye, My Party, Good-bye," was credited with having won the Populist victories in Kansas.[30]

[30] *Cincinnati Enquirer*, May 19, 1891; C. S. White, *Alliance and Labor Songster*, 20. B. O. Fowler, "Songs of the People," *Arena*, 6:xlvi–l, reprints several Alliance campaign songs used in Nebraska. Among the frequent contributions of Mrs. J. T. Kellie to the columns of the *Farmers' Alliance* "The Independent Man," adapted to the tune of "The Girl I Left Behind Me," is typical:

> "I was a party man one time
>   The party would not mind me
> So now I'm working for myself,
>   The party's left behind me.
>
> "A true and independent man
>   You ever more shall find me—
> I work and vote, and ne'er regret
>   The party left behind me."
>
> *Farmers' Alliance*, September 27, 1890.

Another contributor, Arthur L. Kellog, presented "The Hayseed," which could be sung to the tune of "Save a Poor Sinner Like Me."

> "I was once a tool of oppression,
>   And as green as a sucker could be
> And monopolies banded together
>   To beat a poor hayseed like me.
>
> "The railroads and old party bosses
>   Together did sweetly agree;
> And they thought there would be little trouble
>   In working a hayseed like me.
>
> . . . .
> "But now I've roused up a little
>   And their greed and corruption I see,
>
> . . . .
> "And the ticket we vote next November
>   Will be made up of hayseeds like me."
>
> *Farmers' Alliance*, October 4, 1890

Good-bye, My Party, Good-bye
Air: "Good-bye, My Lover, Good-bye."

It was no more than a year ago,
　　　　Good-bye, my party, good-bye
That I was in love with my party so,
　　　　Good-bye, my party, good-bye
To hear aught else I never would go;
　　　　Good-bye, my party, good-bye.
Like the rest I made a great blow;
　　　　Good-bye, my party, good-bye.

　Chorus:
　Bye, party, bye, lo; bye, party, bye, lo;
　Bye, party, bye, lo; good-bye, my party, good-bye.

I was often scourged with the party lash,
　　　　Good-bye, my party, good-bye.
The bosses laid on with demands for cash;
　　　　Good-bye, my party, good-bye
To do aught else I deemed it rash,
　　　　Good-bye, my party, good-bye.
So I had to take it or lose my hash,
　　　　Good-bye, my party, good-bye.

　Chorus:

I was raised up in the kind of school,
　　　　Good-bye, my party, good-bye.
That taught to bow to money rule,
　　　　Good-bye, my party, good-bye.
And it made of me a " Kansas Fool,"
　　　　Good-bye, my party, good-bye.
When they found I was a willing tool,
　　　　Good-bye, my party, good-bye.

　Chorus:

The old party is on the downward track,
　　　　Good-bye, my party, good-bye.
Picking its teeth with a tariff tack,
　　　　Good-bye, my party, good-bye.

With a placard pinned upon his back,
>Good-bye, my party, good-bye.
That plainly states, "I will never go back ";
>Good-bye, my party, good-bye.
Chorus:

Meantime in the South events had been pursuing a somewhat different course. There third-party action was scarcely thought of; instead, the control of the Democratic party by the Alliance was the goal. Following the St. Louis convention of December, 1889, the leaders of the Southern Alliance apparently decided to throw the full weight of their influence in this direction; indeed, they chose as their president Colonel L. L. Polk of Raleigh, North Carolina, who had been actively engaged in the work of lining up the farmer vote of his state solidly against the old Bourbon machine and in favor of Alliance principles and candidates. Polk's powerful position — for the president of the Southern Alliance was no mere figurehead — made it possible for him to bring pressure elsewhere than in North Carolina that would lead to an aggressive political stand by the Alliance.[31]

Great pains were taken to make it clear that the Alliance did not contemplate the organization of a third party to divide the white vote of the South. The *National Economist,* which for the time being reflected accurately the official position, promised positively that "a third political party will not be formed." The Alliance movement, according to this journal, was still a strictly nonpartisan affair, and each member might remain true to his party. But each member was urged, nevertheless, to "see to it that this party continues true to him."[32] The farmers were never allowed to forget that by acting in concert their numbers were "sufficient to control both or either of the existing parties "; moreover, it was fully understood that no one cared much what happened

---

[31] *Progressive Farmer* (Raleigh), April 29, May 27, 1890; Hamilton, *North Carolina since 1860,* 232.
[32] *National Economist,* 2:264 (January 11, 1890).

to the Republican party. It was control of the Democratic party that would count and that must be achieved without delay.[33]

In South Carolina steady progress towards this goal under the leadership of Tillman had been registered prior to the election of 1890, and Tillman easily persuaded the outside leaders of the Alliance to allow him a free hand to go ahead in his own way. The political activities of the year were inaugurated in the famous "Shell Manifesto," written by Tillman but issued in the name of G. W. Shell, president of the Farmers' Association, which had been organized in South Carolina before the coming of the Alliance. This document called for a delegate convention of farmers to meet well before the date of the regular Democratic convention and to outline a program that the farmers might follow to the end that "the aristocratic coterie" in control of the state Senate and the state executive offices should be driven from power. The manifesto made it clear, however, that whatever action the farmers might take would stop short of bolting the Democratic party. Should the regularly elected Democratic state convention turn down their proposals they would abide by its decision.[34]

The convention thus called met on March 27, 1890, with two hundred and thirty-three delegates present, a great majority of whom were farmers. After a heated debate it was decided by a majority vote that a full slate of nominees for the state executive offices should be selected, and Tillman himself was promptly chosen to head it. For the other executive offices prominent farmers or farmer sympathizers were named, and with the active support of the Alliance, to which Tillman and most of his adherents belonged, the battle for control of the Democratic party was joined. The forces opposed to Tillman were by no means idle. Having taken the stand that pre-convention nominations such as the farmers

[33] *Ibid.*, 1:145 (May 25, 1889).   [34] Simkins, *Tillman Movement*, 68, 104.

had made were reprehensible, they could not name an opposing slate; but most of the conservatives agreed that Joseph H. Earle was the logical man for governor, although there was some sentiment in favor of General James Bratton. Earle and Tillman debated the issues of the day before excited throngs in various parts of the state. Earle's language and bearing showed clearly that he represented the well-to-do and better educated classes of the state. This made him anathema to the back-country farmers, who interrupted his speeches with rude outbursts and shouted their faith in Tillman. The contest was in very truth, as Earle asserted, a class movement; but it was not wholly inspired, as he thought, by the selfish ambitions of Tillman. Back of it lay the feeling on the part of the less fortunate of the rural whites that the governing classes had consistently betrayed the interests of the common people and had left them a ready prey to hostile economic forces from which they could never hope to escape without political aid.[35]

The excitement of the campaign for the Democratic nomination in South Carolina equaled that roused in the northwestern states by the appearance of third-party tickets. The debates between Earle and Tillman attracted crowds of turbulent citizens from far and near, some of whom were so eager for the fray that they arrived on the scene the day before the speaking was to take place. At Laurens a pro-Tillman audience refused to listen to the farmer's opponents, shouting "Give us Tillman" and "Bring out the one-eyed plowboy" until the conservative orators gave up. At Columbia the anti-Tillmanites were in the majority, and Tillman spoke to a hostile crowd, perhaps with danger to his life. But usually the majority was with Tillman, and the county-to-county canvass that he made demonstrated clearly the strength of the movement for which he stood.[36]

[35] Simkins, *Tillman Movement*, ch. 5.
[36] *Ibid.*, 119–125.

The fiery Tillman was a fit leader for such a cause. He was himself a farmer and looked the part; he talked in a coarse, vituperative manner that the none-too-squeamish rural classes could fully appreciate; he had suffered enough in his lifetime through physical injuries, family misfortunes, and now economic reverses to feel strongly the grievances he voiced. After Reconstruction Tillman had stuck to his farm, and up to 1881, according to his own acount, he had made money and " bought land and mules right along. In that year, I ran thirty plows, bought guano, rations, etc., as usual, and the devil tempted me to buy a steam engine and other machinery, amounting to two thousand dollars, all on credit." [37]

Then his troubles began. Drought and crop failure, " sickening " prices paid to country merchants, the forced sale of much of his land, drove him to the conviction that there was something radically wrong with the practice of agriculture in South Carolina. He thought at first that the remedy might be found in agricultural education and made some noteworthy efforts along that line, but he soon concluded that the state, by countenancing such devices as the crop mortgage, was largely responsible for the " decay of that sturdy independence of character, which was once so marked in our people." [38] And so into politics he came, maintaining that the farmers, constituting as they did some seventy-six per cent of the population of South Carolina, might " justly claim that they constitute the State, although they do not govern."

The farmers of South Carolina now had it in their power to demonstrate by the nomination and election of Tillman that the majority could really govern.[39] So effective was the Tillman canvass and so genuine was the enthusiasm of the farmers that not only Tillman but the entire Tillman ticket won the nominations overwhelmingly. More quietly the

[37] *Ibid.*, 51.   [38] *Ibid.*, 57.   [39] *Ibid.*, 64.

farmers picked up nominations for the state legislature and for Congress, took over to a very great extent the party machinery of the state, and forced the conservative minority into an undignified and unavailing opposition.[40]

Unfortunately for the farmers of North Carolina, who were almost as well organized for the struggle as those of South Carolina, the year 1890 was an off year in state elections, and a governor was not to be chosen. Members of the legislature were to be nominated, however, and in most cases the Farmers' Alliance saw to it that men in sympathy with Alliance views were placed on the Democratic ticket. Alliancemen dominated the state convention, which nominated candidates approved by the Alliance and adopted a platform that expressed sympathy with " the efforts of the farmers to throw off the yoke of Bourbonism." [41] Members of Congress who refused to pledge themselves to support the Alliance demands were denied Alliance support; and on this account two unsatisfactory candidates were forced to withdraw. Four out of nine Democratic candidates for Congress from the state were Alliancemen, and only two out of the nine dared express themselves as hostile in any degree to the Alliance and its aims.[42]

Colonel L. L. Polk furnished in North Carolina the leadership for the farmers' movement that Tillman furnished in South Carolina. Through his paper, the *Progressive Farmer*, he speedily came to advocate not only better methods of farming but also political reforms that might help the farmer. Polk's movement for reform in North Carolina, like Tillman's in South Carolina, anticipated the Alliance and was well under way before Alliance organizers entered the state. But whereas Tillman merely tolerated the Alliance, joining

[40] *Ibid.*, 127–134; *Appletons' Annual Cyclopaedia, 1890,* 778–779.
[41] *Appletons' Annual Cyclopaedia, 1890,* 625.
[42] Hamilton, *North Carolina since 1860,* 230; *Progressive Farmer,* April 29, 1890.

it and using it to advance his own program but showing it
no real devotion, Polk threw himself heart and soul into the
Alliance movement. His paper became the official Alliance
organ in the state, and through its columns ably written
editorials proclaimed the farmer's woes and advocated enthusiastically Alliance reforms. Polk's ardor led him to take
a strong interest in the national Alliance as well as the state
organization, and his reward came at St. Louis in December,
1889, when he was made national president. Polk not only
wrote ably on behalf of the farmer, but he was an orator of
no mean ability also; his newly attained national prominence
led him to spend much of his time outside the state, and
during the campaign he went as far afield as Kansas and
Dakota.[43] Such was Polk's prominence and influence that it
was to him that Senator Zebulon Vance addressed an urgent
appeal " to prevent this popular cry for redress from becoming a clamor for revenge." [44]

The spirit of revolt was strong in Georgia, where, as elsewhere in the South, the Alliance had determined to overthrow the ruling Bourbon aristocracy. So bent were the
farmers on their course and so overwhelming were their
numbers that opposition to the entrance of the Alliance into
politics was scarcely ever voiced — Alliance supremacy was
taken for granted. Indeed, the two chief candidates for the
Democratic nomination for governor, Leonidas F. Livingston and William J. Northen, were both Alliancemen, and
when the contest threatened to develop rancor, Livingston
was persuaded to run for Congress, and the nomination for
governor went to Northen practically without opposition.
Early in the campaign the Alliance announced that to secure
its support every candidate must pledge himself to a long
list of Alliance demands, and candidate after candidate

<hr />

[43] Morgan, *Wheel and Alliance*, 285–288; *National Economist*, 3:285 (July 19,
1890).
[44] *National Economist*, 3:260 (July 12, 1890).

either bowed to the will of the Alliance or lost his place on the party ticket. Three-fourths of the candidates nominated for the state Senate and four-fifths of those nominated for the House were named by the Alliance. The Democratic state nominating convention, which was controlled by the Alliance, ratified the Alliance choice for governor and adopted the Alliance platform. In six of the ten congressional districts the sitting Bourbon congressmen were refused renominations, and Alliance candidates were selected in their stead. The other four made common cause with the farmers.[45]

Among the new nominees for Congress in Georgia the most spectacular was Thomas E. Watson of the tenth district, in which lay the city of Augusta. Watson at the time of this campaign was only thirty-four years old — one of the youngest of the Alliance leaders. His background was that of the poor southern farmer boy who had worked his way upward against almost insuperable odds, aided by some talent and a great unwillingness to be downed. He had had two years at Mercer College, had taught school and read law, had been admitted to the bar, and had won notoriety, if not fame, as a criminal lawyer whose oratory jurors found irresistible. But this struggle to achieve had told on his character. He suspected of the worst motives all those who opposed him; he was combative and injudicious; he had little ability to control his temper. A natural defender of the " under dog " politically as well as otherwise, he had taken no very great part in politics before the farmers' crusade began. In 1882 he was elected to the state legislature, but he declined to stand for re-election; in 1888, as a low tariff reformer, he supported Cleveland for the presidency with considerable ardor, only to be convinced shortly thereafter that the money question far outweighed the tariff question in importance. But with the coming of the Alliance he found

[45] *Appletons' Annual Cyclopaedia, 1890*, 366; Arnett, *Populist Movement in Georgia*, 103–116.

his "cause." He made joyful war against the jute trust, and in 1890 he carried on a campaign for the congressional nomination from his district that was "hot as Nebuchadnezzar's furnace."

In this fight Watson accepted in full the demands of the Farmers' Alliance as his platform and spoke vehemently in every county in the district against the tariff, the national banks, and the currency system. He indulged also in vicious personal thrusts at the congressman whom he was attempting to displace and at his principal supporter, a certain Judge Twiggs. When Watson characterized a speech delivered by the latter as "the vaporing of a soured outlaw who is so accustomed to abusing everything and everybody that the restraints of truth have no power over him," Twiggs challenged his calumniator to a duel. Watson declined the honor but named a day during the fair at Atlanta when he would be on the streets of that city prepared to defend himself. He was not attacked.[46]

The tactics pursued by the Alliance in the Carolinas and Georgia were emulated to a greater or less degree throughout the South, but not always with the same success. In Tennessee the president of the state Alliance won the nomination for governor after a determined fight that ran to twenty-six ballots in the convention.[47] In Texas the attorney-general, H. H. Hogg, sought Alliance support of his candidacy for the Democratic nomination for governor by advocating an effective railway commission; and Alliance votes gave him an easy victory.[48] In Arkansas the Democrats vied with the Union Laborites in their loyalty to Alliance views and made a better showing against the Union Labor–Republican combination than they had made two years before.[49] In Alabama an Alliance leader, R. F. Kolb,

[46] Arnett, op. cit., 113–114; William W. Brewton, The Life of Thomas E. Watson, ch. 4.
[47] Appletons' Annual Cyclopaedia, 1890, 796.
[48] Ibid., 800; Charles Seymour (ed.), The Intimate Papers of Colonel House, 1:28.
[49] Appletons' Annual Cyclopaedia, 1890, 24.

sought the Democratic nomination and with the hearty support of the Alliance almost won it in spite of a powerful
Black Belt opposition, led by the *Montgomery Advertiser*.
But Kolb could muster only a plurality of votes in the
nominating convention, and the opposition combined to beat
him.[50] It chanced that in several southern states — Louisiana, Virginia, and Kentucky — no state elections were held
in 1890, while in several others — Mississippi, Florida, and
Missouri — interest was less feverish because governors were
not to be chosen. The Alliance in these states centered its
chief attention on the selection of legislators, where legislatures were to be chosen, and on the selection of congressmen.[51]

Throughout the South the election results were highly
gratifying to the Alliance. It elected its candidates for governor in South Carolina, Georgia, and Tennessee; and the
candidate it supported won in Texas. Numerous Alliance
candidates for minor state offices were successful in these
and other states. The legislatures of no less than eight southern states — Alabama, Florida, Georgia, Missouri, Mississippi, North Carolina, South Carolina, and Tennessee —
were counted as safely within the Alliance grasp. Perhaps as
many as forty-four of its candidates — men who had
definitely committed themselves to the support of Alliance
principles — had won seats in Congress; and two or three
United States senators professed Alliance views. In Mississippi and Kentucky, where elections for constitutional conventions had been held, a goodly sprinkling of Alliancemen
had won places as delegates. The claim of one southern
enthusiast that "being Democrats and in the majority, we
took possession of the Democratic party," was broader than
the facts entirely justified, although for some states such a
statement was certainly not far from the literal truth. When-

ever the Alliance candidates failed of nomination, however, the normal procedure was to give hearty support to the Democratic nominee, whoever he might be. The threat of a third party was thus studiously minimized.[52]

The Alliance in the Northwest, working through hastily devised third parties, won comparable victories. In Kansas ninety-one legislators elected were members of the new party, which now had an overwhelming majority in the lower house, although hold-over senators kept the state Senate safely Republican. The governorship and all but one of the other state offices went to the Republican candidates, but the margin of safety had been only a few thousand votes. Five congressmen were elected by the revolting farmers, and the legislature, which on joint ballot was safely under the control of the new People's party, unseated John J. Ingalls as United States senator and sent to take his place one of its own men, William A. Peffer of Topeka, editor of the *Kansas Farmer* and from this time on one of the most resourceful leaders in the third-party movement.

The newly elected senator from Kansas had had a varied career. He was born in Cumberland County, Pennsylvania, on September 10, 1831, and received an elementary education in the local country schools. At fifteen he became himself a country school teacher, began to collect a library, and learned to express in debating societies and through the papers his strong antislavery and protemperance views. After his marriage in 1852 he acquired a farm in Indiana, but the hard times that began in 1857 drove him on west to another farm in southwestern Missouri. When the war broke out, his Missouri neighbors, Confederate sympathizers for the most part, ran him out, and again he began life anew, this time in Warren County, Illinois. In 1862 he enlisted in the Union army and served all through the war, most of the

[52] Drew, in the *Political Science Quarterly*, 6:307–308; Haynes, *Third Party Movements*, 237–239; Arnett, *Populist Movement in Georgia*, 122–123; *National Economist*, 4:133 (November 15, 1890).

180   THE POPULIST REVOLT

time holding the rank of first lieutenant. After the war he opened a law office in Clarksburg, Tennessee, and participated actively as a conservative Union man in the reconstruction of that state. In 1870 he moved to Kansas, took a claim in Wilson County, opened a law office, and began to publish a county paper. He was soon active in Kansas politics, at first as a Republican but from 1888 on as an enthusiastic supporter of Alliance views. This is the impression he made on Hamlin Garland in 1891:

His general appearance is that of a clergyman. He wears a trimly fitting Prince Albert coat, a broad hat, and glasses. His beard is almost as long as the newspapers have reported, is nicely combed, and his whole appearance neat. His habitual expression is grave and introspective. He is a man of wide experience in law and newspaper work, and is self-possessed and dignified in his bearing on the floor of the Senate. . . . He has written a great deal upon the money question, and has done a great work in promoting the political revolution of Kansas. He is considered by his Kansas colleagues to be astute, well versed in political history, and an indefatigable worker. He made a peculiar impression upon me; something Hebraic — something intense, narrow, and fanatical. His face is inscrutable, his manner as peculiar as his face. He did not impress me as a small man who can be handled. He seems to me to be one whose purposes are inflexibly carried out, and the Senate of the United States will be sure to find him a powerful advocate.[53]

In Nebraska the successes of the Independents were hardly less startling than in Kansas. On the face of the returns the Democratic candidate for governor won, and although the Independents, whose candidate appeared to have run second, contested the place, the Democrat ultimately was permitted to hold his seat. The campaign was complicated somewhat by the prohibition issue, and many of the

[53] Garland, in the *Arena*, 5:455–456; Haynes, *Third Party Movements*, 240, 251; Miller, in the *Mississippi Valley Historical Review*, 11:369; Barr, in *Kansas and Kansans*, 2:1153–1156. An estimate of Peffer and of the significance of his election is given in the *Nation*, 52:104 (February 5, 1891).

alleged frauds were supposed to have been committed by overardent " wets," who hoped for Democratic victory. Except for the governorship the Republicans won all the state offices, but the Independents had an out-and-out majority in both houses of the legislature. In the congressional elections one Democrat, one Independent, and one fusionist won — not a single Republican representative from an erstwhile solidly Republican state was elected. Unfortunately for the Independents, Nebraska was not at this juncture entitled to elect a United States senator.[54]

Elsewhere in the Northwest the third-party gains were not so startling. In South Dakota and in Minnesota the new party held the balance of power and in the former state used its votes in the legislature to such advantage as to secure a United States senator of Independent views, one Reverend James H. Kyle, a Congregational minister not yet forty years of age.[55] In North Dakota, Michigan, Iowa, Illinois, Indiana, and Colorado the third-party candidacies ate more or less deeply into the customary Republican vote and resulted in the selection of an unusual number of Democrats, particularly to seats in Congress. Altogether the showing made by the Alliance in this first essay at independent political action was by no means discouraging.[56]

In the states where the more notable Alliance victories had been scored much was expected of the newly elected legislatures, but, curiously enough, not a great deal was done by them. Even in South Carolina, where the radical Governor Tillman took charge, the results were meager. Tillman advocated a reapportionment of the legislative districts that would be more to the liking of the back-country whites, a series of changes in the state educational system that would

[54] Barnhart, " Farmers' Alliance and People's Party," 240–241; Haynes, *Third Party Movements*, 241.
[55] Haynes, *op. cit.*, 252.
[56] See the table of results compiled by Drew. in the *Political Science Quarterly*, 6:307.

improve the rural schools and would emphasize the importance of agriculture and the mechanic arts in higher education, a reorganization of the railway commission with
"honest men" in charge, and sundry minor reforms. The
legislature did not hesitate to reapportion seats in the House
in accordance with the governor's wishes, and the state educational system came in for drastic reform; but the railway
commission bill that finally passed was so unsatisfactory to
the governor that he vetoed it.[57]

Elsewhere in the South the results were much the same.
The railway bill that became law in North Carolina was not
unlike the measure that Governor Tillman so wrathfully
vetoed in South Carolina. It established a commission of
three members elected by the legislature for six-year terms,
it forbade rebates and unjust discriminations, and it en-
joined upon the roads the "long and short haul" principle
in rate-making; but the authority over rates given to the
commission was left subject to the usual limitations of judicial review.[58] In Georgia the right of the railway commission
to fix railway rates was asserted, and the jurisdiction of the
commission was extended to include express and telegraph
companies; banking corporations were also more closely cir-
cumscribed in their operations; blacklists were forbidden and
a twelve-hour day for railway employes was established; but
here, as elsewhere in the South, the pressing problem of crop
mortgages was left untouched, the country merchants and
the "trusts" were left free to continue their various depre-
dations on the earnings of the helpless cropper, and hard
times were not abolished.[59]

In Kansas the demands for remedial legislation were
varied and drastic. The state Alliance, in session at Topeka,
insisted that the shrinkage in farm values should be appor-

[57] Simkins, *Tillman Movement*, ch. 6.
[58] *Laws and Resolutions of the State of North Carolina*, legislative session of
1891, ch. 320, pp. 275–288.
[59] Arnett, *Populist Movement in Georgia*, 120–122.

tioned between the mortgagor and the mortgagee — in other words, that the face value of the mortgage should be scaled down by legislative act.[60] The Kansas county Alliance presidents, in convention assembled, urged the establishment of a graduated income tax to insure that the holders of bonds and mortgages would pay their share of the burden of government, but the county presidents felt that homesteads occupied by their owners should be freed, in whole or in part, from taxation. With a vigor that would have pleased Henry George himself they advocated also that lands held for specutive purposes by nonresidents, aliens, and corporations should be taxed.[61] A bill, which passed the lower house of the legislature, but failed in the Senate, proposed to relieve the mortgaged farmers by allowing them two years in which to redeem their homes after foreclosure and by prohibiting the courts from adding a personal judgment when the property value proved inadequate. Another, which also failed, would have prohibited municipalities from subscribing for railway stock or voting bonds in aid of railroads. Another would have reduced freight rates in Kansas by thirty per cent and passenger rates to two and one-half cents a mile. Still another proposed to lower the rate of interest charged on unpaid taxes. A few of the bills sponsored by the new People's party became law. One such prohibited the alien ownership of land in Kansas; another forbade dealers in live stock to prevent competition by any combination of forces; another attempted the regulation of warehouses and the inspecting, grading, weighing, and handling of grain. But even so, the way of the pioneer farmer in Kansas was still hard.[62]

In Nebraska the third-party reformers centered their activities mainly upon the railway problem, introducing and finally passing through both houses a bill known as the Newberry Bill. This measure proposed an extensive revision

[60] *Nation*, 50:269–270 (April 3, 1890).
[61] Barr, in *Kansas and Kansans*, 2:1146.
[62] *Ibid.*, 2:1156–1157.

of freight classifications and prescribed a detailed schedule of maximum charges that would have reduced the local rates in force from forty to sixty per cent. The existing Board of Transportation was to be continued, but with increased regulative powers. Defenders of the measure held that it merely asked for Nebraska what the railroads had already been compelled to concede for Iowa; opponents, that its passage would mean bankruptcy for the roads. The bill reached the governor only to receive his veto; and an attempt at repassage over the governor's veto failed.[63]

The failure of the Newberry Bill was extremely disheartening to the Nebraska Independents, but with a number of minor measures they had better luck. They took the lead in securing the passage of an Australian ballot act; they voted a hundred thousand dollars for the relief of drought-stricken inhabitants in the western part of the state; they subjected elevators and warehouses to state regulation and inspection; they repealed an existing law that granted a bounty to the makers of beet sugar; they legalized mutual insurance companies; and they urged Congress to foreclose the national mortgage on the Union Pacific Railroad, claiming that this mortgage was being used as a pretext to keep rates high. Considering the inexperience of the farmer members, the strength of the opposing lobbies, the effectiveness of the old-party machines, and the hostility of the governor, this was a fair record; but the Independents could not conceal their chagrin over the failure of their pet measure, the Newberry Bill. Jay Burrows in the *Farmers' Alliance* wrote wildly of impending revolution:

We send the plutocrats a grim warning. . . . The twin of this oppression is rebellion — rebellion that will seek revenge with justice, that will bring in its Pandora's box fire, rapine and blood. Unless there is a change and a remedy found, this day is as inevitable as that God reigns, and it will be soon. . . .[64]

[63] Dixon, in the *Political Science Quarterly*, 13:634.
[64] *Farmers' Alliance*, March 28, 1891; Barnhart, " Farmers' Alliance and People's Party," 267–273.

In Minnesota, where the Alliance forces were led by Ignatius Donnelly, a member of the state Senate, the Alliancemen joined with the Democrats to organize both houses against the Republicans. At first Donnelly held a tight rein over the farmer members, and with cordial support from the Democrats apparently assured, he should have scored a great success. He finally decided that the chief Alliance measure should be an elaborate amendment to the constitution, through which the state would certainly acquire adequate regulatory authority over railways and elevators. But the amendment did not even pass the Senate, nor did any other important item of the Alliance program succeed. "No man on the floor of either house," wrote the hostile *St. Paul Globe*, "did as much as Ignatius Donnelly to prevent the enactment of wise and needed laws. He was not the author of a single constructive measure of any value before that body, and by his endless chattering, his buffoonery, his appeals to the gallery, he frittered away the golden hours in which something might have been achieved." But this was not the whole explanation of the failure of the Alliance. The third-party movement had the same insurmountable obstacles to meet in Minnesota as in Kansas and Nebraska. Only by a miracle could it have carried its program through. At the end of the session an Alliance manifesto, written in Donnelly's best style and signed by many of the Alliance legislators, confessed freely, "We are defeated but not disheartened." [65]

It was painfully apparent from these failures that neither the southern formula of working through the Democratic party nor the northwestern formula of local parties organized along state lines had proved entirely adequate. The signs pointed unmistakably in the direction of a national third-party organization, which might supplement the state organizations and take the lead in securing national measures of reform that the states were powerless to effect.

[65] *St. Paul Globe*, January 7, 10, 14, April 22, 29, 1891. See also Donnelly Scrapbooks, Vol. 8.

CHAPTER SEVEN

THE SUBTREASURY PLAN[1]

AMONG the subjects discussed by the farmers during the campaign and election of 1890 was an interesting plan for the relief of agriculture known usually as the "subtreasury." This scheme was first presented at the December, 1889, meeting of the Southern Alliance in St. Louis. On the last day of that session, after the regular demands of the Alliance had been read and approved, a report was received from the "Committee of the Monetary System." This committee consisted of the leading members of the order: C. W. Macune, past president of the Southern Alliance and editor of its official journal, the *National Economist;* Colonel L. L. Polk of Raleigh, North Carolina, newly elected president and editor of the prosperous *Progressive Farmer;* L. F. Livingston, president of the Georgia State Alliance and a power in Georgia politics; W. S. Morgan, author of *A History of the Wheel and Alliance,* which the convention had just branded as "official"; and H. S. P. Ashby, a prominent member of the Texas Alliance.

The personnel of this committee was bound to attract attention to its report, and undoubtedly that was a part of the plan. Just why the report was so long delayed is not fully apparent. Perhaps it is true, as was repeatedly charged, that the men back of the scheme sought to avoid all risk of

[1] This chapter follows in the main an article by the author entitled "The Subtreasury: A Forgotten Plan for the Relief of Agriculture," in the *Mississippi Valley Historical Review,* 15:355–373.

defeat by waiting until many of the delegates had gone home and the convention was too disorderly to know exactly what was happening.[2] Perhaps the leaders of the Knights and of the Northern Alliance had been consulted previously, had refused to approve the plan, and had thus kept it out of the regular "demands." Perhaps Macune, who was the proponent of the idea, had had a hard time convincing the other members of the committee of its wisdom. Some explanation of the tardiness with which the report was made is certainly due, for it could hardly have been the work of only a few hours' conference and must have been given some consideration in advance of the session.

In presenting the report Macune argued that heretofore the Alliance had "scattered too much and tried to cover too much ground" and that it should now "concentrate upon the one most essential thing and force it through as an entering wedge to secure our rights."[3] The subtreasury was that "thing." Apparently, as an unsympathetic observer declared, the southern leaders hoped to shelve all other reforms and to produce "a simultaneous demand of the farmers all along the line, from Dakota to Texas and from the Atlantic to the Pacific, for this warehouse or sub-treasury scheme."[4]

The Macune report, after making an effort to conciliate the free-silverites by urging again "the free and unlimited coinage of silver or the issue of silver certificates against an unlimited deposit of bullion," got down to business as follows:

That the system of using certain banks as United States depositories be abolished, and in place of said system, establish in every county in each of the States that offers for sale during the one year five hundred thousand dollars worth of farm products; including wheat, corn, oats, barley, rye, rice, tobacco, cotton, wool and sugar, all together; a sub-treasury office, which shall

[2] *National Economist* (Washington), 2:215; 3:301; 4:14 (December 21, 1889; July 26, September 20, 1890).
[3] *Ibid.*, 2:199 (December 14, 1889).
[4] *Farmers' Alliance* (Lincoln), January 19, 1890, quoting the *Iowa Homestead*.

have in connection with it such warehouses or elevators as are necessary for carefully storing and preserving such agricultural products as are offered it for storage, and it should be the duty of such sub-treasury department to receive such agricultural products as are offered for storage and make a careful examination of such products and class same as to quality and give a certificate of the deposit showing the amount and quality, and that United States legal-tender paper money equal to eighty per cent of the local current value of the products deposited has been advanced on same on interest at the rate of one per cent per annum, on condition that the owner or such other person as he may authorize will redeem the agricultural product within twelve months from date of the certificate or the trustee will sell same at public auction to the highest bidder for the purpose of satisfying the debt. Besides the one per cent interest the sub-treasurer should be allowed to charge a trifle for handling and storage, and a reasonable amount for insurance, but the premises necessary for conducting the business should be secured by the various counties donating to the general government the land and the government building the very best modern buildings, fireproof and substantial. With this method in vogue the farmer, when his produce was harvested, would place it in storage where it would be perfectly safe and he would secure four-fifths of its value to supply his pressing necessity for money at one per cent per annum. He would negotiate and sell his warehouse or elevator certificates whenever the current price suited him, receiving from the person to whom he sold, only the difference between the price agreed upon and the amount already paid by the sub-treasurer. When, however, these storage certificates reached the hand of the miller or factory, or other consumer, he to get the product would have to return to the sub-treasurer the sum of money advanced, together with the interest on same and the storage and insurance charges on the product.[5]

The authorship of this astounding scheme was quite naturally, and quite properly, attributed to the man who had presented it to the convention. Macune was known to be

[5] *National Economist*, 2:216–217 (December 21, 1889). The subtreasury plan is also printed in full in Bryan, *Farmers' Alliance*, 93–94.

fertile in ideas and unafraid of innovations. It was he who by a daring venture had saved the Texas Alliance from dissolution; it was he who had initiated the policy of expansion that had brought the Southern Alliance into being; it was he who had introduced the system of business agencies that was even then spreading like wildfire over the entire South. Until recently he had opposed the use of the Alliance for political ends, speaking of it as " primarily a business organization "; but by the time of the St. Louis gathering he had come to the conclusion that the government must help the farmer and that the farmer must see to it that the government did whatever was needed.[6] The germ of the subtreasury idea Macune possibly found in an article in *Frank Leslie's Newspaper* for November 30, 1889, by a North Carolinian named Harry Skinner. Skinner's plan was designed as a balance for the protective tariff which the people had authorized by their votes in the election of 1888, and which, it was assumed, Congress would soon enact. In lieu of the tariff protection to be accorded the manufacturer, the farmer, or at least the southern farmer, should receive assistance from the government in the shape of warehouses wherein to deposit his produce, and loans of paper money with these deposits as security. But Skinner ruled grain out of the scheme as " perishable," and made the southern farmer, whose cotton was " non-perishable," the chief beneficiary. What Macune did was to broaden the scope of Skinner's scheme. After all, what should be regarded as " perishable " was merely a matter of definition, and Macune chose to revise Skinner's definition to include all those products, western and southern, that could be stored successfully.[7]

According to the official Alliance historian, the introduction of the subtreasury plan in St. Louis aroused an " ani-

[6] *National Economist*, 1:8; 2:198 (March 4, 1889; December 13, 1890); Garvin and Daws, *Farmers' Alliance*, 68; Morgan, *Wheel and Alliance*, 298-300.
[7] After the St. Louis meeting Macune reprinted Skinner's article in the *National Economist*, 2:228 (December 28, 1899), to show " the writer's idea of the sub-treasury system." See also the *Southern Mercury*, December 1, 1892.

mated discussion," after which it "was adopted by a large majority."[8] Following the convention the men responsible for the measure set about to popularize it. Macune in the *National Economist* devoted columns and pages to expounding it and proclaiming its virtues. Polk in the *Progressive Farmer* was not far behind him; nor were the editors of most of the Southern Alliance papers. Lecturers also devoted themselves to the subject; and numerous state Alliance meetings were induced to indorse the plan in terms of glowing commendation.[9] The idea attracted the attention of editorial writers the country over, and while outside Alliance circles the criticism was caustic, the advertising the plan received thereby was enormous. When the next meeting of the Supreme Council of the Southern Alliance occurred in Ocala, Florida, in December, 1890, the subtreasury was again approved, and it persisted long enough as a popular plan of farm relief to find a place in the early platforms of the People's party.[10]

Defenders of the subtreasury were quick to point out that their scheme was merely an adaptation of the system under which the national banks operated. The banks had the privilege of depositing with the United States treasurer government bonds, upon which as security they were privileged to issue paper money in quantities up to ninety per cent of the value of the bonds deposited. The farmers asked for themselves only the same privilege that had already been accorded to the bankers. They would deposit not bonds but agricultural produce, and they would ask of the government paper only up to a maximum of eighty per cent of the value of the deposits.[11] "On the face of the national bank note these words are printed: 'This note is secured by bonds of the

[8] Morgan, *Wheel and Alliance*, 175.
[9] *National Economist*, 5:376; 6:120 (August 29, November 7, 1891); *Southern Mercury*, December 1, 1892; Dodge, in the *Century* (new series), 21:455.
[10] *National Economist*, 3:1; 4:284; 6:161; 7:257–258 (March 22, 1890; January 17, November 28, 1891; July 2, 1892).
[11] *Ibid.*, 2:239 (January 4, 1890).

United States deposited with the U. S. Treasurer at Washington.' What better is that indorsement than this — ' This note is secured by wheat deposited in the Government warehouse at Washington.' " [12] The national-bank system was clearly intended to expand the currency beyond the amount of gold, silver, and greenbacks in circulation. The subtreasury would "merely push the system of supplementing the national stock of currency a little further to insure expansion and contraction as needed." [13] Macune's committee claimed that theirs was " no new or untried scheme; it is safe and conservative, it harmonizes and carries out the system in vogue on a really safer plan, because the products of the country that must be consumed every year are really the best security in the world, and with more justice to society at large." [14]

It was claimed, too, that there was a certain analogy between what the government was already doing for the distillers and what the farmers were asking it to do for the agricultural classes. The idea was current that the government built warehouses in which distillers of whiskey were privileged to store their liquor rent free and without insurance charges, pending payment of the ninety cents per gallon then collected on their product; that the liquor so stored was guarded by government officials to whom large salaries were paid; that upon it certificates of deposit were issued that were " good collateral security at any bank." If the government could do this for the distillers, why not also for the farmers? [15] A little investigation brought out the fact that the warehouses were built at the expense of the distillers, not the government; that the government custodians were there only to see that the tax was not evaded;

[12] Peffer, *The Farmer's Side*, 245.
[13] Bryan, *Farmers' Alliance*, 92.
[14] *Ibid.*, 94–95.
[15] Chamberlain, in the *Chautauquan*, 13:340. Cf. *National Economist*, 2:383 (March 1, 1890).

and that the distillers who borrowed upon their bonded liquor borrowed from private sources and not from the United States Treasury. Still, there was enough in the analogy to keep it alive, and the defenders of the subtreasury used it persistently. When cornered by the facts they were apt to retort irrelevantly that at least the government built post offices and customhouses, and that even if it had not lent money to the distillers it had once lent freely to the Pacific railroads.[16]

Proponents of the subtreasury never tired of reiterating that their scheme, if adopted, would "emancipate productive labor from the power of money to oppress."[17] According to Macune it was a "waste of energy for all the farmers in this great land to combine and cooperate to raise the prices of a given product . . . twenty-five per cent, while at the same time . . . another class of citizens . . . would depress prices fifty per cent."[18] The second class of citizens referred to was, of course, that of the national bankers, who could at will contract the currency and, according to the quantity theory of money to which the farmer economists pinned their faith, could beat down in this fashion the price of farm products. The subtreasury, it was estimated, would put into circulation about $550,000,000 (some said more), which the banks would have no power to withdraw. It would, moreover, make provision for an elastic currency, expanding in volume when the rush of business demanded and contracting when business was slack. "It is not an average adequate amount that is needed," said Macune, "because under it the greatest abuses may prevail, but a certain adequate amount

Farmers' Alliance, July 19, 1890; Greensboro Daily Record, (Greensboro, North Carolina), September 10, 1891; National Economist, 2:372–373; 3:188 (March 7, June 7, 1890). The subtreasury plan is discussed favorably by C. C. Post, "The Sub-treasury Plan," Arena, 5:342–353. It is vigorously assailed in a report by the House Ways and Means Committee, House Report No. 2143, 52 Congress, Session 1, 1892.

National Economist, 2:355 (February 22, 1890); Bryan, Farmers' Alliance, 98.

National Economist, 2:197 (December 14, 1889).

that adjusts itself to the wants of the country at all seasons." [19] Especially at the time when the crops had to be moved and the farmer had to sell in order to meet his obligations the subtreasury would help. Instead of a dearth of currency with extremely low prices that would rebound as soon as the rush of marketing was over, the subtreasury would make an abundance of funds available, and prices would be stabilized. Afterwards, when the marketing season was over and the produce stored in the subtreasury was taken out for use, the money issued upon it would automatically be withdrawn from circulation. A safe and sound, but flexible, currency, beneficial not only to the farmer but also to society as a whole, would thus be achieved.[20]

An added attraction of the subtreasury was the scheme of short-time rural credits that it provided. If the system ever went into effect, the farmer could borrow money to the extent of eighty per cent of the value of his crop at an interest rate of one per cent. This loan might not last for more than one year, but in that time he could watch for a favorable time to market and could get for himself the profits that had hitherto gone to the speculators, whose practice it was to buy at low prices during the annual orgy of selling and to dispose of their purchases a little later when prices had gone up. Western farmers could pay the interest on their debts, and eventually the debts themselves. Southern farmers could hope for gradual emancipation from the toils of the country merchant.[21]

There seemed, indeed, no end to the good things that might be hoped from the subtreasury. Economy in large storage (and, it might have been added, in storage largely at government expense) would reduce the profits of the middleman and save money for the farmer.[22] The subtreasury

[19] Ibid., 2:227 (December 28, 1889); Bryan, Farmers' Alliance, 92, 102.
[20] Blood, Handbook, 40; Bryan, Farmers' Alliance, 94–99.
[21] Farmers' Alliance, July 19, 1890.
[22] National Economist, 3:97–102 (May 3, 1890).

"would effectually put an end to all gambling in futures and wipe out the illegitimate methods of the various boards of trade that now curse the country. It would also transfer a continuous market for cotton, wheat, etc., from the large business centers to the smaller ones of the country, and thereby give to them the profits which the larger markets now receive." [23] Southern cotton factories "would spring up in every village of the South," and the ideal of Henry Clay, a home market and a home demand, would be achieved at last. [24] With Macune, and with the multitudes who followed his leadership, the subtreasury became an obsession. "Without it," Macune declared after the election of 1892, "both political parties would have given the Alliance all it asked long ago. But without it there was no relief for the people." [25]

Promptly on the adjournment of the St. Louis conventions, bills embodying the subtreasury idea were drawn by Alliance leaders and at their request were presented to Congress, in the House of Representatives by John A. Pickler, Republican, of South Dakota, and in the Senate by Zebulon Vance, Democrat, of North Carolina. In the House the bill was referred to the Ways and Means Committee, as was to be expected; but in the Senate it was referred to the Committee on Agriculture and Forestry, from which Alliance leaders hoped for better treatment than would have been accorded the measure by the Committee on Finance. [26] Petitions and memorials, literally by the hundreds, then began to descend upon Congress, praying for the enactment of the bill into law. From the state of North Carolina alone, there came during this session fifty-seven. [27] Congressmen who had proclaimed their friendship for the farmers were asked now to

[23] Ibid., 5:233 (July 4, 1891).
[24] People's Party Paper (Atlanta), January 14, 1892.
[25] Southern Mercury (Dallas), December 1, 1892.
[26] National Economist, 2:371, 392; 3:259 (March 7, 15, July 12, 1890); Congressional Record, 51 Congress, Session 1, February 18, 1890, p. 1468; February 24, 1890, p. 1645.
[27] Congressional Record, 51 Congress, Session 1, Index, 667; National Economist, 4:133 (November 15, 1890).

THE SUBTREASURY PLAN 195

take a stand, and many of them were visibly embarrassed. "The sudden advent of the subtreasury upon Congress," wrote one observer, "had about the effect of the unexpected descent of a hawk in the barn yard — great commotion and no little flutter." [28] Probably no one was more distressed than Senator Vance, who, after having introduced the bill, decided that he could not support it.[29] The committees to which the bill was referred were careful, however, to give the Alliance leaders an opportunity to present their case; Macune himself, among others, appeared before both committees and spoke eloquently for the measure. But neither committee could be induced to report out a subtreasury bill of any kind, either favorably or unfavorably.[30]

Even the severe defeat administered to the majority party in the election of 1890 — a defeat accomplished in no small part by Alliance votes — failed to arouse Congress to action. In the short session of 1890–91 the subtreasury reappeared, but nothing was done about it, except that the Senate Committee on Agriculture and Forestry washed its hands of the whole matter by reporting back the proposition with the recommendation that it be referred to the Senate Committee on Finance; this was done.[31]

In the first session of the next Congress, the subtreasury measures were similarly buried until Thomas E. Watson, newly elected representative from Georgia, took it upon himself to force the House Ways and Means Committee to report back its findings. One morning in May, 1892, he asked unanimous consent for the passage of a resolution directing the Committee to bring in a report.[32] Objection was of course raised, but thereafter Watson objected to every request for unanimous consent to anything else until his own

[28] *National Economist*, 3:245 (July 5, 1890).
[29] *Ibid.*, 3:259, 278 (July 12, 19, 1890).
[30] *Ibid.*, 3:97, 145, 153, 184, 190 (May 3, 24, June 7, 1890); *Progressive Farmer* (Raleigh), May 13, 1890.
[31] *Congressional Record*, 51 Congress, Session 1, December 9, 1890, p. 237.
[32] *Congressional Record*, 52 Congress, Session 1, May 19, 1892, p. 4432.

request was granted. A few days later his resolution was agreed to, and the Committee was ordered to report; but the report was delayed until the last day of the session, when it was too late to debate it and to secure a record vote.[33]

Meantime the critics of the subtreasury in and out of Congress had not been idle. Papers that before the plan was proposed had scarcely recognized the existence of the Farmers' Alliance now discovered the organization and set out to discredit, principally through ridicule, its favorite reform. The subtreasury was branded by the *New York Times* as " one of the wildest and most fantastic projects ever seriously proposed by sober man," and in one form or another the leading metropolitan dailies echoed these sentiments.[34] The sagacity of the Alliance leaders was called into question. They were spoken of as "hayseed socialists," and projectors of "potato banks."[35] Criticism came even from those who were regarded as the friends of the farmer. The subtreasury, according to R. F. Pettigrew, was "impracticable in all its features, and the most absurd piece of legislation I have ever seen presented."[36] Colonel Cockrell advised " that instead of crying for a sub-treasury we should go to subsoiling."[37]

Humorists did their bit. One self-styled poet, declaring the price of poetry to be low and the market crowded, wanted a literary bureau in the warehouse so that he might realize eighty per cent on unsold poetry, and then " set down an' wait for poetry to go up."[38] A farmer who had great luck plowing up snakes thought it was a pity that there was

[33] *National Economist*, 7:168 (May 28, 1892); *Congressional Record*, 52 Congress, Session 1, May 20, 1892, p. 4480; May 21, 1892, p. 4515; May 23, 1892, p. 4563; *House Report* No. 2143, 52 Congress, Session 1.
[34] *New York Times*, December 12, 1890; *National Economist*, 3:1; 5:193 (March 22, 1890; June 13, 1891).
[35] *National Economist*, 4:230 (December 27, 1889); Bryan, *Farmers' Alliance*, 109.
[36] *National Economist*, 3:203 (June 21, 1890).
[37] *Southern Mercury*, June 30, 1892.
[38] *Greensboro Daily Record*, September 16, 1891.

no subtreasury warehouse in which he could deposit them and draw eighty per cent of their value in currency.[39] But the persistent way in which the opposition harped on the subject seems to indicate that, after all, the subtreasury was not regarded as a joking matter. Under the ridicule was a note of desperate earnestness and dread lest the supporters of the measure might conceivably have their way.[40]

When opponents of the subtreasury were willing to argue the question seriously they found it fairly easy to pick flaws in the plan. They called attention to the possibility of money inflation on an alarming scale — if it took seven billions of dollars to move the crops, why might not eighty per cent of that sum be demanded in paper? Or even if the sum were less, would it still not be enough to derange utterly the nation's finances?[41] Furthermore, could the existing inelasticity of the currency be worse than the rapid expansion and contraction that would be possible, indeed inevitable, should the subtreasury go into operation? Supposing that, under the subtreasury proposal, a billion dollars worth of farm products were stored, and on them eight hundred million dollars' worth of paper was issued. If only half the stored products were removed within six months, the currency would be contracted by four hundred million dollars. Such fluctuations in the money supply would stimulate rather than hamper speculation and would be "the most fruitful source of panics conceivable."[42]

Some argued that the system might give the farmers an unreasonable power to demand high prices and thus seriously to injure the consumer; others wondered if the subtreasury might not operate in such a way as to be an actual detriment to the farmers. It might, for example, encourage Americans

[39] *Ibid.*, October 2, 1891.
[40] *National Economist*, 5:353 (August 22, 1891).
[41] Bryan, *Farmers' Alliance*, 102.
[42] *House Report* No. 2143, 52 Congress, Session 1, p. 3; *National Economist*, 4:3 (September 6, 1890).

to hold their grain too long at home, allowing the foreign demand to be supplied by India and Russia instead of the United States and stimulating wheat-growing in those countries to such an extent that within "a few years the foreign wheat market will be lost to America, and the condition of the American wheat grower worse than ever."[43]

The charges of "paternalism" and "class legislation" were persistently hurled at the subtreasury. Said the House Committee: "No good could come from making every citizen feel that from the time he plants his crop till it is harvested that to the Government he is to look for a banker and broker, and for safe storage and custody. . . . It will be an unfortunate day, when in a simple republic, Government takes charge of all the farm products, stores them in granaries, and becomes the main agent for the transaction of the business of the citizen."[44] "Is it to be supposed," inquired the *New York Times*, "that the Government will collect money from the great body of the people for the purpose of lending it to a favored class among them at a specially low rate of interest?"[45] Such favors, the supporters of the subtreasury contended, had been conferred freely upon other classes, particularly the national bankers, the tariff-protected manufacturers, and the subsidized railroads; but these arguments were lost on their opponents. The subtreasury, they declared, not only discriminated in favor of the farmers as a class; it discriminated in favor of certain classes of farmers. The men who formulated the plan purposely selected a few main staples, such as could be safely stored and were always in demand. But what good would the subtreasury do the live-stock farmer or the fruit grower?[46] And of what use would the government warehouse be

[43] *Farmers' Alliance*, July 19, 1890.
[44] *House Report* No. 2143, 52 Congress, Session 1, pp. 2–3. Cf. *National Economist*, 3:1, 111 (March 22, May 3, 1890).
[45] December 10, 1890.
[46] T. C. Jory, *What is Populism? An Exposition of the Principles of the Omaha Platform, Adopted by the People's Party*, 9; *National Economist*, 2:328 (February 15, 1890).

even to the farmer whose products were receivable, in case he should live at a distance of twenty-five to fifty miles from the nearest warehouse? And this condition must inevitably occur, for even the most favorable estimates presumed that not more than one out of four counties would have a warehouse. Obviously, too, these counties would be the more wealthy and more populous counties — the ones least in need of help. Perhaps the cotton farmers, whose crop had less bulk in comparison with its value than the crops of the northwestern grain growers, could make the long hauls profitably; but the grain growers could not.[47]

And, granting even that the subtreasury was all that its advocates said, was it not clearly unconstitutional? Most of its critics were convinced that this was the case.[48] The House Ways and Means Committee in its adverse report on the bill so held,[49] and members of Congress who feared to express their doubts as to the wisdom of the measure frequently dodged the issue by asserting that there was no point to passing a law that would promptly be set aside by the Supreme Court. The Court, in fact, had held that it was unconstitutional to lend money to citizens from the Treasury for the carrying on of private business,[50] and the subtreasury plan proposed something closely akin to that. But the advocates of the subtreasury probably were justified in their assumption that any bill that proposed farm relief and money inflation would have been met with numerous legal and constitutional objections. There was occasion for the comment that " a man who says he recognizes the justice of the demands of the farmers, yet can conceive no means by which they are to be granted, is not a capable representative at a time when relief must be had or liberty perish." [51] More-

[47] *Ibid.*, 3:72–73; 4:380 (April 19, 1890; March 24, 1891).
[48] See, for example, the *National Economist*, 3:135 (May 17, 1890), and the *Southern Mercury*, March 10, 1892.
[49] *House Report* No. 2143. 52 Congress, Session 1, p. 5.
[50] *Ibid.*, 5–7.
[51] *National Economist*, 3:266, 278 (July 12, 19, 1890).

over, it was held with some show of reason that the Sherman
Silver Purchase Act, which Congress did not scruple to pass,
really " warehoused *silver* and issued notes against it " just
as the Alliance proposed to do for farm products. The analogy was not complete, for the government actually bought
the silver, but it was close enough to make an effective argument.[52]

Finally, however, the persistent opposition to the subtreasury began to tell, even upon the farmers. Indeed, there
was a disposition from the very first on the part of the
Northern Alliance to agree with the critics of the subtreasury plan.[53] Its adaptability to the needs of the southern cotton grower might be conceded, but in the North its provisions hardly seemed to apply. One northern farmer declared
it " a measure more infamous than the national banks," [54]
and a prominent Alliance editor said of it: " We heard this
scheme presented, on a celebrated occasion, under an injunction of secrecy. We had fondly hoped that the secrecy was
to be eternal; but we were doomed to disappointment." [55]

At Ocala an attempt was made to satisfy the northern
critics by stating that the government should also lend
money upon real estate, " with proper limitations upon the
quantity of land and amount of money," [56] and this modification Senator Peffer, in a book published the following year,
attempted to elaborate and popularize. The subtreasury, as
he saw it, should furnish the short-time loans; but the long-
time loans, which the western farmer must have if he were
to succeed in holding his land, would come from a loan bureau to be created in the Treasury Department in Washing-

[52] *Ibid.*, 2:401; 3:195, 211, 292 (March 15, June 14, June 21, July 19, 1890).
[53] See, for example, the attack on the subtreasury in Ashby, *Riddle of the
Sphinx*, 309–316. Ashby was a lecturer in the Northern Alliance, and his book
is a compound of Alliance views and history.
[54] Drew, in the *Political Science Quarterly*, 6:297.
[55] *Farmers' Alliance*, January 4, 1890.
[56] *Ocala Proceedings*, 32. Senator Leland Stanford of California had proposed
such a measure in Congress the preceding spring. The text of Senator Stanford's
bill is printed in part in Ashby, *Riddle of the Sphinx*, 322–324.

ton. This bureau would lend money to the people at one per
cent on real estate security.[57] Even so, the northern farmers continued generally skeptical. When they agreed to support the subtreasury in the early Populist platforms, as they
did, it was more or less as part of a bargain by which the
South reluctantly conceded the principle of government
ownership of the railways in return for equally reluctant
northern support of the subtreasury; and thereafter many
northern Populists took comfort in the fact that their platforms always expressed a willingness to accept any "better
plan" than the subtreasury that could be devised.[58]

In the Southern Alliance, where the attitude of the national officers was usually sufficient to sway the ordinary
farmers one way or another, the subtreasury was generally
supported, and in the election of 1890 there was a tendency
to measure all candidates for office by the "yard-stick" of
the St. Louis platform, subtreasury and all.[59] Formidable
opposition, however, did exist. Democratic leaders who
viewed with some misgivings the entrance of the Alliance
into politics were quick to seize upon the subtreasury as a
means of discrediting the whole farmers' movement.[60] One
editor insisted that the scheme had been foisted upon the
masses by a few men at the top, and he warned the country
and subordinate lodges to discuss fully, before taking action,
all matters that were sent down to them.[61] Wade Hampton
of South Carolina branded the measure as "so palpably
wrong on its face as to make it absurd to all who have the
prosperity and welfare of the country at heart. This feature
of the Alliance has never been fully accepted in the South,
and I have too much confidence in our people to think that

[57] Peffer, The Farmers' Side, 247–256.
[58] Southern Mercury, February 23, 1893; National Economist, 3:184, 190;
7:184 (June 7, 1890; June 4, 1892); Walker, in the Andover Review, 14:139.
[59] Progressive Farmer, January 20, May 19, 1891; Arnett, Populist Movement
in Georgia, 102.
[60] National Economist, 3:135; 6:130 (May 17, 1890; November 14, 1891);
Southern Mercury, March 10, 1892.
[61] Greensboro Daily Record, August 10, 1891.

it ever will be." [62] Naturally, such wholesale criticism of the subtreasury bore fruit. Even an editorial writer on the *Progressive Farmer* resigned because he found the dominant sentiment in the Alliance on the subtreasury question "in irreconcilable conflict" with his views.[63]

In fact, the dissension in the Southern Alliance over the subtreasury went almost the length of an open split. President Hall of the Missouri State Alliance was violently against the proposal; the official organ of the Alliance in Missouri refused to present the subject for discussion; and an Alliance state convention held by the Missourians in 1891 escaped complete disruption only by agreeing to leave the subtreasury plan out of the resolutions altogether.[64] In Texas the opposition went so far as to establish an anti-Alliance paper and convene a convention to oppose the demands of the order.[65] In Mississippi, Tennessee, and Louisiana, also, "retrograde" movements were reluctantly conceded to exist.[66] When the Supreme Council of the Southern Alliance, meeting in Indianapolis in November, 1891, continued the indorsement of the subtreasury that had been given a year before in Ocala and two years before in St. Louis, the anti-subtreasury leaders issued a call for the formation of a new Alliance that would be free from the heresy.[67] This proposed rival order never got on its feet, but the dissension in Alliance circles that its would-be founders were able to create had a most depressing effect. "Everybody has left the Alliance except the few that believe in the sub-treasury heresy," declared one of the come-outers, a Mississippian. "Not only in Mississippi, but throughout the South, lodge after lodge of the order has disbanded, until in some sections there is not a corporal's guard left." [68] This was a

[62] *Ibid.*, September 21, 1891.   [63] *Ibid.*, June 26, 1891.
[64] *New York Times*, December 9, 1890; *National Economist*, 5:233; 6:21, 39 (July 4, September 26, October 3, 1891).
[65] *National Economist*, 5:296, 377; 6:130 (August 1, November 14, 1891).
[66] *Ibid.*, 5:72, 87, 233; 6:2 (April 18, 25, July 4, September 19, 1891).
[67] Stewart, in the *Indiana Magazine of History*, 14:355.
[68] *Greensboro Daily Record*, December 10, 1891.

gross exaggeration, but the subtreasury did put the Alliance leaders definitely on the defensive, and the membership of the Alliance did decline.[69]

Realizing, apparently, that their arguments were not wholly convincing, the advocates of the subtreasury more and more frequently expressed themselves as being willing to accept any alternative plan that would " stop the present discrimination against the farmer." [70] Opponents of the measure were challenged to propose some better way of securing a flexible currency that would "keep up a true ratio between the volume of money and the demand for its use." [71] A prominent Alliance leader, appearing before the Senate Committee on Agriculture and Forestry, confessed that while the subtreasury was the " very best thing we farmers can make, . . . you can do better. . . . You have all the advantages, and what we ask is not that you give us the sub-treasury bill as we have framed it, but give us that or a better bill." [72] The Populist platform of 1892 called only for " loans to producers under the sub-treasury or some better system."

But satisfactory alternative propositions were not forthcoming — at least not from the opposition. Watson of Georgia suggested modifications that were fairly sensible. He proposed that the amount of money advanced should be limited to one hundred million dollars, and that only wheat and cotton be accepted for storage. Moreover, instead of government warehouses, suitable private elevators were to serve as depositories. But Watson's proposition came too late.[73] By January, 1893, when he advocated this change, the subtreasury idea had lost its appeal; and a few months later it could be said that " the cry of the sub-treasury cannot now be heard even in the silence of the night." [74]

[69] *Ibid.*, July 24, 1891; *People's Party Paper*, December 3, 1891.

[70] *National Economist*, 3:184 (June 7, 1890).

[71] *Ibid.*, 7:184 (April 2, 1892).

[72] *Ibid.*, 3:190 (June 7, 1890). Cf. Walker, in the *Andover Review*, 14:139.

[73] *Congressional Record*, 52 Congress, Session 2, January 9, 1893, p. 460; *People's Party Paper*, January 20, May 12, 1893.

[74] C. S. Gleed, in the *Forum*, 16:256; Tracy, in the *Forum*, 16:248.

What had happened to it? Part of the answer is to be found in the dissension, already described, that the subtreasury wrought in reform ranks. There was little to be gained by insisting on a measure that threatened to destroy utterly all hope of farmer solidarity. And, because of the injection of the third-party issue during the campaign of 1892, that hope was already dim in the South, where the Populist party, as the political legatee of the Alliance, had hardly half the strength of its predecessor. It was good politics to forget the subtreasury. Moreover, Macune, admitted author and chief advocate of the subtreasury idea, utterly lost his reputation with reformers by his conduct during the campaign of 1892. Although as editor of the *National Economist* he seemed to support the Populists, he was actually in close touch with the Democratic campaign managers, whom he aided in the printing and distribution of documents designed to induce Alliancemen to vote the Democratic ticket — a type of activity that seemingly paid him well.[75] For this offense Macune was forced to resign from the Alliance, and the official character of his paper was denied.[76] Naturally the fall of the man whose name was so closely identified with the subtreasury must have affected the confidence that people placed in that measure. But doubtless quite as important as all these things together was the growing favor with which free silver was viewed. To many this appeared to be the "better plan" for which they had sought.

> The dollar of our daddies,
> Of silver coinage free,
> Will make us rich and happy
> Will bring prosperity.[77]

[75] A full disclosure was printed in the *St. Louis Globe-Democrat*, November 4, 1892. See *post*, 272–273.

[76] *Southern Mercury*, December 1, 1892; *People's Party Paper*, April 28, 1893; Blood, *Handbook*, 43.

[77] From an undated manuscript in the Weller Papers in the possession of the State Historical Society of Wisconsin. The authorship is credited to M. H. Daley.

CHAPTER EIGHT

THE BIRTH OF A PARTY

IN addition to the formulation of the subtreasury plan, two other features of the St. Louis conference of 1889 are notable. First, the conference gave the various farm orders substantially one and the same platform of political demands; secondly, it made perfectly obvious the impossibility of gathering all such organizations together in one great Alliance. Both considerations pointed vaguely in the same direction. The orders had been able to agree upon a platform, and precisely the sort of platform that a new political party would need should such a party be formed. Moreover, since the Alliances could not unite as such, did it not behoove their members to create a separate organization through which they could cooperate as individuals to put their principles into effect?

These notions would not be downed, although the idea of forming a third party was frowned upon by nearly all the Alliance leaders. The *National Economist* was moved to say editorially: "A third political party will not be formed by these organizations. It is a nonpartisan movement in which each member may remain true to his party, but each one will see to it that this party continues true to him." A little later, however, the same journal could not refrain from the comment that if the farmers should "take it into their heads to act with solidarity in politics, there may be, in the next

year or two, some of the liveliest and most surprising politics
ever known in these United States." [1]

The farmers did, indeed, " take it into their heads " so to
act. In greater and greater numbers they came to realize that
their nonpartisan and bipartisan efforts were mainly wasted.
As Ignatius Donnelly of Minnesota had once said to the
Grangers, this creation of a nonpolitical organization was
like making a gun "that will do everything but shoot." [2]
The men whom they chose on old-party tickets to represent
them in the legislature or in Congress almost invariably
bowed before the demands of the party machine.[3] The legislation that they desired failed to materialize. As the election
of 1890 approached, signs multiplied that the farmers were
on the verge of political revolution. In several northwestern
states Alliance conventions met to nominate full state and
local tickets and even candidates for Congress. In the South,
where the need of a solid white vote was still keenly felt, the
farmers sought to capture completely the Democratic party
machine. The results were startling. In Kansas and Nebraska the Alliance lost the governorships but elected majorities in one or both houses of the legislatures and some members of Congress. In South Dakota, in Minnesota, and even
in Indiana the Alliance showed amazing strength. Throughout the South the old guard of the Democratic party was put
to shame — completely routed, as in Georgia, South Carolina, and Tennessee, or thoroughly frightened, as in Alabama,
North Carolina, and Missouri.[4] What would the future
bring forth?

[1] *National Economist* (Washington), 2:264; 3:1 (January 11, March 22,
1890). This chapter follows in large part an article published in *Minnesota History*, 9:219–247, and is reprinted by permission.
[2] Ignatius Donnelly, *Facts for the Granges*, 18. A copy of this pamphlet,
which is made up of extracts from his speeches before a number of county
granges in southern Minnesota in the spring of 1873, is in the library of the
Minnesota Historical Society.
[3] Walker, in the *Annals of the American Academy of Political and Social
Science*, 4:796.
[4] Haynes, *Third Party Movements*, 236–252; Arnett, *Populist Movement in
Georgia*, 116, 122–124.

The successes scored by the farmers in the election of 1890 greatly stimulated the agitation, already under way, for the organization of a third party along national lines. The expectation that an Alliance group, composed of nominal Democrats and Republicans as well as Independents, but acting as a unit on all matters pertaining to agriculture, would now appear in Congress, suggested to some the possibility of a farm bloc to occupy seats " on either side of the center aisle in the House of Representatives . . . and to take the place of the Center in the French Assembly." [5] But to others it suggested the immediate necessity of forming a new political party. The opportunity to air these radical views was soon to be vouchsafed, for the Supreme Council of the Southern Alliance was under call to meet at Ocala, Florida, in December, 1890.

This convention became the mecca of all the leading advocates of the third-party idea. Perhaps they were intrigued somewhat by the prospect of attending at the same time the "Semi-Tropical Exposition" arranged for the entertainment of the visitors by the local state Alliance; and once they had arrived in Florida they enjoyed such a round of drives, receptions, and demonstrations, with free special trains to points of interest, free accommodations at hotels, and free use of orange and lemon groves as to suggest a Florida of much later date. Nevertheless, for a certain busy few the chief work of the Ocala convention was to press for action looking in the direction of a new party.[6]

Among these few none were more interested and active than the delegates from Kansas. By virtue of the fact that the Kansas State Alliance had left the northern order the year before to join its southern rival, the Sunflower State was officially represented in the Supreme Council; and the Kansans made it their chief concern to pledge the whole Alliance organization to the support of the third-party move-

[5] Gladden, in the *Forum*, 10:321.
[6] *New York Times*, December 3, 30, 1890; Blood, *Handbook*, 41.

ment.[7] But they found the average Southerner definitely
opposed to the project. To him the lesson of the election
of 1890 seemed to be that the capture of the Democratic
party, nationally as well as locally, was not out of the question. Moreover, anything that would threaten the southern
one-party system, by which the political ambitions of the
colored population could be permanently suppressed, would
provoke unlimited criticism. Should the Alliance sponsor
any such program, doubtless it would lose heavily in membership and prestige.[8]

To promote harmony Dr. C. W. Macune, of whom it was
well said that "in him beats the heart and in him the brains
of our body," proposed a compromise.[9] It was clear enough,
he argued, that there was a strong demand in the North for
third-party action; but it was equally clear that consent to
form such a party could not now be obtained in the South.
Then let the matter rest for a time. On the eve of the next
national campaign, about February, 1892, let there be held
a delegate convention chosen by all "organizations of producers upon a fair basis of representation." Let this convention draw up a joint set of demands and a plan to enforce
them. "If the people by delegates coming direct from them
agree that a third party move is necessary, it need not be
feared." [10]

Macune's plan offered a way out, and the convention
adopted it. The work of promoting such a convention as
was proposed was turned over to a committee on confederation, which held an informal meeting at Ocala and agreed to
meet again the following month in Washington together with
such similar committees as might be selected by other organi-

[7] Barr, in *Kansas and Kansans*, 2:1159; Miller, in the *Mississippi Valley Historical Review*, 11:469.
[8] *Greensboro Daily Record*, December 6, 1890; Chamberlain, in the *Chautauquan*, 13:341.
[9] *National Economist*, 4:252 (January 3, 1891).
[10] Blood, *Handbook*, 66; *Ocala Proceedings*, 25.

zations. The joint committees might then issue the formal call.[11]

But the Macune compromise failed to satisfy the extremists among the third-party men, who believed that the inauguration of third-party action should not be so long delayed, and they decided to call a convention in the immediate future regardless of the Alliance decision. The alternative plan might, indeed, have special merits of its own. The call might be so worded as to make it appear that the third-party movement was broader than the Alliance, broader even than the farmers' organizations; and in the final draft delegates were invited from the Independent party, the People's party, the Union Labor party, organizations of former Union and Confederate soldiers, the Farmers' Alliance — Northern and Southern — the Farmers' Mutual Benefit Association, the Citizens' Alliance, the Knights of Labor, the Colored Farmers' Alliance, and all others who agreed to the St. Louis demands in December, 1889.[12] According to Congressman John Davis of Kansas, who claimed to have been consulted in the matter, the call was drawn up by three Vincent brothers from Winfield, Kansas, two of whom were editors of the radical *Non-Conformist*. They were aided by Captain C. A. Power of Indiana and by General J. H. Rice of Kansas.[13] Individuals present at Ocala and others were asked to sign the call, which at first proposed a convention at Cincinnati on the twenty-third of the following February, but later, when the chairman of the state committee of the Kansas People's party pointed out that the date set would conflict with the meeting of the Kansas legislature, the date was changed to May 19, 1891.[14]

[11] *Ibid.*, 25, 37; Blood, *Handbook*, 67; Drew, in the *Political Science Quarterly*, 6:309.
[12] *New York Times*, December 5, 1890; *Cincinnati Enquirer*, May 20, 1891.
[13] Power is not to be confused with J. H. Powers, for a time president of the National Farmers' Alliance.
[14] *New York Times*, December 5, 1890; *Greensboro Daily Record*, December 6, 1890; *National Economist*, 5:106 (May 2, 1891); Haynes, *Third Party Movements*, 246.

In general, northern Alliancemen were favorable to the idea of a third party, although there was much criticism of the laxness and haste involved in the Cincinnati call.[15] The annual meeting of the Northern Alliance was held at Omaha in January, 1891, and while the sentiment of the gathering strongly favored the Alliance taking " no part as partisans in a political struggle by affiliating with Republicans or Democrats,"[16] a plan differing from that embodied in the Cincinnati call was announced. Six fundamental principles were set forth: (1) free silver; (2) abolition of national banks and substitution therefor of direct issues of legal tender notes; (3) government ownership of all railroads and telegraphs; (4) prohibition of alien landownership and of gambling in stocks, options, and futures; (5) a constitutional amendment requiring the election of president, vice president, and senators by direct vote of the people; and (6) the Australian ballot system.

A petition stating these principles and calling for a convention to nominate candidates for president and vice president in 1892 upon this platform was to be circulated by means of the executive officers of each industrial organization in every state and territory. Whenever five million signers were reported throughout the United States, all such officers in each state, acting together, were to select a state representative upon a provisional national committee. The committee thus constituted should meet in Cincinnati on February 22, 1892, to fix a ratio of representation based on the number of signatures in each state and to determine the time and place for the meeting of the nominating convention.[17]

Meanwhile the plan that Macune had proposed at Ocala for a great industrial conference early in 1892 was being worked out. On January 22, 1891, a few representatives

---

[15] *Farmers' Alliance* (Lincoln), April 4, 1891.
[16] *National Economist*, 4:333 (February 17, 1891).
[17] *Farmers' Alliance*, February 21, 1891.

from the Southern Alliance, the Knights of Labor, the Farmers' Mutual Benefit Association, the Colored Farmers' Alliance, and the Citizens' Alliance met at Washington, D. C.
and organized what they were pleased to call the "Confederation of Industrial Organizations." February 22, 1892,
was fixed upon as the date for the proposed conference of
all the orders, and an executive committee was named to
decide the place of meeting and all other details. This committee first planned the meeting for Washington but later
chose St. Louis.[18]

Neither the Omaha plan nor the Washington plan promised speedy enough action to satisfy the extremists, however,
and preparations for the Cincinnati convention went on. In
the states of the Northwest, especially Kansas, where local
third parties had scored successes in the election of 1890 and
it was supposed could count on even greater successes if supported by a national party, the early convention was popular.[19] Why take chances on what the Alliance might do later?
Better decide the matter at once. There was, moreover,
throughout the entire country, a type of professional third-
party politician who fairly doted on this sort of meeting and
would not let the idea die. A strictly Alliance gathering
might exclude many such, but the Cincinnati call was broad
enough to take them all in.[20]

As the delegates gathered it became increasingly clear that
the convention was to consist of hundreds of determined
farmers from the West and of other hundreds of habitual reformers. One member admitted that this was the fifth national convention that he had attended with the sole object
in view of founding a third party, "two in Chicago, two in
Cincinnati, and now another in Cincinnati."[21] Delegates
came who still called themselves Greenbackers; others were

[18] National Economist, 4:310 (January 31, 1891); Blood, Handbook, 67.
[19] Cincinnati Enquirer, May 19, 1891.
[20] Farmers' Alliance, April 11, 1891; National Economist, 5:199 (June 13, 1891).
[21] Farmers' Alliance, May 14, 1891.

followers of Edward Bellamy and took the name Nationalists; still others pinned their faith to Henry George and were proud to be called Single-Taxers. "A large majority," commented a former third-party man, "are honest, well-intentioned men, a few are dead-beats, and too many . . . don't know what they want and will never be satisfied until they get it." According to one reporter, "all the second and third class hotels are crowded to overflowing." [22]

Known officially as the "National Union Conference," the convention was called to order in Music Hall on the afternoon of May 19, 1891, by Judge W. F. Rightmire of Kansas. Three great inscriptions, "United we stand; divided we fall," "Opposition to all monopolies," and "Nine million mortgaged homes," looked down from the walls upon the assembly, which was seated in state delegations as in national political conventions. [23] Captain Power, who had worked actively to promote the conference, read the call from the original document. As he mentioned the name of each organization invited to participate, he asked its representatives to rise. Members of the Farmers' Alliances were clearly more numerous than members of any other orders, but it was evident that many of the delegates were "joiners" who belonged to several orders.

A credentials committee solved the difficult problem of the allotment of seats by giving to practically everyone with any sort of papers the right to a place in the convention. Said one observer, "I think that if anyone would sprinkle a few hayseeds on his coat he would be admitted to the floor and have a right to vote." Over a hundred members of the National Reform Press Association, which was meeting in Cincinnati at the time, were allowed seats in the convention, and some southerners who were present without any credentials whatever were allowed the privilege of the floor with the right to participate in debate. Altogether more than four-

[22] *Cincinnati Enquirer*, May 19, 20, 1891.
[23] *Farmers' Alliance*, May 28, 1891.

teen hundred delegates were recognized, representing some thirty-three states and territories; but more than four hundred of them were from Kansas, more than three hundred from Ohio, about a hundred and fifty each from Illinois and Nebraska, and the rest mainly from other northwestern states. Few southerners attended. The credentials committee ruled that delegates representing more than one order could have only one vote, but a proposition that each state have one vote and one additional for each fifty delegates was voted down. The gathering might as well have been in name what it was in fact, a mass convention of self-appointed delegates.[24]

After effecting an organization the convention authorized the various state delegations to select members of the customary committees, including the all-important committee on resolutions. At this juncture Donnelly of Minnesota raised a tumult by suggesting that members of a national executive committee should be chosen at the same time, thus assuming that there would be a third party. This aroused James B. Weaver of Iowa, and " amid much confusion he strided [sic] down the middle aisle. He shook his finger angrily at Donnelly, and denounced him for endeavoring to pledge the convention on the sly without there being a word of discussion to the most vital question it had to consider." [25]

Donnelly's motion was not brought to a vote, and order was restored. The incident, however, was significant because it revealed the two points of view held by delegates. The vast majority, including nearly all the Kansans, were with Donnelly and were ready to form a third party on the spot. A more conservative group, headed by Weaver, were for drawing up resolutions and perhaps for suggesting the advisability of forming a third party, but the actual launching of the party they would postpone until the election year.[26]

[24] *Cincinnati Enquirer*, May 19, 20, 21, 1891; *Times-Star* (Cincinnati), May 18, 20, 1891.
[25] *Cincinnati Enquirer*, May 20, 1891.
[26] *Ibid.*, May 18, 19, 20, 1891.

Doubtless the conservatives hoped that by biding their time they might win greater support from the South. President L. L. Polk of the Southern Alliance sent a letter to the convention counseling delay. He thought that the coming year might more properly be used for " education in the principles of reform " and if then, on full reflection, the third party seemed necessary, let it come. But according to one account Polk's letter was "received with painful silence, which was broken at the conclusion by a delegate from Arkansas moving that 'we sit down on that communication as hard as we can.'" This remark occasioned great applause, but a motion to refer the letter to the committee on resolutions was made and carried.[27]

The work of reconciling the divergent opinions expressed by Donnelly and Weaver, if it could be done, fell naturally to the committee on resolutions, of which Donnelly, whose facile tongue and pen were everywhere known and recognized, was made chairman. Donnelly seems, indeed, to have been mainly responsible for the invention of a formula that would suit both factions.[28] The committee, he explained later to the convention, had before it two alternatives: (1) to form a new party at once without regard to anyone else or (2) in the interest of harmony to concentrate on the convention to be held on February 22, 1892. The latter course was finally decided upon, with important reservations. The resolutions announced the immediate formation of the People's party with a national executive committee to consist of a chairman, elected by the convention in general session, and three members from each state represented, elected by the delegations of the respective states. This committee was directed to attend the proposed St. Louis conference and "if possible unite with that and all other reform organizations

---

[27] *National Economist,* 5:34 (April 4, 1891); *Cincinnati Enquirer,* May 21, 1891.
[28] The Donnelly Scrapbooks, especially Volume 8, contain many clippings on this subject.

there assembled. If no satisfactory arrangement can be effected this committee shall call a national convention not later than June 1, 1892, to name a presidential ticket."[29] A third party was thus assured. If the St. Louis conference did not agree to it, the national executive committee emanating from the Cincinnati convention was authorized to go ahead. When the astute plan that the Donnelly committee had devised, together with the platform upon which the new party was to stand, was announced, the convention broke forth into prolonged applause. "It took nearly half an hour for the excited delegates to cool their pent-up enthusiasm."[30]

The platform adopted at Cincinnati contained little that was new. Rather, it sought to codify and restate the demands previously adopted at St. Louis, at Ocala, and at Omaha. The Prohibitionists, who were present at Cincinnati under the leadership of John P. St. John, sought in vain to secure the inclusion of their pet project, and the woman suffragists fared only a little better.[31] "We apologize," said Donnelly, in explaining the report of the committee, "because we have not been able to cover all the interests in the minds of men here to-day. . . . We believe that the party that, in such a crisis as this, shortens its platform, lengthens its muster roll. . . . We feel that we are not here so much to proclaim a creed as to erect a banner around which the swarming hosts of reform could rally." One paper commented that a banner was a rather poor substitute for a creed, but as a matter of fact, the creed was fairly complete.[32]

[29] *Farmers' Alliance*, May 28, 1891.
[30] *Cincinnati Enquirer*, May 21, 1891.
[31] *Times-Star*, May 21, 1891; *Cincinnati Enquirer*, May 21, 1891.
[32] Clippings in the Donnelly Scrapbooks, Vol. 8, dated May 21, 1891. The Cincinnati demands differed from the St. Louis demands (see *post*, Appendixes A and D) mainly by including the subtreasury plan along with the other demands for financial reform, by favoring the direct election of president, vice president, and United States senators, and by urging government control of the means of transportation and communication (favored at Ocala over government ownership), but "if this control and supervision does not remove the abuses now existing . . . government ownership."

Most of the state delegations present at Cincinnati promptly elected their three national committeemen, and the convention chose H. E. Taubeneck of Illinois as national chairman. The newly formed committee then met, chose the other necessary officers, encouraged third-party men in each state to proceed in their own way with the selection of a state executive committee, promised to this end the help of the national committee, and discussed plans for the future.[33]

The course pursued at Cincinnati won approval on all sides. Radicals everywhere rejoiced immoderately that the new party was an actual fact. One Kansas delegate said that had anything less been accomplished, the representatives from his state "wouldn't have dared to go home."[34] Such northerners as had counseled delay — Weaver and "Sockless" Jerry Simpson, for example — felt also that their advice had been taken. Weaver admitted that a new party had been formed, a fact that Simpson at first refused to concede, but the two agreed that the main action was now postponed until the February conference of 1892.[35] The Omaha plan for the formation of a third party, favored by the Northern Alliance at its last national meeting, but never especially popular, was now definitely dropped. Officials of the Southern Alliance, who had in general opposed holding the Cincinnati convention, professed complete satisfaction with the results. The *National Economist* thought the decision "wise and conservative" and well calculated to supply "the link that will unite the farmers with all other occupations in the great approaching conflict."[36]

One significant result of the Cincinnati convention that seems generally to have been overlooked at the time was that

[33] *Cincinnati Enquirer*, May 21, 1891; *Times-Star*, May 21, 1891; *Farmers' Alliance*, May 28, 1891.
[34] *Cincinnati Enquirer*, May 21, 1891.
[35] *Times-Star*, May 21, 1891.
[36] *National Economist*, 5:161 (May 30, 1891). See also the *Farmers' Alliance*, April 4, 1891.

the professional third-party men insured for themselves, quite apart from what the farmer organizations might do later, a prominent place in the councils of the new party. They were " in on the ground floor."

As for the southerners, there was a growing disposition to concede that the attempt to work through the Democratic party was a failure. Local successes were offset by the fact that concessions from the national organization were practically unobtainable. Leaders of the Southern Alliance were particularly aggrieved that the subtreasury plan found no greater favor with the Democratic politicians than with the Republican and was scornfully rejected by Congress.[37] President Polk of the Southern Alliance voiced a general sentiment through the columns of his paper, the *Progressive Farmer*, when he said: " The new party has adopted the Alliance demands into its platform. Does anyone suppose intelligent Alliancemen will vote against a party that adopts those demands, and in favor of a party that not only fails to adopt, but resists those demands? "[38] Polk repeatedly made it clear that the southern farmers preferred to remain in the Democratic party; but he never failed to threaten their secession from it in case the Alliance program of reform, national as well as local, were not adopted as the Democratic program of reform. During the summer of 1891 a delegation of third-party men from Kansas visited the South, presumably to impress upon southern Alliancemen the necessity of independent political action; and a little later Polk made three addresses in Kansas to encourage the western revolters in their hope of southern sympathy and help.[39]

With the southern leaders wavering or coming over, and with the fires of revolt still bright in the West, the third-

[37] *National Economist*, 4:133 (November 15, 1890); *Farmers' Alliance*, July 19, 1890; Bryan, *Farmers' Alliance*, 99.

[38] *Progressive Farmer* (Raleigh), June 3, 1891. Compare the same paper for February 2, 1892.

[39] *Greensboro Daily Record*, August 27, 28, September 2, 15, 21, 1891.

party advocates bent every effort to increase the popularity
of their cause. In June five members of the executive committee of the newly formed People's party met in St. Louis
and directed the secretary, Robert Schilling, to establish a
literary bureau from which a steady stream of third-party
propaganda should flow to the reform press. Likewise, a call
for funds was sounded, evidently with good effect, for an
army of lecturers issued forth to convince the country of the
necessity of third-party action. Mighty men and mightier
women from Kansas — Jerry Simpson, Senator Peffer, John
H. Willits, Mrs. Lease, and Mrs. Diggs — led the van-
guard, and a host of lesser lights followed in their train. The
picturesque features of the campaign of 1890 were reproduced, and orgies of speech-making went hand in hand with
picnics, barbecues, and similar festive occasions. All the various reform organizations were urged to find a place under
the third-party banner. "The Alliance, the F. M. B. A., the
Grange, the Knights of Labor," wrote Taubeneck, " are nothing more or less than industrial schools whose one object is
to teach the industrial masses the principles of economic
government that we call for. All of these organizations are
working for a common end and the People's Party reaches
out and takes them all in." [40]

While this campaign of education waxed warm throughout the West and the South and pressed even into the East,
in only a few states were elections of any considerable consequence held in 1891. In these states the third-party propagandists redoubled their efforts, but with surprisingly slender results, as far as the election of their adherents to office
was concerned.

In Iowa the People's party perfected a state organization
and nominated a full state ticket with A. J. Westfall at its
head. Westfall had made a good showing in 1890 as an Inde-

[40] Stewart, in the *Indiana Magazine of History*, 14:349; 15:73; Barr, in
Kansas and Kansans, 2:1154; Chamberlain, in the *Chautauquan*, 13:341;
Haynes, *Third Party Movements*, 249.

pendent candidate for Congress from the eleventh Iowa district, and he again staged a vigorous campaign, which doubtless greatly increased interest in the new party, even though the third-party vote was light. Governor Horace Boies, who won re-election on the Democratic ticket as a friend of free silver and an enemy of prohibition doubtless drew many votes that under other circumstances might have been cast for Westfall.[41]

In Nebraska a judge of the supreme court was to be chosen. Ordinarily such an election would have attracted little attention, but in this particular year, because of an unusual set of circumstances, there was an exciting contest. The results of the election of 1890, which were apparently favorable to the Democratic candidate for governor, James E. Boyd, were contested by both the Independents and the Republicans. The former claimed that their candidate, who had finished second, had been defeated by fraud and they prepared to contest Boyd's right to office. The latter claimed that Boyd was ineligible, not because of fraud in the election but because he was not a legally naturalized citizen of the United States, and that therefore the Republican governor who had held office the preceding term, John M. Thayer, should hold over until a duly qualified successor could be chosen.

Faced by this situation, the Republican state supreme court had first thrown such obstacles in the way of the contest that Boyd was able to take and hold his office for several weeks and then later had declared Boyd ineligible and had paved the way for the return of Thayer.[42] Under such circumstances the court was clearly open to the charge of rank

---

[41] Haynes, *op. cit.*, ch. 19.

[42] Helen Storms, "The Nebraska State Election of 1890," master's thesis, University of Nebraska, 1924, pp. 92–111. Boyd was actually compelled to turn over his office to Thayer from May, 1891, to February, 1892, but when the United States Supreme Court held that he was a citizen, he resumed office and Thayer went out. The legal aspects of the case are fully reviewed by R. D. Rowley in "The Judicial Career of Samuel Maxwell," master's thesis, University of Nebraska, 1928, pp. 36–44.

partisanship, to which the Independent leaders added the charge of railway domination. The Democrats made no nominations, in general supporting the Independent candidate; but in spite of the excitement of the campaign the Republicans won easily. The Independent candidate was not notably well fitted for his place; furthermore, in much of Nebraska the crops were good in 1891, and the edge was thus taken off the farmers' discontent.[43]

In Kansas many county and judicial offices were to be filled, and such was the dread felt in Republican and Democratic circles that the new party might triumph that there was a strong tendency for the two old parties to get together to prevent it; at least the Democrats refused to cooperate with the Populists as they had done in 1890. By such tactics the old parties usually managed to win; but the People's party men congratulated themselves upon the selection of a goodly number of district judges and county officials "in spite of 90,000 stay-at-homes, mostly farmers."[44] In Kentucky an independent People's party organization contested the state elections held there that year, and while the vote for the third party was light, it was sufficient to elect about a dozen members of the legislature and to make plausible higher hopes for the future.[45] In Louisiana the Alliance forces joined hands with the anti-lottery people, nominated a joint ticket, and conducted a vigorous campaign. The election did not come until April, 1892, but at that time the lottery ticket was decisively defeated. This could scarcely be called a third-party triumph, however; it was more an anti-lottery triumph.[46]

Meager as the election showings seemed to be, the third-

---

[43] Barnhart, "Farmers' Alliance and People's Party," 286.

[44] A Kansas correspondent to the *People's Party Paper* (Atlanta), January 7, 1892; *National Economist*, 6:215; Barr, in *Kansas and Kansans*, 2:1155; Miller, "Populist Party in Kansas," 210.

[45] *Appletons' Annual Cyclopaedia, 1891,* 408.

[46] *Ibid., 1891,* 444–445; *1892,* 425; Melvin J. White, "Populism in Louisiana during the Nineties," *Mississippi Valley Historical Review,* 5:1–19.

party leaders professed to be satisfied. Their main concern
for the moment, they contended, was not so much to elect
candidates as to perfect their organization and to prepare
for the campaign of 1892. If the enemies of reform harbored
the delusion that the third-party movement was over, so
much the better.[47] It was with confidence that the state central committee of Kansas, just after the election, issued its
famous manifesto:

Every branch of business is depressed. The merchant fails
for want of trade and the banker from depreciation of values.
Labor is unemployed and inadequately paid. Our cities are the
abode of poverty and want and consequent crime, while the
country is overrun with tramps. Starvation stalks abroad amid
an overproduction of food and illy clad men and women and
helpless children are freezing amid an overproduction of clothing. We hold these conditions are the legitimate result of vicious
legislation in the interests of the favored classes and adverse
to the masses of American citizens, and we appeal to the great
body of the people, irrespective of occupation or calling, to rise
above the partisan prejudices engendered by political contests,
and calmly and dispassionately examine the facts which we are
prepared to submit in support of our claims. We appeal to reason and not to prejudice, and if the facts and arguments we
present can be refuted we neither ask nor expect your support.[48]

When the next meeting of the Supreme Council of the
Southern Alliance was held in November, 1891, this time at
Indianapolis, the adherence of that body to the third-party
movement seemed assured. At this gathering the executive
committee of the new party put in its appearance, bent on
obtaining the full cooperation of the Southern Alliance in
the forthcoming campaign; here also appeared the executive
committee of the Confederation of Industrial Organizations
to repeat the call for the February, 1892, conference at St.
Louis and to preach a type of joint action by all reform

[47] *People's Party Paper*, December 3, 1891.
[48] Barr, in *Kansas and Kansans*, 2:1162.

orders that logically could end only in support of the Independent party.[49] While numerous southerners voiced their distress at the thought of dividing the Democratic vote of the South and while some of them even went the length of withdrawing entirely from the Alliance, evidently a majority of the delegates were ready to concede that the third party had come to stay and that they might as well throw the strength of their organization to it. They did what they could to insure the nomination of their president, Colonel Polk, to head a third-party ticket in the coming campaign, and they voted with enthusiasm to request all congressmen elected "by the aid of Alliance constituencies . . . to decline to enter into any party caucus called to designate a candidate for Speaker, unless adherence to the principles of the Ocala platform are made a test of admission to said caucus."[50]

This latter bit of advice was not fully heeded. Shortly before the Democratic caucus was held twenty-five Alliance congressmen met in Washington to determine what their attitude towards it should be. The majority of the group strongly favored entering the caucus as usual and abiding by its results as loyal party men should. But a small minority, led by Thomas E. Watson of Georgia and "Sockless Jerry" Simpson of Kansas, adhered steadfastly to the doctrine stated at Indianapolis, and they finally carried their revolt to the point of making Watson himself their candidate for speaker. Besides Watson, five representatives from Kansas, one from Minnesota, and two from Nebraska thus declared their independence of the old-party machines; "so was formed," so Watson's paper declared, "the first distinctive political body known as the People's party."[51]

[49] *National Economist*, 6:233 (December 26, 1891).
[50] *National Economist*, 6:181 (November 13, 1891). This convention is fully covered in Stewart, "The Populist Party in Indiana," in the *Indiana Magazine of History*, 14:354.
[51] *People's Party Paper*, December 17, 1891; November 25, 1892; *National*

Those Alliance congressmen who entered the Democratic caucus did not admit, however, that they had deserted their principles. They joined in a successful movement to select as speaker Charles F. Crisp of Georgia, a suave, conservative politician of the old school but reputed to be a more outspoken friend of free silver than his leading opponent, Roger Q. Mills, of Texas, whose interest lay mainly in the tariff. "The victory is not in securing Crisp," the *National Economist* explained, " but in whipping the bosses of both parties, and securing supremacy for the wishes of the people by making the money question the great issue in 1892." [52] This was cold comfort to the followers of Watson, who by this time were unwilling to be appeased by the prospect of a slightly regenerated Democracy and hoped rather for the development of a powerful third party. Crisp, Watson declared, was one of the first to come out in opposition to the subtreasury plan, and he had defended the occasional instances of Democratic and Republican fusion against the People's party in Kansas. Why rejoice at the victory of such a man? [53]

While these events held the center of the stage in Washington, the scene was being set for the St. Louis conference, which was to determine finally and formally the attitude of the great "industrial organizations" of the country towards the third-party movement. To this gathering the well-known farm orders were all invited to send delegates, as were also such organizations of manual laborers as cared to participate; for while the agricultural societies took the lead, they were exceedingly anxious to have strong labor support. Practically all who were invited, and many besides, responded to the call. Taking advantage of the special railway rates offered, thousands of farmers and their political friends

Economist, 6:193-194 (November 20, 1891); Arnett, *Populist Movement in Georgia*, 130.
[52] *National Economist*, 6:194 (November 20, 1891).
[53] *People's Party Paper*, December 3, 1891.

flocked to the Missouri metropolis. The attendance, including delegates and interested observers, " went far beyond the most sanguine hopes." [54] According to one reporter, those who came " were mostly gray-haired, sunburned and roughly clothed men. . . . The ' ward-bummer,' the political ' boss,' and the ' worker at the polls ' were conspicuously absent."

Held in Exposition Music Hall, the convention turned out to be a " singing session " and in that respect different from " any other political meeting ever witnessed in St. Louis." True, the Populist Marseillaise had not yet been discovered, but it was confidently expected and numerous aspiring airs were given a trial. Like all large audiences, this one was a thrilling spectacle.

The banners of the different States rose above the delegates throughout the hall, fluttering like the flags over an army encamped. The great stage, brilliant and vivid with the national colors, was filled with the leaders of the Alliance, the Knights of Labor, the single tax people, the Prohibitionists, the Anti-Monopolists, the People's party, the Reform Press, and the Women's Alliance. To the right of the stage was stretched a broad poster of bunting which bore the words: " We do not ask for sympathy or pity. We ask for justice." [55]

According to the call each organization invited to St. Louis was entitled to twenty-five delegates at large from the United States and one additional delegate for every ten thousand members. In the selection of delegates, however, not much attention was paid to this rule; indeed, it made little difference whether an organization had been invited to send delegates or not, for any that wished to be represented sent them regardless of formality. It thus became a matter of considerable importance to have a credentials committee pass upon the merits of those clamoring to par-

[54] *Southern Mercury*, February 11, 1892; *National Economist*, 6:380 (February 27, 1892); Blood, *Handbook*, 67.
[55] *National Economist*, 6:394 (March 5, 1892).

ticipate in the work of the convention. Until this was accomplished little else could be done. Ben Terrell, president of the " Confederation of Industrial Organizations," called the meeting to order; a temporary organization was effected, and the eight organizations included in the call were authorized each to choose three representatives to serve upon the credentials committee.[56]

Pending the report of this committee, the convention yielded itself to the ministrations of its favorite orators. Donnelly was "historical, classical, eloquent, amusingly exaggerative." Weaver was called upon for a speech and declined "until there was something before the convention for him to speak on," but Colonel Polk, Mrs. Mary E. Lease, Simpson, Powderly, and many others were undeterred by such irrelevancies. The feast of oratory continued until well into the second day, for the credentials committee, in spite of the all-night session of a subcommittee, was even then unprepared to report.[57]

When the committee did report, the reasons for delay became clear. Organizations "the existence of which none of the old delegates had ever heard of before" clamored for recognition. Some of these orders were suspected of "mushroom growth and doubtful purposes," but they were all given a hearing, and some eight hundred delegates, representing twenty-one different orders, were awarded seats. On motion of Donnelly, Miss Frances E. Willard and two other Women's Christian Temperance Union workers were given places in the convention, raising the number of orders represented to twenty-two. Thus amended, the work of the credentials committee was accepted by the convention, although there was much exhibition of temper on the part of

[56] St. Louis Republic, February 23, 1892; National Economist, 6:380 (February 27, 1892).
[57] St. Louis Globe-Democrat, February 23, 1892; National Economist, 6:394 (March 5, 1892).

some who were not received. One tempestuous would-be delegate had to be "unceremoniously removed."[58] On one contest the credentials committee refused to rule. Some Georgia seats were fought over so fiercely by delegates favoring and delegates opposing a third party that the matter was left for the convention itself to decide, the committee recommending, however, that the third-party men be seated. When this recommendation was promptly followed by the convention, it became apparent that the conservative element was in the minority and that in all probability action favorable to the third party would be taken.[59] The election of Polk as permanent chairman likewise scored a victory for the third-party men, for his willingness to carry the Alliance into the new political party was now well known. He was elected over Ben Terrell, who was reputed still to be hoping for reform through Democratic channels.[60] Nevertheless, there was a strong undercurrent of opposition to placing the convention on record as favorable to the third party. Led by Leonidas F. Livingston of Georgia, a number of southern delegates made it perfectly plain that they would never consent to any program that would threaten the unity of the white vote in the South and they promised to bolt the convention should such action be taken. To avoid disruption,

---

[58] *National Economist*, 6:380, 395, 396 (February 27, March 5, 1892); *St. Louis Republic*, February 23, 1892; *Southern Mercury* (Dallas), February 25, 1892. The number of seats allotted to each of the eight orders included in the call was as follows:

National Farmers' Alliance and Industrial Union...................... 246
Farmers' Mutual Benefit Association.................................   53
Knights of Labor...   82
National Citizens' Industrial Alliance..............................   27
Patrons of Industry..   75
National Citizens' Alliance...   25
National Colored Alliance and Co-operative Union....................   97
National Farmers' Alliance..   97

On the history of some of the lesser farm orders see Allen, *Labor and Capital*, chs. 14–21, and Drew, in the *Political Science Quarterly*, 6:282–310.

[59] *National Economist*, 6:395 (March 5, 1892); *St. Louis Republic*, February 24, 1892; Arnett, *Populist Movement in Georgia*, 131.

[60] *St. Louis Republic*, February 24, 1892; *People's Party Paper*, March 17, 1892.

therefore, the third-party decision was waived and the convention devoted itself to the business of drawing up a satisfactory list of demands.[61]

The work of the platform committee thus became the major concern of the convention. This committee, consisting of one member from each state and one for every twenty-five delegates from each organization given representation in the convention, drew up the usual demands. On only one matter of consequence was there anything new about them. The Ocala and Cincinnati conventions had swung away from the uncompromising stand of the first St. Louis convention on the matter of government ownership of the railway, telegraph, and telephone systems of the country; but now, under pressure from the labor leaders and the antirailroad delegates of the Northwest, government ownership was again indorsed. This, it was well understood, would do some violence to opinion in the South, where such measures were traditionally regarded as "the essence of paternalism and centralization," but the northern delegates gave in on the matter of the subtreasury, for which they cared little, and the railway plank was essentially the price of the concession.[62]

But the platform committee reported a preamble as well as a platform, the former far exceeding the latter in length and in the richness of its rhetoric. The preamble, written by Donnelly, drew heavily upon the language of the convention call and also upon that of the "Populist Manifesto" issued by the Kansas state central committee in November, 1891.[63] It was none the less a unique and startling document, which not only carried with it a ringing denunciation of the existing ills of society but also, inferentially, the promise of a

[61] St. Louis Republic, February 23, 24, 1892; New York Tribune, February 24, 1892.
[62] National Economist, 6:395, 402 (March 5, 12, 1892); Southern Mercury, February 23, 1893.
[63] Southern Mercury, January 14, 1892; National Economist, 6:233 (December 26, 1891); Barr, in Kansas and Kansans, 2:1162.

third party to remedy these ills. Donnelly read the preamble to the convention and was followed by Hugh Kavanaugh, the chairman of the platform committee, who read the actual demands.

At the conclusion, as if by magic, everyone was upon his feet in an instant and thundering cheers from 10,000 throats greeted these demands as the road to liberty. Hats, papers, handkerchiefs, etc., were thrown into the air; wraps, umbrellas and parasols waved; cheer after cheer thundered and reverberated through the vast hall, reaching the outside of the building where thousands [who] had been waiting the outcome, joined in the applause till for blocks in every direction the exultation made the din indescribable. For fully ten minutes the cheering continued, reminding one of the lashing of the ocean against a rocky beach during a hurricane. . . . Shouts for Donnelly went up all over the hall . . . and people crowded around him and grasped his hands reaching up from the orchestra to greet him.

Livingston, who was opposed to third-party action, perhaps sought to avoid any appearance of committing the conference to the new party by moving the adoption of the *platform*. The motion was carried with enthusiasm, but " some delegate saw through the ruse, got the ear of Donnelly, and rushed through a motion to include the preamble." [64]

With the platform and some minor resolutions adopted, the convention adjourned *sine die,* but by preconcerted plan the delegates, or at least the great majority of them, remained in their seats. Thereupon Dr. Macune promptly took the chair and began the organization of a mass meeting of " individual and independent citizens who love their country." General Weaver was made the presiding officer and to him was delegated the one important task the adjourned session performed. This was the appointment of a committee of fifteen to confer with the executive committee of the

[64] *Southern Mercury,* March 3, 1892; *St. Louis Republic,* February 25, 1892; *People's Party Paper,* March 3, 1892; *National Economist,* 6:396 (March 5, 1892).

People's party with regard to the calling of a nominating convention.[65]

The executive committee of the People's party was, of course, on hand, although it had taken no part as such in the St. Louis conference. It now met, absorbed the committee appointed by Weaver, and proceeded to the business in hand. The matter of greatest immediate concern proved to be the date of the proposed nominating convention. Donnelly was eager to have the convention meet before either of the old parties could have a chance to prepare for the People's party onslaught, and he debated the matter earnestly with his old antagonist, Weaver. Weaver and others who agreed with him held that it would help the People's party cause if when the nominating convention met definite information could be on hand to show that the St. Louis platform had been rejected by Democrats and Republicans alike. Voters who agreed to the St. Louis demands would then feel that they had been turned out of their old-party home because of their principles and they would have no choice but to join the new party. Weaver's policy, which Donnelly declared "suicidal," was adopted, and July 4, 1892, was fixed upon as the date of meeting.[66]

Other necessary preliminaries were taken care of by subcommittees. To a group of ten, appointed by Chairman H. E. Taubeneck, was intrusted the task of selecting the meeting place and it chose Omaha over Kansas City, St. Louis, and Indianapolis.[67] To a group of five was given the more important duty of drawing up the convention call. This subcommittee discharged its obligations promptly by inviting all those who approved of the preamble and platform adopted at St. Louis to hold mass meetings in their

[65] *National Economist*, 6:385, 397 (March 5, 1892); *Farmers' Alliance*, March 3, 1892.

[66] *St. Louis Republic*, February 25, 1892; *Farmers' Alliance*, March 3, 1892; *National Economist*, 6:385, 397 (March 5, 1892).

[67] *St. Louis Republic*, February 26, 1892; *Farmers' Alliance*, March 3, 1892.

respective towns and villages on the last Saturday in March (March 26) to ratify the St. Louis demands and to take the initial steps toward the election of delegates to the Omaha convention.[68]

As later elaborated the plan of organization was as follows: Each of the March meetings was urged to form a local organization and to appoint a committee of three to meet at the county seat not later than April 16. The business of this April gathering was to fix the time, place, and basis of representation for county conventions and to appoint a committee of three to confer with like committees from all other counties in the same legislative and congressional districts to fix the time and place and basis of representation for legislative and congressional district conventions. The executive committees of each state, already organized or to be organized in conformity with the plan adopted at Cincinnati, were asked to meet as early as convenient and to fix dates for state nominating conventions, designating how the delegates from their state to the national convention were to be chosen. Eight delegates at large from each state and four delegates from each congressional district were authorized — the total number for the Fourth of July convention thus being set by accident or intent at 1776. The names of delegates were to be sent as fast as they were chosen to Robert Schilling, secretary of the national committee, and donations to the campaign treasury, which were earnestly solicited, to M. C. Rankin, the national treasurer.[69]

Building thus from the very foundation, the actual construction of the party edifice went on. Not everywhere could a complete organization be worked out, for not everywhere was there sufficient third-party sentiment to justify it, but when the Omaha convention met in July, between thirteen and fourteen hundred accredited delegates were on hand, in

[68] *National Economist*, 6:385, 397 (March 5, 1892).
[69] *Farmers' Alliance*, March 3, 1892; *National Economist*, 6:385, 397; 7:41 (March 5, April 2, 1892).

spite of the fact that "probably through some oversight" many railroads had failed to grant the usual convention rates to the third-party delegates. Marion Cannon of California declared that "it was not by accident that the Pacific coast delegates have been overlooked. Our request for the customary courtesy was denied deliberately and with insolence." When the convention got around to the matter, sentiment seemed to oppose asking "any privilege whatever" from the railroads; but it was voted to refer to the Interstate Commerce Commission the propriety of railroads discriminating against one, and in favor of other, political conventions. Thousands of observers, not participants, also attended the Omaha meeting, and the thrifty management sold season tickets to the sessions at ten dollars a ticket.[70]

Because of the desire to make nominations on the Fourth of July, the delegates met for temporary organization on Saturday, July 2. The procedure followed in the main the well-known rules of the older parties. A platform committee was appointed to report before the nominations were made, which on the following Monday presented the results of its deliberations. But the Omaha platform was no hastily assembled document; it contained little that had not been adopted by other conventions — at St. Louis in 1889, Ocala in 1890, Cincinnati and Indianapolis in 1891, Omaha and St. Louis earlier in 1892. The preamble that Donnelly had presented so dramatically at the Washington's Birthday convention in St. Louis was repeated to make a perfect Fourth of July in Omaha. Reforms that had to do with land, transportation, and finance were still the chief concern.[71]

The reception accorded these well-worn demands showed how admirably they fitted the temper of the crowd. For the antirailroad plank there was a "tumultuous ovation," exceeding in volume the applause for the free-silver plank. The

[70] *Omaha World-Herald*, July 1–5, 1892; *National Economist*, 7:279 (July 16, 1892); *People's Party Paper*, July 8, 1892.
[71] *National Economist*, 7:257 (July 9, 1892).

land plank was greeted by a "regular Baptist camp meeting chorus." And finally on the adoption of the platform "the convention broke over all restraint and went wild in a demonstration that," if we may believe a none too grammatical reporter,

had a likeness of the enthusiastic Bastile demonstration in France, the whole convention, audience and delegates, rose to their feet and the first platform of the People's party was ushered into the world with a scene of enthusiasm that in its intensity and earnestness surpassed the cyclonic ovation which greeted the mention of the name of James G. Blaine at Minneapolis. The crowd broke forth time and again in applause until the leaders finally concluded to stem the tide, and after vigorous efforts secured it. The band played "Yankee Doodle" and it lasted twenty minutes.

Little wonder that a platform so christened should come to have among Populists a sort of religious sanction. These demands were not like ordinary political demands — they were a sacred creed.[72] As one who saw what went on at Omaha observed, "this dramatic and historical scene must have told every quiet, thoughtful witness that there was something at the back of all this turmoil more than the failure of crops or the scarcity of ready cash."[73]

Among those who had been prominently mentioned for the leading place on the Populist ticket was Colonel L. L. Polk, of North Carolina, president of the Southern Alliance. Shortly after the Cincinnati convention Polk began to express his friendly interest in the newly formed party, and well before the end of the year he could be counted definitely as within its ranks.[74] Since he was a southerner and an ex-Confederate soldier, his candidacy was bound to attract northern opposition, but this he apparently attempted to

---

[72] *National Economist*, 7:279 (July 16, 1892); *Southern Mercury*, November 14, 1895; *Appletons' Annual Cyclopaedia, 1901*, 421.
[73] *National Economist*, 7:279, 293 (July 16, 23, 1892).
[74] Hamilton, *North Carolina since 1860*, 236; Stewart, in the *Indiana Magazine of History*, 14:355; *Progressive Farmer*, June 6, 1891. See *post*, Appendix F.

offset by an address, delivered in Kansas during the fall of
1891, in which he declared that he had always favored the
preservation of the Union, and had fought against his country most reluctantly — a statement hardly calculated to increase his popularity in the South.[75] But Polk's power and
fame were great. Had he failed to achieve the first place on
the ticket, without a doubt he would have been offered
second place. In the spring of 1892, when he was already
planning to descend upon Omaha with several trainloads of
Confederate veterans, death suddenly cut short his career.[76]

The passing of Polk eliminated the only serious southern
contender for the nomination and increased the pre-convention talk that Judge Walter Q. Gresham, formerly of Indiana, but now a federal circuit judge with headquarters at
Chicago, would be tendered and would accept the Populist
nomination. Rumor had it that Judge Gresham's flirtation
with Populist doctrines began with his reading of Ignatius
Donnelly's *Caesar's Column;* at any rate, he was thought
to be a convinced believer in free silver, an enemy of railroad
extortion, and a friend of organized labor.[77] Gresham's failure adequately to deny his candidacy and to denounce those
who were using his name seemed to indicate a willingness on
his part to consider the honor, and many leading Populists
felt that their cause could best be served by naming him. His
acceptance could then be used as evidence that respectable
men were becoming Populists — even men of high standing in the old parties. Undoubtedly the Gresham candidacy

[75] *Greensboro Daily Record*, September 15, 1891; *Greensboro Patriot*, September 16, 1891.

[76] *Greensboro Daily Record*, May 30, 1892. Perhaps Polk had a premonition
of the fact that his days were almost numbered, for he was widely quoted as
having said on July 4, 1890: "I am standing just behind the curtains, and in
full glow of the coming sunset. Behind me are the shadows on the track, before
me lies the dark valley and the river. When I mingle with the dark waters
I want to cast one lingering look upon a country whose government is of the
people, for the people, and by the people." *National Economist*, 7:281 (July 16,
1892).

[77] Haynes, *Third Party Movements*, 261; Arnett, *Populist Movement in
Georgia*, 141; Stewart, in the *Indiana Magazine of History*, 14:363.

was feared by old-party leaders, and they must have breathed a sigh of relief when they learned that on the Saturday before nominations were to be made he had sent a telegram to one of his friends declining to allow the convention to consider his name.

Some of the delegates may have joined in the rejoicing. Only a few days before, a spokesman for the Georgia delegation had declared: "We do not want a man to come to us at the last hour of the day or to come with a single idea." And earlier in the campaign the editor of the *Great West* (St. Paul) had written: "We do not believe in going outside the ranks of the workers-in-the-vineyard to find men who ought to be of us and are not, to execute our will."[78] But probably a far greater number were disappointed. Members of the Indiana, Illinois, and Iowa delegations caucused and decided to send a committee to Chicago to win over the judge; and four prominent delegates were found to undertake the task. But they were doomed to disappointment. Taubeneck of Illinois read a telegram to the convention that seemed to show Gresham in a mood to accept under certain conditions, and on the strength of it the convention might possibly have been induced to vote for Gresham; but unfortunately the telegram was not from Gresham himself, and when authentic news did arrive, it was to the effect that the committee had failed and that Gresham had refused unconditionally.[79]

The declination of Gresham narrowed the field down to the orthodox third-party leaders of the Northwest. Indeed, as the *Chicago Herald* put it, there was little else left to do but to nominate Weaver. A few months before, Senator

[78] *Omaha World-Herald*, July 1-5, 1892; *Great West*, May 13, 1892; *People's Party Paper*, July 8, 1892; Matilda Gresham, *Life of Walter Quintin Gresham*, 2:659–663.
[79] *Chicago Daily Tribune*, July 5, 1892; Stewart, in the *Indiana Magazine of History*, 14:362–363.

Leland Stanford of California, whose views on financial matters were somewhat alluring, had received frequent mention, but the California delegation to the St. Louis convention had denounced Stanford in bitter resolutions as an unprincipled monopolist; and that had ended his boom.[80] Ignatius Donnelly of Minnesota had likewise attracted much attention. He was beyond all doubt the greatest orator of Populism, and at the opening of the convention he had not neglected to make the most of his talent. Moreover, it was Donnelly who had written the famous preamble to the platform just adopted, and the vivid rhetoric in which this document was couched made it, in the language of Simpson of Kansas, "one of the most vigorous and at the same time, most classic productions of modern literature."[81] But Donnelly for president on any ticket would have been a joke, and the delegates knew it. Once some one had suggested the following epitaph for use, should Donnelly be named:

> Forbear, good friend, to touch these bones,
> For underneath these piled stones
> Lies party third, ne'er to awaken,
> Killed by Ignatius Donnelly Bacon.[82]

Of the other candidates there was none so eligible as General James B. Weaver of Iowa. He was a seasoned campaigner — had even run for the presidency once before back in 1880 as a Greenbacker — and he had friends and supporters in every section of the country. His long-standing connection with a disgruntled minority party, however, was regarded by some of the newcomers in the party as a liability rather than an asset. These men were sensitive to the charge, sometimes made, that their party "did not represent new ideas, but old ones relabelled" and that it "was not

composed of new men, but of old time agitators with new
hopes." They proposed, therefore, as an alternative to
Weaver, with whom defeat was a habit, one of the younger
men in the party, Senator Kyle of South Dakota. Kyle was
inclined at first to refuse permission to use his name, but he
finally consented, and the opposition to Weaver centered on
him. But these efforts were unavailing, and Weaver won
handily, the vote standing 995 to 275.[83]

The head of the ticket being from the North, it followed
almost inevitably that second place should go to the South,
and two well-known southerners, General James G. Field
of Virginia and Ben Terrell of Texas, were proposed. The
convention preferred Field to Terrell by a vote of 733 to
554, perhaps for no better reason than that an ex-Union
general and an ex-Confederate general on the same ticket
had an irresistibly dramatic appeal.[84]

Among the most curious of the acts of the convention was
the adoption of a "self-denying ordinance," so styled, which
established it as " a fundamental party law " that " no person holding office . . . federal, state or municipal . . . including senators, congressmen and members of legislatures
. . . shall be eligible to sit or vote in any convention of this
party. . . ."[85] The convention of 1892 of course could not
bind its successors in this regard, nor did the precedent it
sought to establish endure. But the effort thus made to keep
the People's party in the hands of the people and out of the
hands of party bosses is worthy of note.

It was remarked, too, that in Omaha "the enthusiasm
was all spent on the platform, while at Minneapolis and
Chicago they spent their enthusiasm upon the candidates."[86]

[83] *Omaha World-Herald*, July 4–5, 1892; *Chicago Daily Tribune*, July 4–5,
1892; *People's Party Paper*, July 8, 1892; Haynes, *Third Party Movements*, 317.
[84] Haynes, *op. cit.*, 262–263; Arnett, *Populist Movement in Georgia*, 142;
Manning, *Fadeout of Populism*, 28.
[85] Jory, *What is Populism?* 32; *National Economist*, 7:279 (July 16, 1892);
People's Party Paper, February 9, 1894.
[86] *National Economist*, 7:279, 293 (July 16, 23, 1892).

Whatever else may be said of it, the People's party was born a party of principle and those who brought it forth were in deadly earnest. Nor did they lack a genuine grievance. Whether they knew it or not, the delegates were beginning the last phase of a long and perhaps a losing struggle — the struggle to save agricultural America from the devouring jaws of industrial America.

CHAPTER NINE

THE ELECTION OF 1892

THE next problem of the third-party leaders, once the convention was over and the ticket decided on, was to persuade reformers generally that they should abandon their old-party allegiance and become "Populists."[1] In the West, where independent parties of state-wide dimensions had existed ever since the campaign of 1890, this was not particularly difficult; indeed, to western reformers evolution into a nation-wide party of the people seemed natural and inevitable. But in the South, where independent action had thus far been studiously avoided and where most Alliancemen had prided themselves upon their party regularity, the change of party banners was not easily accomplished.

According to Watson, who for all his froth was sometimes right, "the argument against the independent political movement in the South may be boiled down into one

[1] This term and another even less pleasing, "Populites," appeared first as derisive epithets early in the nineties. The former was finally accepted by People's party adherents; the latter, which for a time was especially common in the South, gradually disappeared.

According to John W. Breidenthal, one of the most prominent of the Kansas third-party leaders, the name "Populist" originated as follows: At a conference of Democratic and People's party leaders to discuss fusion plans the late W. F. Rightmire complained about the difficulty he had in using the name "People's party" in ordinary conversation. For example, he could easily refer to a man as a Republican, or a Democrat, or a Prohibitionist, but he could not call him a People's. He had to use a whole sentence in referring to a member of the People's party.

"We need a shorter name for everyday use," said Rightmire. Then, turning

word — *nigger.*" [2] There were other ways of expressing this idea, however, not so concise but perhaps more comprehensible to the uninitiated. One editor wrote:

The Democratic party at the South is something more than a mere political organization striving to enforce an administrative policy. It is a white man's party, organized to maintain white supremacy and prevent a repetition of the destructive rule of ignorant negroes and unscrupulous whites. . . . The safety of the South . . . as well as the conservation of free institutions on these shores, depend upon the strength, unity and perpetuity of the Democratic party.[3]

Furthermore, the Alliance had acquired control, or was in a fair way towards acquiring control, of the Democratic party in every southern state. What good could come from " deserting a party of which we have entire control and going into one that, at best, can only divide the white vote and open wide the door for an irresponsible race to take within their inexperienced hands the reins of government? " [4] Thus Representative Everett of Georgia inquired, and he went on to predict that

the installation of a third party is the death-knell to the Alliance. The strength of the Alliance lies in its unity. Destroy that and you destroy the whole organization. It is absolutely certain that all the Alliance will not go into the third party; hence there will necessarily be division, which will result in final disintegration and ruin.[5]

to Overmyer, one of the Democratic leaders, he said: " Dave, you are long on words — give us a nickname for our party."

Overmyer scratched his head and brought forth from his recollection of Latin the word Populist.

" I'm afraid of that, because the newspapers would be calling us Pops within a week," said the judge.

" So much the better," Overmyer replied. " You want a short name. You can't find one much shorter than ' Pops.' "

As predicted, the word " Pops " appeared next day in the *Kansas City Star.* See the *Memphis Appeal-Avalanche,* October 9, 1892; Stewart, in the *Indiana Magazine of History,* 14:361; *Kansas City Star,* April 12, 1931.

[3] *People's Party Paper* (Atlanta), August 26, 1892.
[3] *Greensboro Daily Record,* March 7, August 19, 1892.
[4] *National Economist* (Washington), 7:117 (May 7, 1892).
[5] *Ibid.*

The seriousness of this situation was by no means lost on the third-party leaders. "The duty of the hour is education and agitation," wrote Watson. "Preach the new doctrine of political regeneration at every opportunity."[6] In general the campaign of education that was instituted took the form of pointing out how hopeless it was to expect to control the Democratic party nationally. Control in the southern states the Alliance might obtain, and state-wide reforms through an Alliance-controlled Democracy might indeed be expected. But of what avail were such reforms when the real root of the southern farmers' difficulty could be reached only through national legislation? Years earlier an Alliance leader had insisted that "the one thing needful in the present condition of the people is a debt-paying system of finance, in comparison with which all other questions sink into utter insignificance."[7] By the year 1892 there were few Alliancemen, southern or northern, who would dispute that statement. And how could a debt-paying system of finance be devised by the individual states? It was the reform of the monetary system through the subtreasury or "some better plan" that alone, according to the Alliance leaders, could be depended on to bring about a real change for the better. And on this matter what hope did either old party hold out? "The demands of the people are unheeded," railed Watson. "Silver is killed, the sub-treasury is denied a hearing. . . ."[8]

Repeatedly President Polk of the Southern Alliance had declared his willingness to hold in line for the Democratic party the farmers under his control if only that party would "come out and take a stand squarely on the Alliance platform";[9] and with this pronouncement in mind many southern Alliancemen awaited the outcome of the Democratic

[6] People's Party Paper, December 31, 1891.
[7] Proceedings of the Fourth Annual Session of the North Carolina State Alliance, 1890, p. 9; Progressive Farmer (Raleigh), December 15, 1887; April 17, August 21, 1888.
[8] People's Party Paper, July 15, 1892.
[9] Greensboro Patriot, September 10, 1891.

nominating convention of 1892 before making their decisions. Should that convention show a willingness to adopt
the Ocala and St. Louis demands, or something similar to
them, and should it nominate a candidate satisfactory to the
reformers, then there might yet be hope of accomplishing
something worth while through Democratic channels. The
defeat of Cleveland for a third nomination seemed particularly essential, for Cleveland's out-and-out hard-money
views were well known. David B. Hill, senator from New
York, was Cleveland's leading opponent, and while Hill's
views on the money question were hardly to be preferred
to Cleveland's, southerners of Alliance leanings felt that,
since the defeat of Cleveland must be achieved at all costs,
perhaps the best thing to do was to support Hill. It was at
least possible that Hill might yield a little on the money
question, whereas certainly Cleveland would not.[10]

When finally the Democrats met in Chicago, however,
they made not the slightest attempt to conciliate the Alliance. Their platform bore no authentic resemblance to the
Alliance demands, and their candidate was Grover Cleveland, the man whose prospective nomination had lately been
characterized by the Tillmanites in South Carolina " as a
prostitution of the principles of Democracy, as a repudiation
of the demands of the Farmers' Alliance, which embody the
true principles of Democracy, and a surrender of the rights
of the people to the financial kings of the country." [11]

The nomination of Cleveland became a rallying point for
the third-party agitators throughout the South.[12] Southern
Democrats could never be expected to vote for a Republican,

---

[10] *Greensboro Daily Record*, April 21, 1892; *National Economist*, 6:210–212
(December 19, 1891); Arnett, *Populist Movement in Georgia*, 148.
    [11] Haynes, *Third Party Movements*, 266. Watson seemed to think that the
Hill-Cleveland contest might prove valuable to the Populists:

> " Little drops of Grover,
> Little grains of Dave,
> Make their busted party
> Mighty hard to save."
> *People's Party Paper*, June 17, 1892.

[12] *National Economist*, 7:241–242 (July 2, 1892).

even though Harrison's stand on the money question was
far less positive than Cleveland's. But there was practically
no chance for anyone to express his opinion on the money
question by voting either old-party ticket. Long before, it
had been suggested that the Democratic and Republican
leaders might agree among themselves to shelve the question
of financial reform in order to stage a mock battle on the
tariff.[13] Was this not precisely what they had done?

The leadership of the Southern Alliance strongly favored
the new party. Colonel Polk had been so disposed in his
time, and by chance the vice president of the order who succeeded to Polk's duties was an ardent third-party man and
a northerner as well, H. L. Loucks of South Dakota. Polk
had embraced Populism reluctantly and as a last resort.
Not so Loucks, who had none of the repugnance to party
irregularity so natural in a southerner. Under his leadership
the Southern Alliance became even more aggressively favorable to the third party than it had been before.[14] Numerous
suballiances voted to fall in line; numerous Alliance newspapers devoted themselves to third-party propaganda; and
state and local third-party tickets made their appearance
nearly everywhere in the South.

But the solidarity of action that had characterized the
work of the Alliance in the campaign of 1890 was now lost
forever. Many Alliancemen protested vigorously against
the attempt to deliver them over to the third party and insisted on their right to remain at once good Democrats and
good Alliancemen. Tillman in South Carolina, for all his
radicalism, stayed with the Democratic party and loyally
supported the Chicago nominees.[15] Elias Carr, formerly

[13] *Southern Mercury* (Dallas), February 11, 1892.
[14] *People's Party Paper*, August 19, 1892; *Memphis Appeal-Avalanche*, November 15, 1892. "The Alliance is non-partisan," declared the *National Economist* on July 2, 1892 (7:241), " and does not interfere with the political affiliations of its members; but these two political parties have made war on the
Alliance and ostracised its members on account of their opinions, and now, for
conscience sake, they are compelled to affiliate together in a new party."
[15] Simkins, *Tillman Movement*, 172.

president of the Farmers' Alliance in North Carolina, not only repudiated the third party but he also accepted the Democratic nomination for governor of the state, and he campaigned earnestly against the third-party ticket.

But in spite of the fact that these and most of the other leading Alliancemen in the South stood by the Democratic party, the third-party men won many adherents. Perhaps only a southerner can realize how keenly these converts to Populism must have felt their grievances. They became in the eyes of their Democratic neighbors not merely political apostates but traitors to civilization itself, more to be reviled even than the Republicans into whose hands they played. They invited upon themselves every stinging epithet, every scandalous remark, that a host of scurrilous editors and orators could devise. No doubt there were a number who joined the movement only to secure place and preferment for themselves, but the rank and file of these third-party men were enlisted like crusaders to battle for a cause, regardless of the difficulties that beset their way. Perhaps half the members of the Alliance gave the new party their support. The rest were unwilling to risk the reality of white supremacy in what might prove to be a vain struggle for a new order. The third-party strength was recruited mainly from the strictly rural districts, the towns and cities generally remaining intensely and almost unanimously Democratic.[16]

Knowing that if only the South could be won for the third party, victory would be in sight, the leaders of the movement did not hesitate to send their ablest campaigners into that section. Of these General Field was possibly the most successful. He had lost a leg while fighting for the Confederate cause; he knew the oratorical language of the South and could speak it well; and he had little of the harshness and rancor in his system that characterized so many of the

[16] Hamilton, *North Carolina since 1860*, 236–237; John D. Hicks, "The Farmers' Alliance in North Carolina," *North Carolina Historical Review*, 2:181–183; Arnett, *Populist Movement in Georgia*, 152.

third-party speakers. His doctrine was set forth in the conclusion of one of his perorations:

> All hail the power of the People's name,
> Let autocrats prostrate fall,
> Bring forth the royal diadem
> And crown the people sovereign, all.[17]

Weaver and Mrs. Lease, who also planned a tour of the
South, were less fortunate. In earlier years the Populist
candidate had visited the South and had been well received,
but in the campaign of 1892 he was given plainly to understand that his presence was not desired. His antisouthern
record of Civil War and Reconstruction days was dragged
forth and painted as black as possible. At Pulaski, Tennessee, where he had once been in command as a Union officer, he was charged with having been guilty of conduct
unbecoming a soldier and a gentleman and was warned not
to appear. He did appear, however, and he spoke, too, in
spite of the threats.[18] He spoke also in North Carolina and
in Georgia, but everywhere he was bitterly assailed in the
opposition press and roughly handled on the stump.

Mrs. Lease received quite as little favorable consideration.
One southern paper found "the sight of a woman traveling
around the country making political speeches . . . simply
disgusting" and declared hotly that "Southern manhood
revolts at the idea of degrading womanhood to the level of
politics."[19] Howling mobs, brass bands, hoodlums with
rocks and eggs sought to prevent the speakers from deliv-
ering their addresses. According to Mrs. Lease, General
Weaver "was made a regular walking omelet by the Southern chivalry of Georgia."[20] If it was of any consequence at
all, this invasion of the South by northern speakers probably

---

[17] Manning, *Fadeout of Populism*, 28–29; *National Econmist*, 7:359 (August 20, 1892).
[18] *Memphis Appeal-Avalanche*, October 7, 9, 1892.
[19] *Greensboro Daily Record*, September 29, 30, 1892.
[20] Barr, in *Kansas and Kansans*, 2:1170.

did more harm than good. Local third-party speakers received a considerable amount of punishment also, but on the whole their activities were less resented than the work of the outsiders.[21]

Had it been possible for the third party to operate as a national organization only, without the encumbrance of state and local tickets, perhaps the opposition to Populism in the South would have been far less violent. But the third-party leaders were unwilling so to compromise — the new organization must be built from the ground up. And no doubt they were right. What hope of reward would there have been for the local workers, on whom such tasks as getting out the vote would rest, if all the minor offices had been left to the Democrats? Places on the ballot and recognition at the polls would likewise have been difficult to obtain for half a party with half a ticket. For a fully organized new party there might indeed be trouble, as there had been in the Alliance before, from " politicians hunting a fat place " and from "unprincipled demagogues," but local leaders might be recruited, also, from " the young, ardent and progressive elements of the white population," who were ready to carry their revolt "against the tyranny and arbitrary methods of the Bourbons " even to this extreme.[22]

The formation of local organizations, however, tempted inevitably to cooperation with the Republicans, who were from the first altogether too eager to promote the welfare of the Populist movement. Many southern Republicans actually joined the new party and worked energetically for it, no doubt looking forward to the day when the two parties opposed to the Democrats would unite.[23] In North Carolina the Republicans hesitated for a long time before they

[21] *People's Party Paper*, August 5, 1892; *National Economist*, 7:359 (August 20, 1892); Haynes, *James Baird Weaver*, 324–325.
[22] *Greensboro Daily Record*, March 9, August 10, 1891; *Greensboro Patriot*, May 11, 1892; *National Economist*, 7:370 (August 27, 1892).
[23] Hamilton, *North Carolina since 1860*, 235.

finally decided to nominate a state ticket of their own, and
undoubtedly they finally took such action only because they
thought that division in the Democratic ranks might give
the Republicans a chance to win alone. The Populists were
frankly disappointed.[24] In Georgia, where the Republicans
had by no means the strength that they possessed in North
Carolina, the Populist candidate for governor was indorsed
by the Republicans, but separate nominations were made
for the other state offices.[25] In South Carolina the situation
was unique. There neither the Republicans nor the Populists
nominated a state ticket, but their combined strength went
to Tillman, who, to the utter anguish of the conservatives,
ran in 1892 as in 1890 on the regular Democratic ticket. The
Populists nominated presidential electors only.[26] In Florida
the Republicans denounced Democratic rule in the state as
"a long-studied plan to rob the majority of its liberty" and
by making no nominations left such of their adherents as
might be permitted to vote free to support the Populist
ticket. In Alabama cordial cooperation at the beginning of
the campaign ended finally in a fusion of forces, fairly complete and thoroughgoing.[27] In Louisiana the Republicans
and the Populists divided the congressional and the electoral
nominations between them and agreed that every honorable
means should be used by each party to elect the combination ticket. In Arkansas there was no open agreement, but
there was a tacit understanding that the two minority parties were not to make war on one another.[28]

Throughout the South, even where entirely separate
tickets were nominated, this tendency of Republicans and
Populists to work together was plainly manifest. Local and

[24] *Greensboro Daily Record*, September 1, 5, 6, 14, 1892; *Greensboro Patriot*,
September 14, 1892.
[25] Arnett, *Populist Movement in Georgia*, 153.
[26] Simkins, *Tillman Movement*, 172; Haynes, *Third Party Movements*, 226.
[27] *Appletons' Annual Cyclopaedia, 1892*, 5, 279; Manning, *Fadeout of Populism*, 23–25.
[28] White, in the *Mississippi Valley Historical Review*, 5:9–11.

state tickets that were avowedly Populist often carried well-known Republican names, and "pepper and salt" combinations, i.e., negroes nominated along with whites, were not unknown.[29] It was freely charged that money and encouragement from the Republican national headquarters found its way into the hands of southern Populist campaigners — money that was originally furnished by the "Gold-Bugs of Wall Street."[30] All this affinity with Republicanism, real or fancied, told heavily against the new party. Prospective Populists were confronted with the argument that the end of Populism would be "just what the end of the Readjuster party in Virginia was. That the whole thing will go over to the Republican party."[31] And, convinced by this argument, thousands of deserting Democrats were driven back to the party fold.

To encourage wavering Democrats still further, the regular Democrats of about half the southern states did not hesitate to come out for sound Populistic doctrines in their state platforms. In Florida, for example, the Democratic state platform demanded the unlimited coinage of silver, the abolition of the national banks, the substitution in the currency of legal tender notes for the existing national bank notes, the increase of the circulating medium to not less than fifty dollars per capita, the elimination by national law of dealing in futures, the control by the government of the railroads, and such other reforms as might be well calculated to "redress the agricultural grievances of the state."[32]

In North Carolina, South Carolina, Georgia, and Texas, the same or similar clauses were likewise inserted in the Democratic state platforms, and the South Carolina Democrats frankly approved by name the Ocala demands. Occa-

---

[29] *Appletons' Annual Cyclopaedia, 1892*, 755; Hicks, in the *North Carolina Historical Review*, 2:184 note.
[30] *Greensboro Daily Record*, September 1, 15, October 11, November 1, 1892.
[31] *Greensboro Patriot*, June 22, 1892. Compare Arnett, *Populist Movement in Georgia*, 150–151.
[32] *Appletons' Annual Cyclopaedia, 1892*, 279.

sionally other more or less genuinely Populistic doctrines would creep in, such as the repeal of the prohibitory tax on state bank notes, the demand for a just and equitable system of graduated income taxes, the enactment of laws prohibiting the alien ownership of land, and the creation of a fractional currency. In several southern states, where there was no prospect whatever of Populist success, the tendency to abandon traditional Democratic views was less evident. But in the states named the Democratic and the Populist demands could almost have been exchanged without doing violence to the spirit of either platform.[33] Under these circumstances wavering Democrats could be the more easily persuaded to stand by their party. Locally, at least, its views were not far different from those of the Populists, and the time might come when the national Democratic party could no longer resist all liberalizing tendencies.

As the campaign wore on it became fully evident that the Populists had no chance to win in most of the southern states. In the states of the upper South, where the Republicans were numerous enough to warrant hopes of victory should the Democrats split, the third-party movement failed utterly to take hold, the Alliance was disrupted, and the Democrats were assured of their usual victories. People's party tickets were very generally nominated, but with the full realization that they stood no chance to win. In Louisiana, Mississippi, and Florida, also, the Democrats succeeded

[33] Most of these platforms may be found, in part at least, in *Appletons' Annual Cyclopaedia, 1892*, under the names of states. Ordinarily, of course, the Democrats would have nothing to do with the subtreasury plan, which was still good Populistic doctrine, and in the press they opposed it gleefully:

> "Why should the farmer delve and ditch,
> Why should the farmer's wife darn and stitch?
> The Government can make 'em rich:
>   And the People's Party knows it.
> So hurrah, hurrah for the great P. P.!
>   1=7 and 0=3,
> A is B and X is Z:
>   And the People's Party knows it!"
> *New York Sun*, quoted in the *Greensboro Patriot*, July 27, 1892.

in preventing any dangerous division of their forces. In North Carolina a different story might have been told had Polk lived, but his death left the Populists without adequate leadership, and they so bungled their nominations and their campaign that they finished finally a poor third.[34] South Carolina would certainly have been debatable territory had Tillman gone over to the new party, but since he refused to abandon the Democratic standard, the men who normally would have favored the new party were technically the regular Democrats, and the conservatives furnished opposition only in the primaries.[35]

In Texas, too, the strength of the reformers was sapped by the Democrats, who renominated Governor Hogg. Hogg had waged relentless warfare against the malpractices of the railroads and had exhibited such hostility to corporations in general that, according to the conservatives, capital was being driven from the state. Indeed, so dangerous did the conservatives deem the governor that they carried their fight against him beyond the primaries and the convention to the election itself, where they supported a ticket headed by Judge George Clark. Under the circumstances the People's party men might well have been pardoned had they supported Hogg, and some of them doubtless did that very thing, but there was a separate Populist ticket, which, although it polled over a hundred thousand votes, finished third. The Republicans supported Clark.[36]

In Alabama the Populists seemed to have a good chance to win. The campaign was a long-drawn-out affair, since the election of state officials came in August and the nominating conventions had to be held early in the year. The Alliance leaders, still working together with fair unanimity when the campaign began, sought again, as in 1890, to bring about the

[34] *Greensboro Daily Record*, August 22, 1892; Hicks, in the *North Carolina Historical Review*, 2:183–184.
[35] Simkins, *Tillman Movement*, 170–172.
[36] *Appletons' Annual Cyclopaedia, 1892*, 741.

nomination of one of their own number, R. F. Kolb, on the regular Democratic ticket. In this, however, they failed — less, they believed, because of lack of votes than because of unfair tactics on the part of their adversaries. Accordingly the defeated faction held a second convention and named Kolb at the head of a separate "Jeffersonian Democratic" ticket. To this ticket the Populists gave enthusiastic support; indeed, they practically adopted it as their own. The Republicans, also, were drawn into the combination, and a general fusion of the forces opposed to the Democratic machine was completed by dividing the county tickets among Republicans and Populists in accordance with their local strength.

The Democratic control of the state was thus effectively challenged, and for a time the outcome seemed in doubt. But the election, held on August 1, showed on the face of the returns the customary Democratic victory for the state ticket and fewer local victories than the fusionists had confidently expected. Jones, the Democratic nominee for governor, was declared elected by a majority of about eleven thousand, having carried most of the "black" counties where the negroes were in the majority. The fusionist forces, whose candidate carried the greater number of the "white" counties, claimed that the negroes had been fraudulently voted for the Democratic ticket, and they produced strong evidence to show that in a fair election Kolb would have won by a majority of not less than thirty-five or forty thousand. But the Democrats had the legislature and the courts, and the attempt of the Kolb forces to bring about a contest was effectively thwarted. After the state election was over the fusion of Jeffersonian Democrats, Populists, and Republicans was continued, joint electoral and congressional tickets receiving the general support of all three factions. But the results in November were not unlike the results in August.[37]

[37] *Appletons' Annual Cyclopaedia, 1892*, 3–5; Manning, *Fadeout of Populism*, 23–25.

In Georgia, as in Alabama, Democratic supremacy was seriously threatened. There the fiery Watson could have had the Populist nomination for governor had he been willing to accept it, but he chose rather to run for re-election to Congress, and a genuine " dirt farmer," W. L. Peek, headed the third-party ticket. The Republicans apparently favored fusion, but the Populists, fearing no doubt the loss of caste that such action on their part would entail, were unwilling to go the full length. The Republicans, accordingly, made separate nominations for the minor state offices, but on Peek's assurance that he would welcome Republican votes they put his name at the head of their ticket, and they made no separate nominations for Congress. Watson, through his speeches in Congress, his *People's Party Paper*, and his *Campaign Book*, which had the entertaining subtitle, *Not a Revolt, It is a Revolution*, set the pace for the lesser lights of Georgia Populism and gave the Democratic leaders much to do. In the press and on the rostrum the battle of words raged, the money question in one form or another usually commanding chief attention. Local issues were all but forgotten. But in spite of their heroic efforts the Populists lost out in Georgia even more completely than in Alabama. Peek was defeated by a two to one vote, and the Weaver electors did only about half as well as Peek. Watson himself went down in the general slaughter.[38]

These unexpected and disastrous defeats in Alabama and Georgia and the generally poor showing of Populism in the South can be fully understood only by taking into consideration what happened to the negro vote. From Reconstruction times until the election of 1892 negro voting in the South, while by no means abolished, existed on sufferance only, it being well understood that the Republicans must have no chance to win. This situation finally commanded the attention of the small Republican majority in Congress that was

[38] Arnett, *Populist Movement in Georgia*, 143–155.

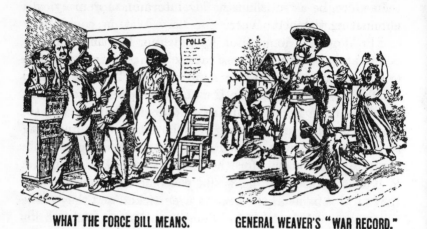

WHAT THE FORCE BILL MEANS.   GENERAL WEAVER'S "WAR RECORD."

Two Cartoons by C. De Grimm Posted on Large Sheets in
the South by the National Democratic Committee
[Reproduced in the *Review of Reviews,* November, 1892.]

returned by the election of 1888, and a Federal Elections
Bill, designed to provide adequate protection to the colored
Republican voter through federal supervision of congressional elections and, if necessary, the exercise of military
power, was forced through the House of Representatives. In
the Senate, however, the bill met defeat, thanks to the fact
that a small group of western silver Republicans joined with
the Democrats to vote the bill out of its place on the calen-
dar and thus make way for the consideration of a silver bill.
Bitter resentment in the South over this attempted northern
interference with "home rule" was freely expressed, and
there was a general feeling of relief when the "Lodge Bill,"
or the "Force Bill," as it was usually called, was definitely
shelved. All this agitation, however, told heavily against the
third party in the South, which in the face of such a menace
proposed political division, and at least one southern state,
Mississippi, promptly adopted into its constitution what

seemed to be a satisfactory legal formula for practically eliminating the negro vote.[39]

The Populists, in spite of the touchiness of this situation, deliberately set out to obtain for the new party all the negro votes they could get. Through the Colored Alliance, always a mere adjunct of the Southern Alliance, they spread their third-party propaganda, and, since the great majority of the colored people were in some way identified with agriculture, it seemed reasonable to hope that they would vote for the party that promised most to agriculture.

Realizing the danger, the Democratic leaders made preparations to meet it. There were other well-known methods of influencing colored voters than the appeal to reason, and these methods would have to be used. Bribery and intimidation, the stuffing of ballot boxes, the falsification of election returns, all these means had been employed in time past to save society from the rule of the ignorant and the vicious; the same high end would have to justify the same low means once again. The Democratic leaders took no chances. In the so-called "black" counties, the planters sometimes herded their employes to the polls and voted them in droves for the Democratic ticket. Barrels of whiskey and beer were provided in the villages for dusky all-night revelers, who the next morning were "marched to the polls by beat of drum, carefully guarded lest some desert in search of another reward."[40]

The total vote of Augusta, in Tom Watson's district, turned out to be double the number of the legal voters. Wholesale ballot-box stuffing and the importation of wagonloads of negroes from South Carolina, Watson claimed, helped to swell the Democratic majority. Third-party repre-

[39] *Greensboro Patriot*, January 8, 1891; Hamilton, *North Carolina since 1860*, 240; *Appletons' Annual Cyclopaedia, 1890*, 238; *1891*, 232–233; *1892*, 472; Fred Wellborn, "The Influence of the Silver-Republican Senators, 1889–1891," *Mississippi Valley Historical Review*, 14:477–478; Lester B. Shippee, *Recent American History*, 177–178.

[40] Arnett, *Populist Movement in Georgia*, 154.

sentation on election boards was freely denied, and votes were counted as they should have been cast, not as they happened to have been cast. "These things," writes Arnett, himself a southerner, "are matters of common knowledge among Georgians, especially in the black belt, who remember the Populist campaigns."[41] "Yes, we counted you out," Manning makes the Democratic leaders in Alabama say. "What are you going to do about it?"[42] Doubtless the Populists were not innocent of all wrongdoing, but the results seem to indicate at least that they were inept as compared with the Democrats. There was some reason in the argument of a Texas Populist that "the democrats rely on the negro to perpetuate themselves in office, and that the people's party is the only white man's party in the South."[43]

Out in the West the campaign of 1892 was really two campaigns. One campaign centered in the territory of the old Northern, or Northwestern, Alliance — the Middle Border. In this region the doctrines of Populism were nothing new — indeed, it was with these same doctrines as a platform that the various independent parties of state-wide dimensions had done battle in the campaigns of 1890 and 1891. Here the whole Populist program was given unstinted devotion; and the reforms that could be instituted by authority of the state governments were still deemed only slightly less important than those that the national government alone could institute. Here such matters as the more effective state regulation of railroads and elevators or laws to alleviate the suffering occasioned by unfair taxation, high legal rates of in-

[41] *Ibid.*, 155 note.
[42] Manning, *Fadeout of Populism*, 25. See also *Appletons' Annual Cyclopaedia, 1892*, 5.
[43] *Southern Mercury*, October 13, 20, December 29, 1892; *People's Party Paper*, December 9, 23, 1892; *Memphis Appeal-Avalanche*, November 16, 1892. As a result of the unfair tactics employed to defeat the Populists in 1892, at least two secret quasi-military orders were formed to employ the methods of the Grand Army of the Republic or if necessary of the Ku Klux Klan to secure honest elections. One of these orders, designed to be national in scope, was known as "The Industrial Legion of the United States" and actually founded some lodges in Texas and Kansas. The other, a "Gideon's Band," was ap-

terest, stern methods of foreclosure, and so forth, were seriously demanded.

The other campaign, in the mountain states of the farther West, was fought along much narrower lines. In this part of the country the Alliance had never acquired great strength, the majority of its doctrines had had no vogue whatever, and interest in the third party arose almost entirely from the fact that it was the only national party openly in favor of free silver. As a party of general reform, concerned with a variety of reforms, state and national, Populism could never have made headway in the so-called " silver states," but as a party that stood ready to redress the wrongs done by the " crime of 1873," the People's party had a well-nigh irresistible appeal.

The genuine Populists, as distinguished from the free-silver Populists, were most numerous in Kansas, Nebraska, the Dakotas, and Minnesota. In each of these states pre-Populist independent parties had already dealt shattering blows to old-party complacency, and in 1892 the new party had high hopes of success. The question of the hour was, Should the Populists avoid all entangling connections with the Democratic party in order to fight a straight-out battle for supremacy, or should it accept Democratic support and with such assistance look forward with greater confidence to victory? In a general way the Democrats were ready to be convinced on the matter of fusion. They sympathized thoroughly with the Populist efforts to end Republican supremacy locally, they were more than half convinced that as far as the silver question went the Populists were right, and they knew full well that such electoral votes as Weaver won in the West would be taken from the Republican

parently confined to North Carolina. Its membership was to be restricted to a select inner circle of Alliance and third-party leaders, and its proposed methods were of such doubtful legality that it became an easy point of attack for the opposition. The " Band " did not last long. See Haynes, *Third Party Movements*, 270–271; *Greensboro Patriot*, October 26, November 2, 1892; Hamilton, *North Carolina since 1860*, 240.

column and would better by just so much Cleveland's chance of success. All this was tempered, however, by the fear of what the Populists might do should they get control of the state governments. J. Sterling Morton, the leader of the Democratic party in Nebraska, was one of a number of Democrats who were desperately opposed to fusion; and he made it his chief concern politically to try to stamp Populism out of existence wherever he could.[44]

In the end the policy of fusion prevailed fully in only two of the border states, Kansas and North Dakota; and in both these states the Democrats practically forced themselves upon the Populists. In the other border states there were occasional instances of combination, or at least of cooperation, but none of thoroughgoing fusion.

In Kansas the Populists and the Democrats supported the same tickets, state, electoral, and congressional; but nearly all the nominees were Populists, and all the electors were pledged to Weaver. In the campaign of 1890 the Democrats had fused with the Populists in voting on a number of nominees with gratifying results, but in 1891, distressed at the refusal of the third party to recognize in any way the help it had been given, the Democrats had again nominated candidates of their own for local offices, or had even cooperated with the Republicans. In 1892, however, they could not resist the argument that Democratic support of the Weaver ticket would take the state safely out of the Republican column. On this subject they were delightfully frank; the same convention that indorsed the Weaver electors also indorsed the Democratic national ticket. As for the state ticket, there was no chance for the Democrats to win that anyway. This "unprincipled" action gave great offense to some of the old-line Democrats, who proceeded forthwith to hold another convention; but their convention

[44] J. Sterling Morton and Albert Watkins, *Illustrated History of Nebraska*, 1:712.

failed to make nominations and contented itself with advising the voters to defeat the Populist ticket at the polls. Inasmuch as the only way this could be done was by voting the Republican ticket, the disloyalty to party of these "come-outers" was quite as marked as that of their fellow Democrats who sought to throw the party vote to the Populists.[45] It was perhaps with an eye to the Democratic vote in the state, and certainly with an eye to the effect that such action would have south of Mason and Dixon's Line, that a delegate to the People's party nominating convention in Kansas — a Union veteran of the Civil War — was permitted to nominate for a place on the state ticket an ex-Confederate. This nomination was promptly seconded by two hundred and sixty-four other Union veterans present, who rose to be counted. In this dramatic way "the bloody shirt was folded decently and laid out of sight."[46]

Fusion in North Dakota was accomplished in much the same way as in Kansas. A third-party convention met and made nominations for electoral, state, and congressional candidates in advance of the meeting of the Democratic convention. All the standard Populistic doctrines were indorsed, but especial emphasis was placed upon the need of state-owned terminal elevators. The Democrats, also following the Kansas example, made no nominations of their own but indorsed the Populist ticket. The electoral ticket included, however, both Cleveland and Weaver electors.[47]

In South Dakota the Democrats successfully resisted the temptation to fuse with the Populists, and three separate tickets made their appearance. But the Populists were confident. Their convention, according to an eyewitness, "was the largest ever held in the state by any political party. It

[45] *Omaha World-Herald*, August 1, 1892; *Appletons' Annual Cyclopaedia, 1892*, 370–371; Miller, "Populist Party in Kansas," ch. 10.

[46] *National Economist*, 7:247 (July 2, 1892).

[47] *Appletons' Annual Cyclopaedia, 1892*, 530; Fossum, *Agrarian Movement in North Dakota*, 37.

lasted one week, and partook of the good old-fashioned camp-meeting order. I never saw men so filled with zeal and inspired by a cause." [48]

In Minnesota, where the versatile Donnelly became the Populist candidate for governor, fusion was spurned. Donnelly claimed that he had tried fusion repeatedly, and that it had always failed him. But the opposition to Donnelly's leadership among Minnesota third-party men was always strong, and resentment at the failure of Donnelly to make overtures to the Democrats was freely expressed. The Republicans nominated for governor Congressman Knute Nelson, whose stand on public questions — particularly the tariff — had been so much in harmony with Alliance views that in 1890 the revolting farmers could scarcely be restrained from adopting him into the fold and nominating him for governor themselves. "If the Alliance doesn't come nearer Knute Nelson's conception of political orthodoxy than the Republican party then his views are not in consonance with his acts," is what one Democratic editor thought about it. Nelson, moreover, was a leader among the Scandinavians and, as the same irreverent observer declared, was "supposed to carry the Norwegian vote of the State in the coat-tail pocket of his trousers." [49] Lawler, the Democratic nominee for governor, was a Catholic whose nomination, Donnelly claimed, was engineered by James J. Hill and other leading Democrats to cut into the heavy vote that Donnelly would normally receive from members of the Catholic church, particularly the Irish members. [50] Eventually the Populists, because they were "short on lawyers," Donnelly explained, did nominate two Democrats for positions in the state supreme court; and the Democrats, whether in return for this favor or for reasons of their own, took down the names of

[48] *National Economist*, 7:346 (August 13, 1892).
[49] *Broad-Axe* (St. Paul), September 24, 1891; February 25, August 25, 1892.
[50] *Minneapolis Tribune*, July 10, 1892; *Duluth Daily News*, September 9, 1892; *Penny Press* (Minneapolis), June 8, 1894.

four out of their nine candidates for presidential electors to make way for the names of four Weaver candidates.[51]

In Nebraska, as in Minnesota, the campaign was in no small part a contest of personalities. The third party made a strong bid for success by nominating as its candidate for governor an able and prominent politician, Charles H. Van Wyck. Before the Civil War Van Wyck had represented a New York district in the lower house of Congress; during the war he had commanded a regiment at the front; and shortly after the war, on coming to Nebraska, he had resumed his interest in politics. Until the formation of the Populist party he had been classified as a Republican, and as such he had represented Nebraska in the United States Senate, but inasmuch as his Republicanism from Granger times on had been strongly tinctured with antirailroad and antimonopolistic views, his eligibility for high office at the hands of his party was slight. There was some talk that the Omaha convention might choose Van Wyck as the Populist presidential standard bearer, but perhaps out of deference to the Nebraska delegates, who wished to hold Van Wyck in reserve to run for governor, the talk did not progress as far as an open candidacy. Van Wyck, as candidate for governor, promised to sign the famous Newberry Railroad Rate Bill, defeated by Governor Boyd's veto the year before, and the plan was to make this the chief Populist appeal of the campaign.[52]

Spurred on by the prospect of the Van Wyck candidacy, the two old parties also made unusually good nominations for governor. Lorenzo Crounse, the Republican candidate, had served as a member of the state supreme court and in Congress, but like Van Wyck he was thoroughly independent of railway control, and he had been rather effectively shelved until the threat of Populism required a reform pose on the

[51] *Farm, Stock and Home* (Minneapolis), May 1, 1893; *Representative* (St. Paul), April 19, 1893.
[52] Marie U. Harmer, "The Life of Charles H. Van Wyck," master's thesis, University of Nebraska, 1929.

part of the Republicans. During the campaign Crounse could and did maintain that he had been quite as fearless in his dealings with the railroads as Van Wyck — sometimes more so. J. Sterling Morton, the Democratic candidate, was not outstanding as a reformer, but he was a far abler man than the Democrats of Nebraska were wont to nominate for governor. Morton was a hard-money man, and according to general belief he harbored no very deep-seated grudge against the railroads; in fact, the main purpose of his candidacy was clearly to hold as many Democrats in line as possible and thus to prevent the election of the radical Van Wyck. All three parties had full tickets of presidential electors, but in two or three of the congressional districts there was fusion, or what practically amounted to fusion.[53]

In these five border states the Populist campaign followed the precedents set in preceding elections — parades five miles long, meetings lasting hours and days, oratory without end. The national executive committee of the People's party was desperately poor — Secretary Schilling reported only fifty dollars in the treasury at the time of the Omaha convention — but subscriptions in small sums were solicited with some success, and campaign speakers took up collections for expense money at every meeting they held. Gifts even from ladies' auxiliaries were gratefully received.[54]

Weaver, Mrs. Lease, and Donnelly showed remarkable endurance. The latter began his fight months before he was nominated. "We are making a tremendous campaign," he wrote to "Calamity" Weller of Iowa. "I start out tomorrow to begin a series of 65 speeches before July. After July I shall be at work all the time."[55] "From Forge and Farm; from Shop and Counter; from Highways and Firesides," ran a Donnelly broadside, "come and hear the 'Great Com-

[53] *Omaha World-Herald*, August 4, 31, 1892; Morton and Watkins, *Illustrated History of Nebraska*, 1:631–634; Barnhart, "Farmers' Alliance and People's Party in Nebraska," 310–315.
[54] Haynes, *Third Party Movements*, 276.
[55] Letter of April 15, 1892, in the Weller Papers.

moner' on the mighty issues which are moving mankind to the ballot box in the great struggle for their rights." And the crowds came.[56] Weaver and Mrs. Lease, speaking at a huge rally in Lincoln, Nebraska, were introduced after a parade with floats had set forth the issues of the day. The Republican party was a young woman kneeling before a golden calf. Tradesmen at work were pulled about, helpless to prevent it, by a representative of the "money power." [57] In general, the Populist meetings were not greatly disturbed by the opposition, and there was distinctly less rowdyism in this region than in the South; but in Kansas, where the campaign was probably hottest, a few notable examples of persecution were recorded.[58]

But in spite of all their earnest labor and their high hopes the Populists won only modest victories on election day. The results were far less satisfactory in this region than they had been in 1890, and according to Donnelly, whose ticket emerged a sorry third in the Minnesota contest, it was clear enough "that the People's party afforded no promise of reward for me or any other man." [59] The best results were obtained in Kansas, where the Populist-Democratic combination threw the vote of the state to Weaver, elected the entire Populist state ticket, and won five out of seven seats in Congress. But the joy that these successes brought turned to grief when the discovery was made that on the face of the returns the lower house of the legislature had gone Republican by a narrow majority. The cry of fraud was inevitably raised, but there seemed to be little chance of altering the result.[60]

In Nebraska, thanks to the fact that several thousands of Democrats abandoned their national ticket to vote for Wea-

[56] Donnelly Scrapbooks, Vol. 11.
[57] Barnhart, " Farmers' Alliance and People's Party in Nebraska," 322.
[58] Barr, in *Kansas and Kansans*, 2:1165.
[59] *Representative*, April 19, 1893.
[60] *Appletons' Annual Cyclopaedia, 1892*, 370–371; Barr, in *Kansas and Kansans*, 2:1163, 1167.

ver, the electoral vote of the state went to Harrison by only
a very slender margin; but the contest for state offices was
not so close, the Republicans winning everything as usual.
In the congressional districts the Populists did better. They
helped the Democrats elect Bryan in the first district, and
the Democrats helped them elect their candidate, Kem, in
the sixth. In the fifth a straight-out fusionist was chosen by
Democratic and Populist votes. The other three districts in
the state went Republican. But the legislature, which as a
result of the preceding election had been Populist in both
branches, was in chaos—no party had a majority in either
house. The chances were that the whole Populist program
of legislation would be blocked.[61]

In North Dakota, as a result of fusion, the contest was
much closer than in Nebraska. The entire fusionist state
ticket, with the exception of the candidate for secretary of
state, was successful; but the Republicans won a majority
in both houses of the legislature and elected the single representative from the state to Congress. The electoral vote
split three ways, one for Weaver, one for Cleveland, one for
Harrison. In Minnesota the Populists held the balance of
power in the state Senate, elected one Populist to a seat in
Congress, but lost everything else, although only the cumbersome character of the Minnesota ballot prevented the
four Populist electors that the Democrats had indorsed from
winning. In South Dakota the Republicans won by a clean
sweep.[62]

These failures to achieve a complete victory in the very
heart of the Populist territory can be accounted for without
much difficulty. In the first place, it was a presidential election year, and old-party loyalty was stronger than in off
years when less was at stake. Wavering Republicans, who
might otherwise have voted the third-party tickets, were

[61] Barnhart, "Farmers' Alliance and People's Party in Nebraska," 325–331.
[62] *Appletons' Annual Cyclopaedia, 1892*, 497, 530, 707–708; Hicks, in the
Minnesota History Bulletin, 5:545–546.

constrained to help save the country from the rule of the Democrats. Moreover, only a moderate number of Democrats could be induced to vote for Weaver, even when they knew that a vote for Weaver would be equivalent to a vote for Cleveland. The *Omaha World-Herald* pointed out the practical wisdom of this policy in Nebraska, as did also ex-Governor Boyd; and had their advice been generally followed the Weaver electors in Nebraska would have won.[63] But the

DISTRIBUTION OF THE POPULAR VOTE FOR WEAVER IN 1892

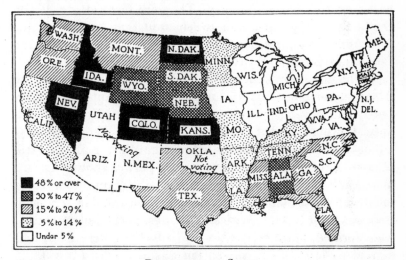

PERCENTAGES BY STATES

Alabama 36.60	Maine 2.04	New York....... 1.20
Arkansas 8.07	Maryland	Ohio 1.74
California 9.38	Massachusetts ... 0.82	Oregon 16.24
Colorado 57.07	Michigan 4.32	Pennsylvania ....
Connecticut	Minnesota 11.35	Rhode Island..... 0.43
Delaware	Mississippi 19.42	South Carolina... 3.42
Florida 16.06	Missouri 7.59	South Dakota....37.58
Georgia 19.17	Montana 16.55	Tennessee 8.92
Idaho 54.66	Nebraska 41.00	Texas 23.64
Illinois 2.54	Nevada 66.76	Vermont 4.17
Indiana 4.00	New Hampshire.. ...	Washington 21.79
Iowa 4.65	New Jersey...... 0.28	West Virginia.... 2.49
Kansas 48.44	North Carolina..15.94	Wisconsin 2.66
Kentucky 6.92	North Dakota....48.96	Wyoming 46.14
Louisiana 5.30		

[63] *Omaha World-Herald*, November 3, 1892.

majority of the Democrats stuck by their party standard.
Only in those states where the leaders saw to it that no
Democratic ticket for electors existed could they be sure
that the rank and file would follow their advice. It was true
also that the harvests of 1891 and 1892 had been fairly adequate in most of the border states. Prices were too low for
the farmers to be making much money, and the hard times
were by no means at an end; but the feeling of utter hope-
lessness that had driven many to revolt in 1890 was not so
much in evidence in 1892. "All the farmers want," one jour-
nalist wrote, "is more money. They are fast getting it, and
the faster they get it the more reluctant they become to ride
forty miles in a lumber-wagon through the rain to hear Mrs.
Lease and General Weaver make speeches." [64]

In the mountain states there had been some interest in the
Alliance movement as an antirailroad and antitrust affair,
but it had never achieved much prominence. Both the Alliance, however, and its political legatee, the People's party,
advocated the free and unlimited coinage of silver at the ratio
of sixteen to one, and when the older parties showed an unmistakable tendency to ignore the silver question, these
mountain states showed an equally unmistakable tendency
to desert their former political moorings. Local " silver clubs "
were formed, state conventions of silverites were held, and
a great national silver convention was called for May, 1892,
in Washington, D. C., to make sure that silver was not forgotten in the coming campaign. In the mountain region Republicans and Democrats, quite as enthusiastically as Populists, adopted free-silver platforms. A state convention of
Colorado Republicans went the length of denouncing President Harrison for his opposition to free silver, and the Democrats of Nevada officially served notice

that in the event of the Chicago National Convention failing
to nominate a candidate who is unequivocally in favor of the

[64] C. S. Gleed, in the *Forum*, 16:257. See also Tracy, *ibid.*, 16:249.

free coinage of silver and upon the free-coinage platform, the Democrats of Nevada are hereby absolved from all obligations to support nominees of the National Democratic Party.[65]

Both old parties being unwilling to go to extremes on the money question, these Democrats of the silver states promptly threw their strength to the Populists. In Nevada a silver convention met at Reno, organized a Silver party, nominated presidential electors, and sent delegates to the national nominating convention of the People's party at Omaha. A later convention of this so-called Silver party instructed the electors already named to vote for Weaver and Field. Even the Republicans in this preeminently silver state failed to hold fast their party lines, and an anti-Harrison faction also went over to the silverites, bringing with it F. G. Newlands, who became the universal silver candidate for Congress. In Colorado the Democrats officially indorsed a Populist ticket headed by Davis H. Waite for governor and supported the Weaver electors. A number of conservative Democrats came out both for Cleveland and for free silver, but they were a hopeless minority and failed to figure seriously in the campaign. In Idaho Republicans, Democrats, and Populists vied with one another in defending the rights of silver, and each party nominated a full ticket; but eventually the Democrats supported the Weaver electors. In Wyoming a fusion ticket of Democrats and Populists was arranged, the candidates for electors being pledged to Weaver. In the other mountain states there was much sympathy for silver, but only in these four was the sentiment strong enough to eliminate any effective Democratic opposition. Except in the case of Nevada the Republicans succeeded in maintaining a lively opposition. In Oregon, where the Democratic state convention actually rejected a free-silver plank in its platform, the state central committee deliberately adopted one of the Populist electors. This was

[65] *Appletons' Annual Cyclopaedia, 1892*, 127–128, 490; Joel F. Vaile, "Colorado's Experiment with Populism," *Forum*, 18:715.

done, obviously, with a view to taking at least one electoral vote out of the Republican column.[66]

All this Democratic show of interest was tremendously heartening to the Populists, who were, for the most part, not too proud to accept aid from any quarter and on any reasonable terms. Mrs. Lease progressed from one triumph to another as she toured the silver states. In Nevada, where she made eight speeches a day, she found that in some localities the old parties had so diminished in strength that they were unable to form the customary party committees.[67] Weaver, who also campaigned in the West before his ill-fated southern tour, found "the West on fire," and telegraphed back to headquarters late in August that "the whole group of States west of the Missouri is with us and the tide is sweeping eastward. The Republican party is completely disorganized and we shall win. Be strong and of good courage." [68]

The election showed, moreover, that Weaver's optimism, so far as it applied to the strictly silver states, was not altogether mistaken. The fusionist state tickets won decisive victories in Wyoming and Colorado; while in Nevada, Colorado, and Idaho the Weaver electors were chosen by similarly impressive majorities. In Oregon the one Populist elector on the Democratic ticket received a greater number of votes in the November election than were cast for any other Oregon elector. In the states where fusion was undertaken it failed to result in the choice of Weaver electors only in Wyoming,

[66] *Appletons' Annual Cyclopaedia, 1892*, 127–128, 340, 490, 614–615, 755, 827–828. The Populist movement in Wyoming was affected by the conflict then raging in that state between the cattlemen and the so-called "rustlers." The state Republican machine seemed to be in league with the cattlemen, and when an "invasion" of the rustlers' chief stronghold, Johnson County, was undertaken, state authorities were blind to what went on until the "war" began to go against the cattlemen. In the election of 1892 the rustler element, comprising in the main the small farmers and stock owners, as distinguished from the great cattle companies, supported the Populist-Democratic combination and helped the fusionists to the victories they won. Osgood, *Day of the Cattleman*, 246–255. See also A. S. Mercer and John Mercer Boots, *Powder River Invasion, War on the Rustlers in 1892*.
[67] Barr, in *Kansas and Kansans*, 2:1165.
[68] *National Economist*, 7:328, 369 (August 6, 27, 1892).

where practically no silver was mined. In many of the other states with three full tickets in the field the Republicans won only by pluralities, not majorities; and in California so many Republicans deserted to the Populists that the Democrats won every electoral vote of the state but one.[69]

Viewing the results as a whole, the Populists had made a good showing for a new party. Their candidate for president had received a total of over a million popular votes and twenty-two electoral votes, the first third party to break into the electoral college since the Civil War. The purely Populist delegation in the lower house was eight or ten strong, and the number who owed their election to deals of one kind or another with the Populists was even greater. Populist governors had been chosen in Kansas, North Dakota, and Colorado; and according to one computation not less than fifty state officials and fifteen hundred county officials and members of state legislatures would owe allegiance to the new party.[70] Weaver expressed himself as profoundly pleased at the " enviable record and surprising success at the polls " of a party at once so young and so badly embarrassed for lack of funds. He professed to believe that the Populists held the balance of power in a majority of the states, and he pointed hopefully to the fact that the new party had " succeeded in arousing a spirit of political independence among the people of the Northwest " that could no longer be disregarded.[71]

But there were certain unpleasant facts that the Populists had also to face. The most painful was that they had failed utterly to break into the " solid South." They had reasoned that, while the southern farmers could never be led to abandon the Democratic party for the Republican party, they

[69] *Appletons' Annual Cyclopaedia, 1892,* 755. See also under the names of states.
[70] Haynes, *Third Party Movements,* 274, 281; McVey, *Populist Movement,* 197.
[71] Haynes, *Third Party Movements,* 270.

might be induced to join an entirely new party.[72] But the
great majority of southern Democrats had refused to budge
from their traditional allegiance to the "white man's party."
Moreover, the victories in the West were not as reassuring as
they might have been.

In the states of the Middle Border, where Populism was
supposedly strongest, there was not a single complete and
thoroughgoing victory, except possibly in Kansas. And even
in Kansas there had been far more reliance upon Democratic
votes to secure results than was flattering to the new party,
while the sad fact remained that the lower house of the state
legislature would probably be in the hands of the Republicans.

In the farther West the silver states supported the Populist ticket enthusiastically and cast half the electoral votes
received by Weaver. But there was none the less a conspicuous fly in the ointment. The silver states were not interested in Populism, they were interested in *silver*, and they
supported the Populist ticket solely because of this one item
in the Populist creed. Had the Populist program not included free coinage it could hardly have appealed seriously
to any of the mountain states.

One other reflection there was to pain the thoughtful
Populist. From Iowa to New England, from the states of
the upper South to the Canadian border, Populism had made
no genuine impression. Eastern farming communities were
older and better established than those in the West — they
suffered less from debts and drought. Moreover, the eastern
farmers, by virtue of the fact that they were closer to market
than the westerners, received prices for their produce which,
had they been obtainable in the West, would have broken

[72] Watson's theory was that "the western farmer was a republican and would
never vote the democratic ticket, while the southern farmer was a democrat
and would never vote a republican ticket. Therefore, what? A new political
party was absolutely necessary that these two great sections could act in harmony. . . ." *Southern Mercury*, September 3, 1896.

the back of frontier Populism in a season. Even as far west as Iowa, the home state of General Weaver, Mrs. Lease spoke only to small crowds, and the presidential candidate of the new party polled less than five per cent of the total vote cast; and in Indiana, despite a picturesque campaign of rallies, barbecues, and picnics, the Populists cast only four per cent of the total vote.[73] As for the Populist appeal to the industrial classes of the East, it, too, fell upon deaf ears. In spite of the earnestness with which Populist platforms dwelt upon the rights of the laboring man and in spite of the dismal pictures of the downtrodden eastern masses that the Populist orators drew, the laborers of the East were content to vote about as they had voted before.

The Alliance, Northern as well as Southern, emerged from the campaign of 1892 sadly shattered. Thousands of members withdrew or dropped out, northern Republicans sharing with southern Democrats deep resentment that a supposedly nonpartisan organization should deliberately support a particular party. "When I joined the Alliance," one southerner protested, "I was expressly informed that neither my religious nor political convictions would be interfered with. When representatives of the Alliance go to a convention, as in St. Louis for example, and there vote to indorse a political party, and at the same time to commit the Alliance to this party, they violate the constitution of the Alliance." [74]

But the foundations of the Alliance were even more effectively undermined by the very existence of the third party. Why have two separate and parallel agencies of reform? If the Populist party could and did express the farmers' protests, why continue the Alliance to do the same thing? Accordingly, as the third party waxed stronger, the Alliance

[73] Haynes, *Third Party Movements,* 325; Stewart, in the *Indiana Magazine of History,* 14:364.
[74] *Greensboro Patriot,* May 11, 1892.

weakened. By sheer inertia the old orders, Northern and
Southern, state and local, lived on for several years, but each
year the number of delegates in attendance at Alliance conventions declined, and each year the attention accorded such
gatherings by the press diminished.

Among those who had worked for the success of the new
party during the campaign, none had been more energetic
and enthusiastic than the president of the Southern Alliance,
H. L. Loucks of South Dakota. He had held that the farmer
organization over which he presided was certainly bound to
support its own principles, and, since its principles were also
the principles of the Populist party, that "no true Allianceman can hesitate which course to pursue."[75] Support of the
Populist party he knew would be sure to cause the Alliance
the loss of many members, but such a test might prove only
"a winnowing process, a separation of the wheat from the
chaff and foul seeds that grow up with the order. The result
will be a reduction in numbers, but a great improvement in
quality."[76] At the Memphis meeting of the Supreme Council of the Southern Alliance, held just after the election, this
"reduction in numbers," was painfully apparent — only the
faithful few put in their appearance, and the question was
freely asked, "Would the order go forward, or had it out-
lived its usefulness, and was it ready to die?"[77]

Among those present there were some to point out how
greatly the order had flourished in its nonpartisan rôle and
to urge that an attempt be made by a return to nonpartisan-
ship to regain the ground that had been lost.[78] The chief of
those advocating such a policy was none other than the celebrated Dr. C. W. Macune, editor of the *National Economist*
and author of the subtreasury plan. Macune had steadfastly

[75] *People's Party Paper*, August 19, 1892.
[76] *Memphis Appeal-Avalanche*, November 16, 1892.
[77] Blood, *Handbook*, 43.
[78] *Memphis Appeal-Avalanche*, November 14, 15, 1892.

maintained that the Alliance and the third party must be kept separate, that "the party and the Order could not both be supreme over the same subjects at the same time."[79] He had long foreseen that for the Alliance to become in fact a class party — a political faction — would alienate old-party men and drive them in numbers from the Alliance ranks,[80] and he now frankly deplored their departure. During the campaign his paper had given lip service to the Populist cause, but there was much reason to believe that his personal and his public opinions did not fully coincide. After the campaign he was unwilling to admit "that the new party is entitled to more fealty than the order or its demands";[81] and he arrived at Memphis with a fine new scheme for the organization of the cotton growers into an "Alliance Cotton Cooperative Board" with himself, preferably as newly chosen president of the Alliance, to furnish the guiding hand.[82]

Would the Supreme Council choose Loucks to succeed himself as president and thus continue the Alliance as a mere adjunct to the People's party? Or would it turn to Macune and attempt through his leadership to restore its former glory? Third-party officials were not uninterested in the outcome of this contest. H. E. Taubeneck, chairman of the People's party executive committee, found it possible to visit Memphis while the convention was on, and in spite of the fact that he was not even a member of the Alliance he seemed to be managing Loucks' campaign for re-election. The struggle for control of the Alliance was relatively short and at no time was the result seriously in doubt; but the fight was hot enough to plunge the convention into the wildest disorder. Ultimately Loucks won by a handsome majority, for a ruling was adopted that each delegate might cast

[79] *National Economist*, 4:373 (February 28, 1891).
[80] *Ibid.*, 6:225 (December 26, 1891).
[81] *Southern Mercury*, December 1, 1892.
[82] *Memphis Appeal-Avalanche*, November 19, 1892.

only his own vote and not the entire vote of his state. This told against the badly depleted southern delegations and gave an easy victory to Loucks. All the other officers were chosen from the "Populite" faction, the same old demands that had now become the Populist platform were reiterated, and the public was left to infer that it was no longer possible to be a good Allianceman unless one were also a good Populist.[83]

As for Macune, his days as an Alliance leader were almost numbered. Grave charges of double dealing during the campaign of 1892 had been brought against him only a few days before the Memphis meeting. It was then revealed that J. F. Tillman, who together with Macune was a member of the Southern Alliance executive board, had issued over his signature a pamphlet repudiating Weaver and urging Alliancemen to support Cleveland. In the new pamphlet Tillman had posed as the "general manager and director of the National Farmers' Alliance and Industrial Union Lecture Bureau"—an office that existed only in his own imagination. Over a hundred thousand copies of his pamphlet had been circulated by the national executive committee of the Democratic party, and from all appearances these pamphlets had been printed in the office of the *National Economist*. Some thought that the literary style of the pamphlets was that of Macune rather than that of Tillman; and the opinion generally held among third-party men was that both Tillman and Macune had deliberately sold out to the Democrats. Macune's financial rating was closely examined, and embarrassing questions were propounded. How could he live in a ten-thousand-dollar house, pay off a thirty-five-hundred-dollar mortgage during the campaign, and spend not less than four thousand dollars a year, when the *National Economist*, of which he was only part owner, had never paid any dividends and never would? Had he not "been in possession

[83] *Ibid.*, November 18, 19, 1892; *Southern Mercury*, May 18, 1893.

of money that he did not and could not have earned through any known legitimate efforts of his own?"[84]

All this talk would have been hard to live down in any event; but with the Alliance now wholly controlled by the third-party men, Macune's position in it became intolerable. Shortly after the Memphis meeting he resigned from the order and issued an open letter to the public, defending his conduct and ascribing the attack on him partly to those who hated the subtreasury plan for which he had fought and partly to those who wished to turn the Alliance over to the People's party and "let it die as such."[85] Subsequently he gave up the editorship of the *National Economist*, sold out his interest in it — according to Watson "to his Wall Street partners" — and disappeared utterly from the ranks of the reformers.[86]

[84] *St. Louis Globe-Democrat*, November 4, 1892; *Memphis Appeal-Avalanche*, November 17, 1892; *People's Party Paper*, July 7, 1893.

[85] *Southern Mercury*, December 1, 1892.

[86] *People's Party Paper*, March 3, April 28, July 7, 21, 1893. Macune is still alive (1930) and resides in Waco, Texas.

CHAPTER TEN

ON TRIAL IN THE WEST

THE elections of 1892 failed to give the Populists complete control of a single state government, South or West; but in the latter section there were several states in which the Populists had elected a large enough number of administrative officers or legislators or both to affect materially the character of the régimes that followed. This was particularly true of Kansas and Nebraska, and to a less extent of North Dakota, Minnesota, and Colorado. These states were therefore carefully watched for signs of what might be expected of the Populists should they ever achieve the power they coveted. In a sense Populism was on trial in the West, and only in proportion as it succeeded there might it hope to be intrusted with greater power in these and other states and in the nation. Small wonder that the Populists were exceedingly eager to give a good account of themselves; and small wonder, too, that the enemies of Populism made ready with every weapon at their disposal so to discredit the new party that it would have no chance to live.

Appropriately it was the state of Kansas, where Populism had won its most notable victories, that first challenged the attention of the country. There the mere process of swearing in the "first People's Party administration on earth" presented a thrilling spectacle. Faithful devotees of "the cause" had flocked to Topeka by the thousands. All the local Populist celebrities were there, including Mrs. Lease and Jerry Simpson, and some outsiders as well, such as A. J. Streeter of Illinois, once candidate for president on the Union Labor ticket. There were witnesses, too, who were not Populists and not friendly, among them a large delegation of newspaper reporters seeking sensational news.[1]

In his inaugural address Governor Lewelling gave a good account of himself, as had been expected. He was a new-comer to Kansas, a Wichita produce man who, as president of the local county Alliance, had delivered a stirring address of welcome to the Populist nominating convention and had garnered in first place on the ticket as his reward. He again spoke winged words — words that aptly summarized the whole philosophy of Populism:

Government is a voluntary union for the common good. It guarantees to the individual life, liberty and pursuit of happiness. The Government must make it possible for the citizen to live by his own labor. The Government must make it possible for the citizen to enjoy liberty and pursuit of happiness. If the Government fails in these things, it fails in its mission. It ceases to be of advantage to the citizen; he is absolved from his allegiance, and is no longer bound by the social compact. What is the State to him who toils if labor is denied him and his children cry for bread? Is the State powerless against these conditions? Then the State has failed and our boasted civil compact is a hollow mockery. But Government is not a failure, and the State has not been constructed in vain. The people are greater than the law or the statutes, and when a nation sets its heart on doing a great or good thing, it can find a legal way to do it.

After the inaugural ceremony was over Jerry Simpson and Mrs. Lease also addressed the crowd, and in the evening a typical Populist celebration was staged. It was the people's day.[2]

But the real excitement began on the afternoon of Janu-

[1] Barr, in *Kansas and Kansans*, 2:1168–1169.
[2] *Ibid.*, 2:1169–1170; *Nation*, 56:43–44 (January 19, 1893).

276   THE POPULIST REVOLT

ary 10 with the attempt to organize the lower house of the legislature. The state Senate had a comfortable Populist majority and was able to organize without difficulty; but the slender majority claimed by the Republicans in the House was vociferously disputed by the Populists, who were determined to unseat a few of their opponents and take possession for themselves. Four seats that the state canvassing board had given to the Republicans, enough to swing the majority to the Populists, had been called into question immediately after the election, and although in one case the clerk through an error had reversed the figures, giving the election to the Republican by certifying to him the vote that the Populist candidate had received, the state supreme court ruled that it could do nothing about any of the returns, the House alone being the judge of its own elections.[3] Under these circumstances the Populists argued that all those whose seats were contested should stand aside, allowing only those to vote whose election was not called into question. The Republicans, on the other hand, with precedent all on their side, insisted that certificates of election were prima facie evidence of election and that the contest should be brought only after the House was fully organized on this basis — a sure way of disposing of the contesting Populists.[4]

Both sides were determined to have their way, and in advance of the opening of the session they had planned what to do. According to law the newly chosen secretary of state, a Populist, appeared with a list of the members-elect, and he offered to act as temporary chairman until the House could select a chairman of its own. This proposal aroused immediate objection from the Republicans, however; where-

[3] Kansas, *Senate Journal, 1893*, 1–6; Kansas, *House Journal, 1893* (Republican), 20–21; (Populist), 369–374. The *Journals* of the rival houses were printed separately. It should be noted that in the particular case cited the Republican acted honorably and refused all claim to the seat.

[4] The story of this legislative session is well set forth by Barr in *Kansas and Kansans*, 2:1169–1188; by J. K. Hudson in *Letters to Governor Lewelling*, 73–95; and by Miller in his unpublished dissertation on " The Populist Party in

upon both Populists and Republicans, as nearly simultaneously as possible, elected two presiding officers, one chosen by the unanimous vote of the Populists, the other by the unanimous vote of the Republicans. Both chairmen took possession of the chair, both attempted to call the House to order, and in any literal sense both failed miserably, for the result, of course, was pandemonium. Undeterred by this difficulty, however, each side recognized the right of its own adherents to seats, elected a full quota of officers, and sent word to the governor and the Senate that organization had been achieved.[5]

There was now nothing left to do but to adjourn; but each contending faction feared that should it leave the hall, it might never get in again; hence it was mutually agreed that the representatives must all stay on all night. Sympathetic observers brought in sandwiches for the hungry; but when the janitors banked the fires and went home, there was considerable suffering from the cold. Unmoved by such minor discomforts, however, the faithful legislators refused to leave, and they carried motions for adjournment only on the following morning just before the time set for opening the next day's session. At this time there were in the hall sixty-five Republicans, fifty-nine Populists, and three Democrats, although the total membership should have been only one hundred and twenty-four. The Democrats, with good reason, abstained from voting with either side. Hostilities were now fully resumed, but by noon the rival factions, tired and hungry, agreed to a truce until ten o'clock the next morning, when they would reassemble *in statu quo*.[6]

The succeeding days are without parallel in legislative annals. When the House next came together the three

Kansas," 234–260. Briefer accounts may be found in *Appletons' Annual Cyclopaedia, 1893*, 420–422, and in Buck, *Agrarian Crusade*, 165–168.

[5] Kansas, *House Journal, 1893* (Republican), 20, 26; (Populist), 7–8; *Appletons' Annual Cyclopaedia, 1893*, 420.

[6] *Appletons' Annual Cyclopaedia, 1893*, 420.

Democrats took a hand in the proceedings and voted with the Republicans. To offset this misfortune the governor and the Senate recognized the Populist House as the legal House.[7] The tension was somewhat lessened by an agreement between the factions to use the hall alternately instead of simultaneously, each side taking turns at passing laws and making speeches, and neither side making any effort to oust the other by force. For days this manner of procedure was maintained, and without undue friction, even though rival chaplains hinted broadly to the Almighty that divine interposition on the right side would not be unwelcome. During this period the Populists in House and Senate met together in joint session to elect a United States senator, and faced by the necessity of commanding Democratic support they chose a Democrat, John Martin, to take the seat lately held by Preston B. Plumb. The Republicans would have produced a rival senator had they found it possible to do so; but on joint ballot they lacked a majority of the votes of both houses and hence were unable to elect. Martin was given his certificate of election and was seated by the Democratic majority in the United States Senate in spite of Republican protests.[8]

Meantime, actively aided and abetted by the completely happy newspaper men, this "legislative war" had aroused the whole state of Kansas to a high pitch of excitement. Advice and advisers poured into the capital, and military precautions were deemed necessary by both sides. A Republican county sheriff swore in sixty Republican deputies, and the Populists countered with a formidable array of "deputy adjutant generals." Determined to get the dispute before the courts, the Republicans announced about the middle of February that after one week they would bar from the House all who refused to recognize the Republican organiza-

[7] Kansas, *House Journal, 1893* (Populist), 12, 16.
[8] Kansas, *Senate Journal, 1893*, 159–162; *Appletons' Annual Cyclopaedia, 1893*, 421; Haynes, *Third Party Movements*, 274.

tion; they also contested in the courts an appropriation bill passed by the Populist House and Senate and approved by the governor; and they ordered the arrest of the Populist clerk on the ground that his boisterous and unseemly conduct was disturbing the proceedings of the regularly constituted House.

The unhappy clerk, who was spirited away to the office of the state auditor, was defended by a hundred determined Populists; and the Republicans having made the mistake of adjourning, other Populists, armed with Winchesters, took possession of the hall. When the Republican House demanded admittance on the morning of February 15, the Populists agreed to admit only those whom they believed entitled to seats, and they pointedly refused admission to a small army of deputy sergeants at arms. By a clever ruse, however, the Republicans distracted the attention of the Populist guards, rushed the entrance, smashed in the doors of the hall with a sledge hammer, and took possession. The Populists retired temporarily but returned a few hours later at the time when the Republicans customarily permitted them to hold forth in their turn. But this time armed men barred them out.[9]

The situation now began to look really serious. Lewelling, the Populist governor, ordered out the state militia and in this way obtained altogether about two hundred and fifty men, poorly equipped. The commander of the militia, Colonel Hughes, turned out to be a Republican, and he refused point blank to obey when ordered by the governor to clear the hall. Undoubtedly Hughes's action was dictated primarily by partisanship, but the colonel might well have hesitated in view of the fact that a force of well over a thousand men — deputy sheriffs and deputy sergeants at arms — was now arrayed against him. In fact, Hughes not only told the Republican House leaders that they need fear nothing from

[9] *Appletons' Annual Cyclopaedia, 1893,* 421–422.

him, but he also threatened to surrender his command to the enemy. He was at once relieved of command, and a siege of the Republicans in the hall was instituted, though no attempt was made to clear them out. For three days the siege was kept up, the hunger of the beleaguered Republicans being appeased somewhat by food smuggled in in mail sacks or drawn up to the windows in blankets. The sympathetic Populists even permitted their adversaries to leave the hall in search of food and guaranteed them a safe return.[10]

Thoroughly alarmed at the state of affairs, Governor Lewelling made frantic efforts to bring about a settlement. His first attempts were unavailing, but with the able assistance of a Kansas blizzard, which deposited ten inches of snow, he ultimately succeeded in patching up a truce. Both the militia and the deputies were to retire; the Republican House was to have possession of Representative Hall, the Populists meeting elsewhere; and the decision as to which was the legal House was to be left to the courts. This agreement constituted in point of fact a surrender by the Populists, for they knew full well that the courts were in the hands of partisan Republican judges who would decide against them. But they were unwilling to prolong the contest further with the chance that it might lead to bloodshed. As had been expected, the Republican majority in the state supreme court decided for the Republicans, although it might well be argued that the question was political rather than constitutional and no affair of the courts. If this latter view had been taken, the Populist House, accorded recognition as it had been by the governor and the Senate, would have won out.[11]

When the court decision was made known the Populists gave up, and such of them as the Republicans chose to admit

[10] *Ibid.* For complete details see Barr, in *Kansas and Kansans*, 2:1184–1187, and Miller, "Populist Party in Kansas," 248–254.

[11] Kansas, *House Journal, 1893* (Populist), 372–373; *Appletons' Annual Cyclopaedia, 1893*, 422; Barr, in *Kansas and Kansans*, 2:1187.

to membership went over to the Republican House. With
the House Republican and the Senate Populist, it was of
course impossible to put through any comprehensive program of reform, although the Republicans did permit the
passage of an Australian ballot law, a foreclosure law allowing a year's time for the redemption of occupied land, an
anti-Pinkerton law, and some others. Nearly all the measures promised by the Populists during the campaign had
been passed by the ill-fated Populist House, but these acts
were invalidated when the courts decided to recognize the
Republican House. Inevitably the violence and disorder that
had characterized the session were laid at the door of the
Populists, who were branded lawbreakers and anarchists.
Thenceforth Populists were on the defensive in state and
nation whenever the "legislative war" in Kansas was mentioned.[12]

Even after the legislature had adjourned, the Populist
misfortunes in Kansas continued. Irregularities enough to
make capital targets for the Republicans were soon discovered. Governor Lewelling was not especially happy in the
appointments he made, and he quarrelled openly with Mrs.
Lease, whom he had made a member of the state board of
charities. Moreover, the Populists soon fell out with their
Democratic allies, fought disgracefully among themselves,
and as a consequence lost heavily in the local and judicial
elections of 1893.[13]

The Nebraska legislature of 1893 also had some difficulty
in effecting an organization, but there was no such excitement as in Kansas. In each house of the Nebraska legislature the Republicans and the Populists had about equal
strength, but in neither house had any party a majority, the

[12] *Appletons' Annual Cyclopaedia, 1893,* 422; The history of the "legislative
struggle of two years ago" was briefly recounted by Governor Lewelling in his
address to the Kansas legislature of 1895. Kansas, *House Journal, 1895,* 44-47.
[13] *Appletons' Annual Cyclopaedia, 1894,* 392. For detailed treatment see
Miller, "Populist Party in Kansas," 261-265.

balance of power being held by a mere handful of Democrats. It was obvious that the Democrats must throw their
support to one side or the other in order to effect organization, and the precedents, so far as the West was concerned,
were mainly for combining with the Populists. In the House
of Representatives this was done without much trouble; but
in the Senate the Democrats allowed the Republicans to
elect a temporary secretary and a president pro tem before
they changed their minds and joined forces with the Populists. Thereafter all officers and committees were chosen by
agreement between the Populists and the Democrats.[14]

Organization being accomplished, the next important item
of business was the election of a United States senator,
which under the circumstances was sure to be a difficult
task. The Republicans voted for Algernon S. Paddock on
the early ballots, and they returned to him on the last ballot, but they put forth their chief efforts in behalf of John
M. Thurston, general solicitor of the Union Pacific. The
Populists at first sought in vain for a candidate satisfactory
both to themselves and to the Democrats, and they succeeded only after the contest had gone on for three weeks,
finally mustering the entire Democratic-Populist vote for
District Judge William V. Allen of Madison.[15]

Judge Allen was accounted on all sides an excellent choice.
He was an honest and successful country lawyer who had
left the Republican party only a few years before and who
had been nominated and elected district judge by the Populists in 1891. By 1893 he was well known throughout his
district and beyond as " a just, upright judge, whose acts of
kindness and charity are legion " and as " a Populist with a
head filled with wrong financial notions" but otherwise without those eccentricities that made so many third-party men

[14] Nebraska, *Senate Journal*, 1893, 3–27; Nebraska, *House Journal*, 1893,
3–15.
[15] Nebraska, *Senate Journal*, 1893, 245, 254, 292; *Appletons' Annual Cyclopaedia*, 1893, 503.

ridiculous.[16] Unadorned with whiskers, over six feet tall
and of corresponding weight, and "straight as a pine tree,"
he had, as Dr. Albert Shaw pointed out, "the physical basis
of greatness."[17] He had, moreover, a fine, well-balanced
mind; he could speak with composure on many subjects; he
was utterly incorruptible. His triumph over the Republican
candidate, a "chosen friend of the monopolists," cost the
judge exactly $74.25, probably, as someone maliciously re-
marked, "the smallest sum by which a seat in the present
United States Senate was secured." The Populists were immensely proud of Allen. They saw in him a kind of symbol
of that unimpeachable respectability they craved for themselves and their party but could scarcely hope to attain.[18]

The Nebraska legislature of 1893 made yet another contribution to Populist history by passing in somewhat modified form the famous Newberry Bill, vetoed two years before
by Governor Boyd. As passed in 1893, this bill reduced the
freight rates in force in Nebraska an average of twenty-nine
and a half per cent. It also conferred an unnatural duty
upon the state supreme court by giving it the right to raise
rates on representation from the roads that the existing rates
were unjust. The state Board of Transportation to which
this power ordinarily would have gone, might lower rates
but it might not raise them.[19] The Newberry Bill received
Republican and Democratic as well as Populist votes, and
it was signed by Governor Crounse; but it seems clear that
the Populist agitation was the main cause of its passage.
Perhaps the old-party men who supported the measure fore-
saw the difficulty it would have in the courts; for after five

[16] Tracy, in the *Forum*, 16:248.
[17] *Review of Reviews*, 10:32 (July, 1894).
[18] Tracy, in the *Forum*, 16:247; *Dictionary of American Biography*, 1:214.
Several other senatorial contests, similarly long drawn out, occurred in that year
in the western states, but in no other case was a Populist elected. In Montana
and Wyoming the legislatures adjourned with deadlocks still unbroken; and in
North Dakota a combination of Republicans, Democrats, and Populists elected
a Democrat to the United States Senate. *Appletons' Annual Cyclopaedia, 1893*,
501, 535, 774.
[19] *Laws of Nebraska*, 23d legislative session, 164–384.

years of litigation and uncertainty an injunction against its operation was sustained by the United States Supreme Court on the ground that the reductions called for were unjust and unreasonable and deprived the carrier of property without due process of law.[20]

Equally unavailing was a convention held at Lincoln in June of 1893, under call of the Nebraska legislature, to make plans for a Gulf and Interstate Railway. Ten states, from Minnesota and North Dakota in the North to Texas and Louisiana in the South, had been invited to participate in the convention, and seven responded with a total of forty delegates, mostly from Kansas and Nebraska. The project discussed was the building of a railroad through the trans-Mississippi region to Galveston, Texas, at the expense of the states through which it would run, the railway when completed to " be owned by the people of the said states and operated as a single line at cost for the benefit of its owners." With a deep-water harbor at Galveston apparently assured by action of the national government, the convention developed much enthusiasm over the projected road. Its cost, distributed among the states, could be met by a tax of five cents an acre on land. Operation at cost and a short route to the seaboard would bring relief from the high rates that neither state nor federal control of the railways seemed able to secure and in ten years would make Nebraska a state of five million people "with many prosperous cities like Lincoln and Omaha." This attractive scheme was quite in accord with the Populist stand on government ownership of the railways, and it seems to have emanated from the Texas Alliance; but there were obvious difficulties in the way of its execution, constitutional and otherwise, and nothing came of it.[21]

[20] Smith v. Ames, in *169 United States Reports*, 466; Dixon, in the *Political Science Quarterly*, 13:644; Barnhart, " Farmers' Alliance and People's Party in Nebraska," 347–349.

[21] *Report of the Proceedings of the Gulf and Inter-State Railway Convention, 1893*, 3–14.

One other episode of Nebraska history during this period is worthy of note. The fall of 1893 was the time regularly set for the election of a justice of the state supreme court, and circumstances conspired to make this election and the campaign that preceded it seem unusually important. The Populists were concerned about the courts, for most reform legislation, including the greatly prized Newberry Bill, must run their gauntlet. Could they be trusted to take a nonpartisan view? In a recent decision by the state supreme court it appeared that that body, at least, could not. The late legislature had brought articles of impeachment against some of the Republican state officers on the grounds that they had misappropriated state funds, had spent excessive amounts, and had mismanaged state institutions. The indictment seemed fully warranted by the facts, but the supreme court, which was charged in Nebraska with the duty of trying impeachment cases, found in favor of the accused.[22]

This decision, concurred in by two Republican judges, was denounced by the third, Chief Justice Samuel Maxwell, as little short of disgraceful. Such a decision as had been given, he held, would open the door to all sorts of dishonesty in office; and he left room for the inference that certain railways, desirous of retaining these officials as members of the state Board of Transportation, had used their influence during the trial. The dissenting judge was a Republican with twenty years' service in the supreme court, but after this opinion he was denied renomination by his party, in spite of his considerable ability and his long term of office.[23]

Had the Populists not already chosen a candidate, District Judge Silas A. Holcomb, they might well have nominated Maxwell themselves, as they did, in fact, two years later. But Maxwell indorsed Holcomb, who next to Allen was perhaps the ablest man the Nebraska Populists had, and the

[22] State v. Hastings, 37 Nebraska, 96.
[23] This has been reviewed at length by Johannes Klotsche in an article entitled "The Political Career of Samuel Maxwell," in the Nebraska Law Bulletin, 6: 439–454. See also Rowley, "Judicial Career of Samuel Maxwell," 44–49.

contest "between the railroads and the people" was thus joined. It was a lively contest, with the real race clearly between the Populist and the Republican candidates, the Democratic contestant scarcely figuring in some portions of the state. According to one interested writer,

The Populists are holding mass meetings all over the State in support of their judicial candidate. At these meetings there are songs sung by the glee clubs, composed generally of pretty country maidens, dressed in the national colors, whose not unmusical voices adjure the horny-handed Populists to "keep in the middle of the road." When the orator takes the platform he awakens the echoes with his thunderous denunciations of the existing order of things. His audiences are always silent and attentive to a degree rarely if ever found in the political gatherings of either of the old parties. After the speaking comes more music. Frequently there is a prayer offered by some earnest disciple of the faith, who, in all probability, is a farmer preacher who has felt himself "called" to the ministry and has not awaited the diplomas of colleges or the edicts of church conferences or assemblies.[24]

Mrs. Lease came over from Kansas to help with the work, but even intervention by "the Patrick Henry in petticoats," as someone called her, was insufficient, for on election day the Republicans won. Had the Democrats been prevailed upon to cooperate with the Populists, however, the result might have been different, and that fact sank deeply into the third-party consciousness. "It is the Democrats we are after," Mrs. Lease remarked during a tour of Iowa taken that same fall, "and we are going to get them, too, don't forget it."[25]

[24] This quotation is from a correspondent writing to Tom Watson's *People's Party Paper*, November 24, 1893.

[25] *Ibid.* This picturesque campaign, fully set forth in the files of the *Alliance Independent* (Lincoln) for the period from June to November, 1893, is also described by Barnhart in his dissertation on "The Farmers' Alliance and People's Party in Nebraska," 357–360, and by William F. Zimmerman in a master's thesis entitled "Legislative History of Nebraska Populism, 1890–1895," 151–157.

In North Dakota the chief demand that the Populist-Democratic forces had made in the campaign of 1892 was for state-owned terminal elevators. For years the farmers of North Dakota, whose northern hard wheat was in great demand for the manufacture of the better grades of flour, felt that they had been defrauded of their rightful profits by the elevator men, through whom, for the most part, the grain had to be shipped. The farmers, therefore, early demanded that the elevators be subjected to stringent regulation to prevent monopolistic practices, discriminations among customers, and dishonesty in the weighing and grading of grain. Not content with these demands, however, they soon began to urge that the state compel the railroads to construct loading platforms at all sidetracks so that the farmers might ship their grain to the terminal markets without paying any toll whatever to the elevator men.

These demands hit hard at powerful vested interests, both the elevators and their allies, the railroads, being directly concerned; and every step in the direction of the desired reforms was hotly contested. In the legislative sessions of 1889 and 1891 the forces opposed to the farmers found themselves usually able to control the governor or the Senate, and in this way most reform legislation was blocked. When this method failed and laws were actually passed, jokers were inserted somewhere along the way that made the laws nugatory in operation. Before the year 1892 the farmers' movement had accomplished practically nothing by way of reform legislation.[26]

It was this situation that led the Alliance and its political friends to insist that the state itself enter the terminal elevator business. Fearful of what the indignant farmers might do if they won complete control of the state government, the legislature was convened by the governor in advance of the campaign and was induced to concede a satisfactory

[26] Fossum, *Agrarian Movement in North Dakota*, 34–36.

platform law. But the farmers were not now to be appeased by such partial remedies, and they very generally gave their support to the Populist-Democratic combination, which promised, if elected, to provide a state-owned terminal elevator. Given a satisfactory platform law, it was believed that such an elevator would practically insure the North Dakota grain-grower a fair price for his grain; and it was planned also that the storage receipts issued by the elevator might be used as collateral for short-time loans.

Although in the election the fusionists failed to achieve an outright majority in either house of the legislature, there were enough Republican legislators who sympathized with the fusionist demands to make possible the passage of the desired measure.

Unfortunately, however, the terminal elevator bill was drawn by a Populist attorney-general known to the opposition as the "hair-brained Populist" and was full of flaws. It provided that the elevator, which would have to be constructed in Duluth, Minnesota, or Superior, Wisconsin, must be maintained for the exclusive use of North Dakota grain-growers and that the state of North Dakota must be given exclusive political jurisdiction over the land on which it stood; further, while the bill appropriated a hundred thousand dollars for the construction of the elevator, it failed to provide the treasury with the necessary funds. Hence the elevator was never built, and once again the farmers saw their plans miscarry. Some minor measures of reform were passed by the legislature, however, and in spite of the fact that the terminal elevator law was inoperative, the producer was now better off than he had been before. The railroads under a Populist administration did have to provide local track platforms for the direct loading of grain, and faced by the danger of this competition, the elevator men were compelled to abjure many of their former unfair practices.[27]

[27] *Appletons' Annual Cyclopaedia, 1893*, 535; Fossum, *Agrarian Movement in North Dakota*, 37–38.

In Minnesota, where the farmers' problems were comparable to those in North Dakota, the nomination of Knute Nelson for governor by the Republicans in 1892 was taken as a sort of guarantee that the legislation demanded would not long be delayed; and after the election the Populists, by cooperating with the Republicans, were able to get what they wanted on a large number of important matters. State inspection of the weighing and grading of grain at the great terminal points, Minneapolis, St. Paul, and Duluth, had been put into effect under a law of 1885, but the farmers still complained that they were defrauded at the country elevators. A law of 1893 extended the benefit of state inspection to those sellers of grain who could not avoid dealing through local grain merchants. Another law increased the punishment meted out to individuals who were responsible for the creation of pools and trusts, providing that in addition to the punishment by fine already assigned there should be also imprisonment of from one to ten years in the penitentiary. Still another law provided for the purchase of a site and the erection by the state of an elevator at Duluth, to be managed and operated by the state warehouse commission. As in North Dakota, this was a pet Populist measure that followed precisely the demands of the Populist platform. But, as also happened in North Dakota, the law was so framed that it never went into effect. The state attorney-general ruled against it, and his opinion was later upheld by the state supreme court, which declared the act unconstitutional and void. An attempt to place upon the tax list all unsold railway land grants likewise failed.[28]

Meantime Ignatius Donnelly, defeated Populist candidate for governor of Minnesota in 1892, had been actively at work upon two projects of investigation, with which he had been intrusted by the Minnesota legislature. One concerned

[28] The work of this legislature is covered more fully by the author in the *Minnesota History Bulletin*, 5:547–548.

the cutting of timber from state lands by lumber companies, who paid little or nothing for the privilege. Sometimes, apparently, the timber was stolen outright.[29] The other exposed the existence of a coal combine, which had operated for several years as a price-fixing concern. According to Governor Nelson, who seems to have accepted Donnelly's findings, only dealers who were members of the "trust" were furnished coal, and members were required to charge the "trust" prices.[30] Donnelly's exposures aroused much interest, and the general agent of the combine was haled before a local grand jury for indictment. But the grand jury failed to indict, holding that the evidence adduced was too purely circumstantial. Donnelly was acutely distressed at the escape of a culprit who he believed had been caught red-handed, and he proclaimed through his paper that "if there is no remedy in a new party then is this nation lost — ruined — damned beyond redemption, dead of dry-rot and universal corruption." [31]

Nelson himself was moved to further action, and on the authority of the legislature he invited other states to join with Minnesota in sending delegates to an antimonopoly convention to meet in Chicago the first Monday in June. The convention was held and was attended by more than a hundred delegates, representing nearly every state in the union. Nelson was on hand to call the convention to order. He pointed out that while the existence of numerous such combines as had been discovered in Minnesota was apparent enough, the remedy was not. The Sherman Antitrust Act, he asserted, was inadequate, and the antitrust laws of a single state were powerless against trusts that were interstate in their operations. The convention should discuss possible remedies.

[29] Pioneer Press, April 12, 1893; St. Paul Globe, April 11, 12, 1893; Report of the Pine Land Investigating Committee to the Governor.
[30] Minneapolis Tribune, March 18, 1893; Chicago Inter-Ocean, June 6, 1893.
[31] Representative (St. Paul), April 17, 1893; Donnelly Scrapbooks, Vol. 15.

Donnelly, Weaver, Henry Demarest Lloyd, and many other well-known radicals — men "with a wheel loose," one Chicago paper called them — were present, and most of them agreed with Donnelly's demand that "should all other measures fail, the enactment of laws to confiscate the real and personal property of all trusts and combinations, to deny them access to the courts to enforce their claims and to withdraw from their property the protection of the law," should be the final resort.[32] The majority of the convention, however, did not agree with Donnelly, and according to the Minnesota statesman, "everything radical and earnest that would break the combines and shield the people was voted down. The convention was a humbug." A radical, earnest minority, composed almost exclusively of Populists, held a rump convention and adopted resolutions of the kind Donnelly had advocated, pointing them particularly at the owners of the anthracite coal lands in Pennsylvania.[33]

Out in Colorado the sole reason for the triumph of Populism in the election of 1892 was the great public interest in free silver, but the officers chosen to administer the affairs of the state were not merely silverites; they were fairly characteristic Populists. This was especially true of the new governor, Davis H. Waite, who had participated in the work of the St. Louis and Omaha conventions and had helped to frame the Populist platform. Waite was a man rather advanced in years, headstrong and obstinate. His whole personal appearance as well as the occasional frenzy of his rhetoric suggested the narrow-minded fanatic.[34]

In all matters that required legislative assistance, Waite's administration was hopelessly crippled, for in the House of Representatives the Republicans had a clear majority and

[32] *Chicago Evening Post*, June 5, 1893; *Chicago Daily News*, June 5, 1893; *Chicago Dispatch*, June 6, 1893.
[33] *Representative*, June 7, 1893.
[34] *National Cyclopaedia of American Biography*, 6:452; *Review of Reviews*, 8:247 (September, 1893).

in the Senate a plurality that could be overcome only by a
precarious alliance of Populists and Democrats. Most of
Waite's radical recommendations, therefore, including one
that would have created a railway commission with power
"to hear and determine complaints, without recourse to the
courts," went by the board.[35] On the railroad issue Waite
had a particularly bitter experience. The legislature not only
refused to create the type of railway commission he desired,
but it also chose to repeal the existing railroad law, which,
in the words of the governor, "was an epitome of the railway
legislation of the state." Waite promptly vetoed the repeal
bill, but the legislature overrode his veto in spite of the fact
that Populist votes had to be obtained to secure the necessary two-thirds vote in both houses. The new law provided
for a railroad commissioner with extensive powers, but to the
immense chagrin of the governor and, as he thought, to the
great satisfaction of the railroads, the state was left without
a line of statute railroad legislation.[36]

Waite also failed to carry the legislature with him in his
advanced views on the silver question. During the summer
of 1893, while the agitation for the repeal of the Sherman
Silver Purchase Act was at its height, he talked wildly of
the imminence of revolution and asserted, "It is better, in-
finitely better, that blood should flow to the horses' bridles
rather than our national liberties should be destroyed."
Thenceforth he was always known as "Bloody Bridles"
Waite.[37] The plan of action that he developed, however,
fell far short of revolution. In December, 1893, he called the
legislature into special session chiefly that it might enact a
law declaring "all silver dollars, domestic and foreign, containing not less than 412½ grains nine-tenths fine silver . . .
legal tender for the payment of debts, public and private,

[35] *Appletons' Annual Cyclopaedia, 1893,* 175–176; Vaile, in the *Forum,* 18:716.
[36] Colorado, *House Journal, 1895,* 48; *Appletons' Annual Cyclopaedia, 1893,*
177.
[37] *Review of Reviews,* 8:247. See also Frederic L. Paxson, *Recent History of
the United States,* 230.

ON TRIAL IN THE WEST 293

collectable within the State of Colorado."[38] This measure
he defended as practical on the ground that the fate of the
debtor class of the nation would be sad indeed if the only
hope of free coinage lay in what Congress might do, and as
constitutional on the ground that article 1, section 10, of the
Constitution of the United States, by forbidding the states
to make *anything but gold and silver coin* a tender in payment of debts, recognized by implication the right of the
states themselves to make *gold and silver* coins legal tender.

Waite felt that he had made a great discovery. Congressional legislation on behalf of silver might never be obtained,
but, thanks to the provision of the Constitution that he had
noted, there was

another remedy within the constitutional power of every State,
by the exercise of which the fruits of free coinage of silver may
be anticipated, money made abundant, the wheels of business
set in active motion, employment given to labor at remunerative wages, and prosperity drive away [sic] grim want and
poverty that have camped down in almost every household.[39]

But Waite's plan of legalizing the Mexican dollar in the
United States by state action (for that was really what he
had in mind) not only failed to impress the Colorado legislature favorably, but it was also generally rejected by the
Populist fraternity. Even so good a friend of silver as Tom
Watson of Georgia doubted the soundness of Waite's interpretation of the Constitution and objected strenuously to
any scheme that would tend to array the states against the
nation. Waite hit back at Watson through the press, and to
the immense delight of anti-Populists and anti-silverites the
argument waxed exceedingly warm.[40]

Waite's ineptitude was again demonstrated in his dealing
with the Denver board of police, which, under the terms of

[38] Colorado, *House Journal, 1895,* 53.
[39] *Ibid.,* 52.
[40] *Appletons' Annual Cyclopaedia, 1894,* 149; *People's Party Paper,* December 22, 1893; January 12, 1894.

the city charter, he was authorized to appoint — "with power of suspension or removal at any time for cause, to be stated in writing, but not for political reasons." In due time Waite made his appointments, but the new board had not long been in office when the governor reached the conclusion that tribute was being levied upon gamblers for police protection, a practice to which the police board appeared to be wholly indifferent. Thereupon he removed two of the commissioners for cause duly stated in accordance with the law, and appointed others in their places. The ousted commissioners refused to acknowledge the right of the governor so to remove them. Waite's new appointees took possession, nevertheless, and on the question of title won a favorable decision from the state supreme court.

In a short time Waite again became convinced that things were not right in the police board, that his new appointees, both Populists, "were sending policemen to the gambling houses, ostensibly to keep the peace, but really to protect the illegal business of gambling, and that the system of blackmail prevalent for so many years had been resumed." [41] Again the governor made use of his power of removal and ordered the offending commissioners to turn over their offices to new appointees. But this time the induction into office of the new commissioners was resisted by a strong show of force. Policemen and firemen, still loyal to the old commissioners, and several hundred deputies appointed by the sympathetic sheriff of Arapahoe County, were stationed in the city hall, "armed with guns, pistols, and even dynamite" to prevent the entrance of Waite's new men.[42]

To the governor this was no less than "open insurrection against the authority of the state" and, determined to meet force with force, he called out the national guard of the city of Denver. The offending commissioners, meantime, alleging

[41] Colorado, *House Journal, 1895,* 50.
[42] *Ibid.,* 51.

that Waite's opposition to them was in reality political, had secured injunctions from two of the district judges restraining the new appointees from interfering with the work of the old or, as Waite put it, "restraining the executive department from executing its duties." The old commissioners held that they had a right to their offices until the courts should rule on the legality of the governor's action, and they asserted their willingness to yield office at once should the courts find against them. The governor, however, pointing to the previous court decision on practically the same question, held that there could be no question as to the legality of his action and saw in the whole procedure a dark conspiracy of gamblers, capitalists, and judges to retain the old commissioners in office until the governor's term expired. On the fourteenth of March, 1894, he ordered the militia to advance on the city hall.

Prominent citizens of Denver, perturbed at the prospect of bloodshed, now took a hand and managed to avert a clash. They conferred with all parties to the dispute, persuaded the militia officers to delay their attack, and planted in the governor's mind the idea that Brigadier General McCook of the United States Army should be asked to send troops from Fort Logan to assist in preserving order and preventing bloodshed. Waite asked for the troops, evidently expecting them to assist the national guard in overcoming the opposition at the city hall, but General McCook held that his sole duty was to preserve the peace and he kept his men in camp at the railroad station.

Inasmuch as Waite and McCook failed distinctly to agree on a plan of procedure, the governor "respectfully withdrew his request for military aid" and notified all the militia throughout the state to be under arms. McCook stayed on, however, as long as he thought the presence of the United States troops was necessary to prevent disorder, and the governor was persuaded to submit the dispute to the state supreme court under a provision of the constitution that permitted the executive in certain exigencies to call upon that body for its opinion. It was unofficially agreed that the opinion of the court should be rendered within five days.

The decision, duly rendered, lectured the governor roundly for his precipitate action, holding that "by no rule of construction can the power and duty imposed upon the governor 'to execute the laws' be held to authorize the forcible induction of an appointee into office," but the right of the governor to remove for cause was again upheld and the new appointees were recognized as legally entitled to their places. The old commissioners therefore yielded their offices with what grace they could muster, and the insurrection was over.[43]

Waite gained further notoriety for himself and unpleasant advertising for his state by his conduct during a strike that broke out early in 1894 in the Cripple Creek gold-mining district.[44] The strikers, determined that the mines should be worked only on such terms as the union might direct, were reputed to have driven away unfriendly officials and to have maltreated miners who were willing to compromise. From the first, Governor Waite manifested his hearty sympathy with the strikers, and perhaps on this account they were the more insistent in demanding what they believed to be their rights.

On the sixteenth of March, 1894, while the crisis over the police board was at its height in Denver, word came from the sheriff of El Paso County that "a riot existed which he was unable to manage with any force at his command." The governor at once dispatched five companies of the national

---

[43] Slightly conflicting accounts of the episode are given in *Appletons' Annual Cyclopaedia, 1894*, 148–149; by Vaile, in the *Forum*, 18:717–718; and in the Colorado *House Journal, 1895*, 51.

[44] Cy Warman, "Story of Cripple Creek," *Review of Reviews*, 13:161–166. The story of this and subsequent Cripple Creek disorders is well told by Benjamin M. Rastall in *The Labor History of the Cripple Creek District: A Study in Industrial Evolution.*

guard to the scene of the conflict, but an investigation conducted by the state adjutant general reported "that no person in the county had been charged with the commission of any offense in regard to the existing labor troubles, that no warrant or other process of court had ever been issued and that neither the sheriff or any of his deputies had ever been resisted in any way. . . ." On receipt of this report the governor withdrew the troops.[45]

County officials and business men of Cripple Creek had told the adjutant general, however, that the civil authorities were unable to keep the peace in the mining district, that armed men patrolled the roads and trails in defiance of the constituted authorities, and that there was no safety for life and property. Evidently all was not as peaceful as the governor and his subordinates maintained. Certainly the local sheriff did not agree that all was well, for with the national guard shirking its duty, he presently recruited an army of deputies, estimated at about twelve hundred men and "divided into infantry, cavalry, and artillery." The miners also gathered arms and ammunition, built fortifications on a spot known as "Bull Hill," and with a force of perhaps a thousand to eighteen hundred men stood ready to resist attack. By the latter part of May it seemed that only a miracle could avert a battle.

"The El Paso County troops," as the governor called them, "concentrated and drew nearer to the mining district," and at length established a rival camp close to "Bull Hill." Waite continued to show sympathy for the miners. He visited their camp and fraternized with them, and he offered no rebuke when his private secretary said:

The majority of the men on Bull Hill are animated by the conviction that they represent a vital principle. They haven't the slightest fear of death. There are scores of these men who be-

[45] Colorado, *House Journal, 1895*, 54; *Appletons' Annual Cyclopaedia, 1894*, 147.

lieve that if they fall in the impending battle, their blood will
be the seed which will redeem the cause of labor to the world.

To the governor himself the expression "D—n capital" was
attributed. Certainly he did not hesitate to brand the sheriff's army as illegal and the sheriff's actions as unconstitutional.[46]

As "sole arbitrator" for the miners Waite twice attempted to bring about a settlement but failed each time.
He then called out all the national guards of the state to preserve the peace. On June 6 the state troops marched between
the opposing forces, and the sheriff was notified that no further advance by the deputies would be permitted. Such an
advance was attempted, the sheriff pleading that "he had
no control over his men," but when the commander of the
state troops threatened to open fire, the deputies returned
to their camp. Finally the miners surrendered to the national guard, the deputies disbanded, and the national guard,
with the exception of a small detachment, was sent back
home. The Cripple Creek War was over. It had cost the
state and El Paso County approximately $135,000 and had
settled no important issue.[47]

Waite had other troubles during his administration, but
these were the most conspicuous. He did, indeed, avoid
bloodshed, but there is good reason to believe that, had he
deported himself more tactfully, there might have been less
danger of bloodshed. His administration was criticized as
partisan in the extreme and concerned with the welfare of
a particular class rather than with the welfare of the people
as a whole. To this charge the governor replied, "Well, what
if it is? Is it not the truth that for thirty years the two old
parties have been legislating for the creditor class? It is true,
and turn about is fair play."[48] His ideas were those of a

[46] Colorado, *House Journal*, 1895, 55; Vaile in the *Forum*, 18:721.
[47] Warman, in the *Review of Reviews*, 13:162 note; Colorado, *House Journal*, 1895, 56; Vaile, in the *Forum*, 18:721.
[48] *Ibid.*

fanatic, and they tended to bring discredit upon the state. His public utterances were often intemperate, indiscreet, and harmful to the causes they were designed to serve. So many members of his own party and administration fell out with him as to make it clear that the governor was usually to blame. His renomination by the Populists in 1894 gave the Republicans a chance to conduct their campaign with "the suppression of the spirit of anarchy, the restoration and maintenance of law and order" as the paramount issue instead of silver. The verdict of the Democratic nominating convention of 1894, which chose independent candidates and abjured fusion, was equally unfavorable. And, according to its resolutions,

Populism . . . has been in power for a year and a half. Its so-called principles . . . have during that time been in full and uncontrolled operation. The miserable consequences are everywhere apparent. The good name of the State is imperiled. Many functions of Government have been perverted to selfish, ignoble, and unlawful ends, and imbecility in high positions has made our career since 1892 a satire on self-government.[49]

Generally speaking, the experiments with Populism in the West had done little to engender confidence in the ability of the new party to rule. Some legislation favored by the Populists had reached the statute books, but usually such of this as stood the test of the courts was as much the work of the other parties as of the Populists. Purely Populist legislation was all too frequently found defective and unconstitutional. In administrative matters the Populists were even less successful. Evidently their genius lay in protest rather than in performance. The third-party men who took office were almost invariably inexperienced as administrators, and it is not surprising that they were baffled by the difficulties that confronted them, difficulties that, incidentally, were made just as baffling as the opponents of Populism knew how to make

[49] *Appletons' Annual Cyclopaedia, 1894*, 150.

them. Members of the new party, moreover, showed small charity for one another's blunders. From the men they had elected to office the Populists expected more than was humanly possible; and when the millenium failed to arrive they turned promptly from ardent supporters to disappointed critics. Mrs. Lease, for example, thus became a thorn in the flesh to Governor Lewelling of Kansas.

In defense of the Populists it should be noted that in no instance did they have complete control of a state government. Always there was at least one house of the legislature in the hands of the opposition, or a hostile governor, or an apparently partisan supreme court. With a free hand the Populists might have done far better. Also, the Populist tenure of office was very short. A single meeting of a legislature or a brief two years in administrative control could not possibly prove conclusively what the new party was capable of doing. But such allowances were seldom made in judging the Populist record, and undoubtedly the general opinion was that on the basis of these western experiments Populism was proved a failure.

CHAPTER ELEVEN

THE RISE OF THE SILVER ISSUE

DURING the campaign of 1892 the Populists had learned that of all the planks in their platform the silver plank had the widest appeal. In some of the mountain states this issue alone had made Populist tickets invincible, and throughout the country converts had flocked to the Populist standard for no other reason than that the new party promised to do for silver what neither of the older parties dared attempt. Populist orators found free silver their best talking point; Republicans and Democrats frequently found the task of diverting attention from silver to the tariff well beyond their ability. Sometimes in their distress old-party campaigners resorted to the most transparent sophistries. Democrats were known to contend that the phrase in their platform, "without charge for mintage," really meant free silver in the popular sense; Republicans, that the bimetallism for which their platform called was the last possible word in helpfulness to silver.[1]

All the silver arguments harked back to the "crime of 1873." In that year the greenback and national bank-note currency of the Civil War still held universal sway, but Congress, looking forward to a time when coins once more might take their place in the nation's currency, had passed a law revising the coinage lists. This law made no provision for

[1] Arnett, *Populist Movement in Georgia*, 149; *Proceedings of the Tenth Republican National Convention*, 1892, pp. 86, 172.

the coinage of a silver dollar, presumably for the reason that silver dollars had long since ceased to exist except as curiosities and would not be in demand. The war had swept all coined money out of circulation, but even before the war, so Dr. Paxson tells us, "it was possible to grow to manhood without ever seeing a silver dollar of American coinage."[2]

The reason for this is clear. The coinage law of 1834 made provision for a silver dollar of 412½ grains, whereas from that time forward until well after the "crime of 1873" was perpetrated 412½ grains of silver, expressed in terms of gold, were worth more than a dollar. In other words, the coinage ratio by weight of silver to gold, usually spoken of as "sixteen to one," put too much silver in the silver dollar, and too little gold in the gold dollar; hence, according to the well-known Gresham's law, silver as the dearer money was driven out of circulation. Perhaps the statute of 1873, instead of dropping the silver dollar, might have tried to make the coinage ratio conform with the commercial ratio; it might have put a little more gold in the gold dollar or a little less silver in the silver dollar. But the framers of the law evidently felt that the silver dollar was not needed and that its coinage might well be discontinued. The existing silver dollar was not actually "demonetized"; but no provision was made for the coinage of new silver dollars.[3]

The law of 1873 was passed in the perfunctory fashion common with routine legislation. "Perhaps I ought to be ashamed to say so," Garfield remarked in the fall of 1877,

but it is the truth to say that I, at that time being chairman of the Committee on Appropriations and having my hands over full . . . never read the bill. I took it upon the faith of a prominent Democrat and a prominent Republican, and I do not know

---

[2] Paxson, *Recent History of the United States*, 92.

[3] A satisfactory brief history of the silver movement in the United States is given in W. Jett Lauck, *The Causes of the Panic of 1893*, ch. 2. Fuller accounts may be found in J. Laurence Laughlin, *The History of Bimetallism in the United States;* A. D. Noyes, *Forty Years of American Finance;* and Wesley C. Mitchell, *Business Cycles.*

that I voted at all. There was no call of the yeas and nays and nobody opposed the bill that I know of. It was put through as dozens of bills are . . . on the faith of the report of the chairman of the committee.[4]

The bill had been before Congress for several sessions, and its terms must have been available to anyone who had a mind to investigate them, but apparently no one was much more interested than was Garfield. Final passage was accomplished on the word of Sherman that the bill had practically passed Congress at a preceding session and was now, with minor changes, exactly what it had been before.[5]

Probably the silver dollar would never have been missed had the price of silver not begun to decline shortly after this coinage law went into effect. During and after the Civil War, however, new silver mines, some of them amazingly rich, were opened up in the West; and with the assistance of a transcontinental railroad after 1869, they began to turn a steadily increasing stream of the white metal into the channels of trade. Not only were new deposits exploited but the discoveries of modern science made possible also the use of low-grade and refractory ores, wasted in an earlier day, and mining engineers found better ways of getting at the ore. With the production of silver soaring to undreamed-of heights by the middle seventies, there came the inevitable drop in price; precisely the sort of thing that would have happened with wheat or any other commodity under similar circumstances.

By a curious coincidence, just as silver was becoming extremely plentiful, the demand for it as money was on the decline. In the transfer of large balances, gold, which was easier to handle and, after the discoveries of the forties and fifties, more plentiful than formerly, tended to supplant entirely the use of silver. Moreover, there was a growing con-

---

[4] *National Economist* (Washington), 3:263 (July 12, 1890).
[5] *Congressional Globe*, 42 Congress, Session 3, pp. 203, 363, 661, 668, 674, 868, 871, 1150, 1214, 1282; *National Economist*, 2:120 (November 9, 1889).

viction in financial circles that a bimetallic standard was
unsatisfactory, if not, indeed, actually impossible. An international monetary conference held at Paris in 1867 strongly
favored the gold standard, and in the next five or six years
the silver currency of most European countries was either
limited strictly or demonetized outright. Thus the "crime
of 1873" in the United States was paralleled in Europe by
another "crime" that actually resulted in the sale of great
quantities of silver bullion, previously used for monetary
purposes. With the annual world production of gold strangely
at a standstill, or even declining, the gold value of silver now
dropped at an alarming rate.[6]

Once the decline in the price of silver became clearly apparent, the "crime" was out, and silver miners of the American West, for whom coinage at the old ratio would now have
been profitable, were quick to demand that silver be restored
to its former status. In this demand they were joined by a
host of others who saw in the action of Congress no mere
accident or oversight but rather a nefarious conspiracy on
behalf of the creditor class to the everlasting detriment of
the debtors. The conspiracy — doubtless international in its
scope — was to throw upon gold the monetary burden that
previously had been borne by gold and silver together. With
silver demonetized, the demand for gold would increase;
with the demand for gold increasing, the purchasing value
of the gold dollar would rise. And with dearer dollars the
debts of the common man, contracted in a period of cheaper
currency, would on their collection net the lender a handsome profit. No wonder the import of the law that perpetrated this crime was concealed from the Congress that
passed it! No wonder Grant was not permitted to know the
contents of the bill he signed! For, as Donnelly put it, "the
demonetization of silver was intended to increase the value
of money at the expense of labor, and to enrich the creditor

[6] Lauck, *Causes of the Panic of 1893*, 16–18.

class at the expense of the debtor class; it is a reversion to barbarism and can only produce the most destructive consequences."[7]

These arguments concerning silver were obviously nothing less than the well-worn arguments of the Greenbackers in a new dress. The appreciation of the greenback dollar, particularly after the Resumption Act of 1875, was in its day held to be the result of a no less nefarious conspiracy, with precisely similar results. But with resumption well started and presently accomplished, free silver, which also promised a depreciated dollar, became the adopted child of the Greenbackers. They asked for it in their party platforms, and they took much credit to themselves for the passage of the Bland-Allison Act of 1878, which once more created a silver dollar, albeit of limited coinage.[8]

The Bland-Allison Act required the secretary of the treasury to buy each month for coinage purposes not less than two, nor more than four, million dollars worth of silver, but it did not require him to take all the silver that might be presented, nor did it permit him to pay more than the market price for the silver he bought. The law was meant as a sop to the silver miners and may have helped them somewhat, for it did increase slightly the demand for silver. But inasmuch as every secretary of the treasury from the time the law was passed to the time it was superseded in 1890 used his discretionary power " against the debtor classes and against silver; in favor of Wall Street and against the workers "[9] — purchasing, that is to say, the minimum amount of silver — the boon to the silver miners was not great. Nor was the cheap silver dollar permitted to become the standard unit, for the secretary of the treasury saw to it all during

[7] *Anti-Monopolist* (St. Paul), March 15, 1877. See also the issue for November 9, 1876, in which Donnelly made his first exposure of the crime, and the *National Economist*, 2:120; 3:263–264 (November 9, 1889; July 12, 1890).

[8] *Anti-Monopolist*, March 28, 1878.

[9] *National Economist*, 2:232 (December 28, 1889).

this time that every dollar of silver was as "good as gold." Thus without a gold standard law on the statute books and purely by the connivance of public officials, Democrats and Republicans alike, the single gold standard was maintained.[10]

Once more, in 1890, with the output of the silver mines still accelerating, it became politically expedient to "do something" for silver. The Republicans in the campaign of 1888 had in their platform a statement that was interpreted to be a "strong silver plank"; but only under considerable pressure did they set out to redeem their pledge, and the so-called Sherman Silver Purchase Act that resulted merely continued the principle of the Bland-Allison Act. Again the government was required to buy silver at the commercial price, but this time the amount was fixed at fifty-four million ounces per year, the estimated annual output of the American silver mines. Legal tender treasury notes, redeemable in "gold or silver coin," were to be issued in payment for the silver.

The new measure took into the treasury twice as much silver as the Bland-Allison Act required, more by seventeen million ounces, so the *National Economist* charged, than the silver mines of the United States were able to produce;[11] but under its terms the way was still left open for the maintenance of the gold standard. For the treasury could, and in practice did, undertake to redeem in gold rather than in gold or silver the new issues of legal tender notes. This act and the Bland-Allison Act that preceded it did so little, however, to check the decline in the price of silver that some were led to question whether even the free and unlimited coinage for which the silverites contended would have accomplished this end.[12] The Bland-Allison Act had taken twenty-four million dollars worth of silver out of the market each year "abso-

[10] Lauck, *Causes of the Panic of 1893*, 18–28.
[11] *National Economist*, 3:124 (May 10, 1890).
[12] *St. Louis Globe-Democrat*, February 26, 1892.

lutely and permanently," and the Sherman Silver Purchase Act in the years succeeding 1890 absorbed the entire domestic product, but still the price of silver went down. By 1893 the value of the bullion in the silver dollar had dropped to sixty cents; by 1894, to forty-nine.

But the Sherman Act had none the less important consequences, for the increased note circulation placed a heavy burden on the gold reserve in the treasury, upon the maintenance of which depended the preservation of the gold standard. This reserve, amounting in the beginning to about a hundred million dollars, had been accumulated by Secretary Sherman during the Hayes administration to make possible the resumption of specie payments, and inasmuch as resumption had been successfully accomplished by relying upon such a sum of gold, the idea prevailed in the financial world that as long as the treasury could show this item in its balance the country had no need to fear a depreciated currency. Down to 1890, in spite of the added strain of the Bland-Allison Act, the receipts of the treasury showed a large annual excess above expenditures, and there was no difficulty in maintaining a satisfactory gold reserve; indeed, the amount of gold in the treasury was steadily augmented, and during the three years that preceded the passage of the Sherman Act it stood ordinarily only a little under the two-hundred-million-dollar mark.[13] But very soon thereafter the gold balance began to decline. Under the McKinley Act the tariff brought in a smaller revenue, while congressional appropriations showed a marked tendency to increase. Confidence that the government could maintain a satisfactory gold reserve in the face of a diminishing income and accelerated expenditures began to wane. Gold was deliberately hoarded, debts to the government were paid increasingly in silver and paper, and large sums of gold were actually drawn directly from the treasury in exchange for the greenbacks

[13] Lauck, *Causes of the Panic of 1893*, 93.

and treasury notes that the government desired to maintain at a parity with gold. By the time of the election of 1892 the gold balance was too close to the one-hundred-million-dollar minimum for the administration to feel comfortable, and as the end of the Harrison administration approached, the president and Secretary Foster, according to the *Nation*, were "watching the dollars in the treasury with unconcealed anxiety, and hoping against hope that March 4 will come without an actual crash." [14]

For Cleveland and his secretary of the treasury the crash could not long be postponed, and within a few weeks after they took office the whole world was aware of the fact that the gold reserve had dropped below the one-hundred-million-dollar mark. It was clear, moreover, that if the decline continued, the gold in the treasury would soon be entirely exhausted, and the government would have no money except silver with which to meet its obligations. This meant the substitution of the silver standard for the gold standard and a depreciation of from one-third to one-half in the purchasing value of the dollar. The silverites, now stronger than ever in the South and the West, could be depended upon to oppose all efforts to avert such a catastrophe; indeed, to most of them such a result was to be regarded rather in the light of a blessing. Deep apprehension in the business world as to what might happen turned suddenly to panic, the New York Stock Exchange registered a sharp fall in values during the first week in May, and the country was presently in the throes of a disastrous financial collapse. Banks, railroads, and mercantile and industrial establishments followed one another into bankruptcy with disheartening rapidity, and a period of hard times was ushered in that was to last for four full years. [15]

While the panic of 1893 was immediately due to "the ap-

[14] *Nation*, 56:151 (March 2, 1893). See also *ibid.*, 56:96–97 (February 9, 1893), and Lauck, *Causes of the Panic of 1893*, 14–15, 82–83, 86, 91–95.
[15] Lauck, *Causes of the Panic of 1893*, ch. 7.

prehension, which was prevalent both in this country and abroad, that the United States would be unable to maintain the gold standard of payments,"[16] it seems clear enough that a period of business depression was already almost overdue. The decade of the eighties had been characterized by a notable expansion of business. Transcontinental railroads had been built, one after another, and over the whole country improvements in railway transportation had been made, sometimes far beyond any genuine need or any hope of immediate returns. A rapidly expanding market tempted manufacturers, also, to expand their establishments, a process facilitated and accelerated by a host of new inventions. Industrial expansion meant larger cities and new cities, nearly all of which spent lavishly on public improvements. During these years the greater portion of the trans-Mississippi West had been developed, almost exclusively on credit. With expenditures and investments rising each year to higher and higher figures, it was inevitable that they should ultimately exhaust the resources available for such purposes. By the year 1893 that time had come: credit was used up, prosperity was at an end.[17]

Hard times had reached the agricultural West and South well in advance of the year of the panic, but with the whole country now involved, the plight of the western farmer and the southern cotton grower became desperate. Prices struck new low levels. With corn at less than fifteen cents a bushel and cotton at less than five cents a pound, debts merely compounded, to be measured, moreover, in ever dearer dollars. Small business men who depended upon the farmers' trade suffered almost as acutely as the farmers themselves. Even the country merchants in the South, who protected themselves by high prices to the last possible limit, failed by the hundreds, and dependent customers with their source of

---

[16] *Ibid.*, 122.

[17] This theory of the causes of recurring hard times is admirably set forth in Paxson, *Recent History of the United States*, 224–226. See also William Allen White, "The End of an Epoch," *Scribner's Magazine*, 79:564, and *ante*, ch. 1.

supply cut off faced the direst want.[18] In the West partial crop failures in 1893 were followed by almost total failures in 1894.[19] "Never in the history of our Republic," a correspondent wrote to Senator Allen, "has there been so much anxiety made manifest in the minds of the whole people as at the present hour, and in the intensity of sentiment men are running hither and thither to catch onto a ray of hope. This is a life or death struggle for our Republic. . . ."[20]

If such a thing could be, the western silver miner was harder hit by the panic and the hard times that followed it than was the farmer. The low prices paid for silver caused the silver mines, one after another, to shut down. Only "the most fortunate producers, the finders of the great prizes,"[21] were able to continue in operation profitably. A mass meeting of the "friends of silver," which was held in Denver on July 11, 1893, and drew delegates from nearly every town and village of the Colorado mining district, pointed out in its resolutions that ninety-nine per cent of the silver mines of the country had shut down and that, inasmuch as the "silver industry is the very heart from which nearly every other industry receives support," these other industries must likewise soon suspend operations. Already there were estimated to be fifteen thousand idle miners in the state

who know not where to turn if work is not resumed. There will soon be added to this idle army of labor 4,000 men from the smelters. The stone quarries are nearly all shut down, the railway companies are laying off train crews by the score, the foundries are nearly all out of order, the farmers and fruit-growers will be barely paid for the cost of saving their crops, and the merchants are countermanding their Eastern orders.

Such conditions existed not only in Colorado but in all the "silver-mining States and Territories, embracing 1,000,000

[18] Arnett, *Populist Movement in Georgia*, 157–158.
[19] Nebraska, *Senate Journal, 1895*, 45; *Appletons' Annual Cyclopaedia, 1894*, 555; Nixon, in the *Iowa Journal of History and Politics*, 21:391.
[20] John Batie to Allen, August 9, 1893, in the Allen Papers.
[21] Hague, in the *Forum*, 15:67.

square miles of the continent, with 2,000,000 Americans inhabiting them." [22] Appalled at the prospect of such unemployed and starving multitudes as these, another correspondent wrote to Senator Allen, "How long will it be if something is not done before they will rise in their might in vast numbers, in bread riots, demanding and taking what they want?" [23]

President Cleveland, faced at the outset of his administration by unparalleled calamity on every hand, did exactly the thing best calculated to antagonize the suffering West and South. He called Congress together in special session to repeal the Sherman Silver Purchase Act of 1890. This was probably the reasonable thing to do, and it was certainly in entire harmony with Cleveland's past record on the silver question. Just before he took office the first time, on February 25, 1885, Cleveland had declared himself against the free coinage of silver and had advocated also the repeal of the Bland-Allison Act.[24] Silver men had not forgotten this, nor had they forgotten how their hopes had been blasted by the "first public utterance Mr. Cleveland made after election. . . . He did not even wait until his inauguration to recommend the stoppage of silver coinage." [25] In Cleveland's demand for the repeal of the Sherman Act they saw new evidence of the president's total disregard for the welfare of the common man. His financial policies, like those of his predecessor, were conceived in Wall Street and born in utter indifference to the miseries they were sure to inflict upon the American people.[26] It was in a paroxysm of rage at this

[22] *Appletons' Annual Cyclopaedia, 1893*, 178, 502.

[23] J. D. Baily to Allen, October 16, 1893, in the Allen Papers.

[24] This is what the Populists called Marble's Cleveland letter, arguing that the letter was written by Manton Marble and revised by Samuel J. Tilden. Gordon Clark, *Handbook of Money*. See also Cleveland's statement of February, 1891, in opposition to silver, printed with comments in the *Nation*, 52:147, 150 (February 19, 1891).

[25] *National Economist*, 2:232 (December 28, 1889).

[26] *Representative* (St. Paul), July 5, 12, 1893; Stewart, in the *Indiana Magazine of History*, 15:53; B. O. Fowler to Senator Allen, April 25, 1894, in the Allen Papers.

wicked war on silver that "Bloody Bridles" Waite won his famous sobriquet.[27]

Cleveland's request for the repeal of the Sherman Act was carried out by a bipartisan majority, consisting mainly of eastern Republicans and eastern Democrats. Against the repeal bill the silver men of all parties maintained a united front, led forensically in the House by young William Jennings Bryan, a Democrat, and in the Senate by William V. Allen, a Populist. Far from maintaining the modest reserve expected of a newly elected senator, Allen threw himself into the thick of the fight and, aided by other prominent silver men, staged an extended filibuster designed to prevent a vote on the repeal bill. Once during this contest Allen spoke continuously for fifteen hours, beginning his remarks at five o'clock in the afternoon of October 11 and closing a little after eight o'clock the following morning — up to that time the longest continuous oratorical effort in Congress on record. Allen's energy and resourcefulness won great admiration even from non-Populistic sources. "If Populism can produce men of Senator Allen's mould," wrote the learned editor of the *Review of Reviews,* "one might be tempted to suggest that an epidemic of this Western malady would prove beneficial to some Eastern communities and have salutary results for the nation at large." [28] As for the Populists themselves, the sting of defeat was almost forgotten in satisfaction at the good showing their representatives had made and the good feeling that had been established among silver men of all parties. "I want to say that your course and that of Hon. W. J. Bryan is approved by a very large majority of not only Populists but of Democrats and Republicans," wrote one of Allen's correspondents; another hazarded the opinion that "the next Congress will be Populist if we do our duty." [29]

[27] See *ante,* 292, and the Colorado *House Journal, 1894,* 46.
[28] *Review of Reviews,* 10:30 (July, 1894). For Allen's long speech see the *Congressional Record,* 53 Congress, Session 1, Appendix, 289–340.
[29] Robert Kittle to Allen, October 23, 1893, and J. C. Roberts to Allen, September 28, 1893, in the Allen Papers.

The hue and cry over the repeal of the Sherman Act tended to focus public attention more closely than ever upon the silver issue, and converts to free silver as a panacea for all the ills that beset American society were counted by the thousands. An organization known as the American Bimetallic League, which had held its first convention at St. Louis in November, 1889, and had been active ever since, accepted the chief burden of winning votes for silver. This organization, strongly backed by the silver miners, had pursued a strictly nonpartisan course, seeking adherents from every party. By 1892 it had been able to line up a majority of the United States Senate for a free-silver bill, thanks in part to the disproportionate representation in that body from the silver-mining West, but it had failed of its purpose in the House of Representatives. Party lines had been shattered by the votes on silver, the Populists alone giving the measure undivided support. "The Democratic House of Representatives could not pass a free silver bill," Tom Watson's paper observed. "A Republican Senate does. Doesn't it show how utterly confused the old parties are?"[30]

Following the election of 1892 the League redoubled its efforts, holding during the year 1893 three great conventions, one at Washington in February, one at Chicago in August, and one at St. Louis in October. The Washington meeting invited every labor and industrial organization in the country to send representatives to future meetings of the League, and when the Chicago convention assembled, it was evident that this invitation had been generally accepted. The Chicago convention was timed to precede by a few days the special session of Congress called to repeal the Sherman Act, and more than eight hundred delegates, including such notables as William Jennings Bryan, Terence V. Powderly, "Bloody Bridles" Waite, and Ignatius Donnelly, found it possible to attend in spite of the hard times. The resolutions

[30] People's Party Paper, July 8, 15, 1892.

of this "Silver Congress," presented by Donnelly and no doubt drafted in part by him, set forth in "eight whereases and five declarations" the doctrine that the prevailing economic distress was in no sense due to the Sherman Silver Purchase Act but rather to the "crime of 1873," which, in the interest of the creditor class, had outlawed silver, had forced upon gold an unwonted burden, and had thereby raised the value of the dollar to intolerable heights.[31]

These sentiments were echoed by the St. Louis convention and by hundreds of journalists, thousands of orators, and uncounted numbers of common men. "Cleveland is being universally d——d by the masses," Allen learned from a correspondent;[32] Secretary Carlisle, once thought to be a friend of silver, was now branded as an apostate;[33] and John Sherman, for his share in the "crime of 1873" and all subsequent betrayals of silver, was made the archfiend. For Sherman the silverites would brook neither defense nor apology. "Their orators thunder denunciations of his name, their singers execrate him in rhyme and meter, and the audience roars itself hoarse whenever a particularly livid picture of the 'gold-bug conspiracy' is brought to a close with an invective of John Sherman for a climax."[34]

Amidst all the enthusiasm for silver that the repeal of the Sherman Act engendered, there was a strong tendency to ignore the fact that on certain fundamentals the "friends of silver" were by no means in complete accord. Silver orators who failed to recognize these differences were apt to find themselves adrift on a sea of confusion, attempting to steer in at least three different directions all at once.

The silver miners of the West and their dependents, including many easterners who owned stock in the mines, fav-

[31] *Chicago Times*, August 3, 1893; *Chicago Daily Tribune*, August 3, 1893; *New York World*, August 3, 1893; *Representative*, August 9, 1893, which gives a full report; Haynes, *Third Party Movements*, 272.
[32] J. C. Roberts to Allen, September 28, 1893, in the Allen Papers.
[33] B. O. Fowler to Allen, April 25, 1894, in the Allen Papers.
[34] *People's Party Paper*, November 24, 1893.

ored the free and unlimited coinage of silver at the ratio of sixteen to one because they believed that the effect of such a policy would be to restore silver to its former value. According to their argument, the only reason that could be given for the depreciation of silver was that in 1873 the coinage right and privilege was taken from it. "In 1890," so a silver brochure proclaimed, " on the strength of a mere rumor that Congress would pass a free-coinage act, and that Harrison would sign it, silver went in London to $1.21 an oz., or within 8¢ of gold. Had the act passed, 16 oz. silver would have been equal to 1 oz. of gold in ten minutes, and in the Bank of England." [35] Halfway measures, such as the Bland-Allison Act and the Sherman Silver Purchase Act, were branded as thinly disguised frauds to conceal the purposes of those who wished " to absolutely demonetize silver, to reduce it to the level of a mere commodity, and to make this a monometallic gold country." [36]

To the debtor classes of the West and South, who also tended increasingly, regardless of party, to embrace the free-silver doctrine, a silver dollar of equal value to the existing gold dollar had small appeal. They favored free silver precisely because in their opinion it opened the way to a cheaper dollar. For years the gold dollar had been steadily appreciating until now its purchasing power was nearly double what it had been twenty years before. In consequence, they argued, gold as a monetary unit had proved a failure. Gold was too dear. There was not enough of it, and, since year after year the product of the gold mines tended to decline, there seemed no hope that the situation would grow better. Silver, on the other hand, was increasingly plentiful, increasingly able to care for the business of the nation and of the world. Why not give it a trial? Perhaps the free coinage of

[35] *National Economist*, 6:10 (September 19, 1891); Clark, *Handbook of Money*, 290.
[36] *National Economist*, 3:156 (May 24, 1890); C. S. Gleed, in the *Forum*, 16:255–256.

silver would pull the gold dollar down from its artificial height towards the lower value of silver, and perhaps the silver dollar would rise slightly to meet the declining value of gold. Thus a double standard might be maintained. But if this could not be, let Gresham's law operate, and let it operate against gold, the unsuccessful measure, rather than against silver. Should a cheaper dollar become the standard, debts would indeed be scaled down, but what dishonesty would there be in scaling down debts that had been artificially appreciated? This was, in fact, precisely what the debtor class wanted. To silverites of this persuasion the repeal of the Sherman Act was the greater calamity because, had it remained indefinitely on the statute books, the government might have been forced to pay its debts in silver, thus making the cheap silver dollar "standard." For the debtor class a "fifty cent dollar" had no terrors.[37]

From the standpoint of the genuine Populist there was a good deal of nonsense about both these arguments. To him money was money not because it possessed any intrinsic value of its own but because of the fiat of government. "It is the cardinal faith of Populism," so one shrewd observer declared, "without which no man can be saved, that money can be created by the Government in any desired quantity, out of any substance, with no basis but itself."[38]

The genuine Populist also accepted without reservation the quantity theory of money, and he insisted that the currency be made thoroughly elastic, capable of being expanded or contracted as the needs of the country might demand. "We believe it possible," said Senator Allen,

so to regulate the issue of money as to make it of approximately the same value at all times. The value of money ought to bear

[37] People's Party Paper, July 22, 1892; J. W. Gleed, in the Review of Reviews, 10:46; James H. Canfield, in "A Bundle of Western Letters," Review of Reviews, 10:42; Davis, in the Forum, 14:381; William Jennings Bryan, The First Battle, 80–85; Arnett, Populist Movement in Georgia, 163–165; Haynes, Third Party Movements, 223–225.
[38] Tracy, in the Forum, 16:245.

as nearly as possible a fixed relation to the value of commodities. If a man should borrow a thousand dollars on five years' time today, when it would take two bushels of wheat to pay each dollar, it is clear that it ought not to take any more wheat to pay that debt at the time of its maturity, except for the accrued interest.[39]

Populist platforms were agreed, however, that the existing amount of money in circulation was hopelessly inadequate and should be increased to not less than fifty dollars per capita. This money might include gold and silver — the more the better — but, as Donnelly put it, "if gold and silver cannot be issued in sufficient volume, by all means let us have paper money."[40]

For the old-school Populists who in this fashion had thought their way through the money question the current rage for free silver was a little beside the point. They were ready enough to support any measures that would tend to undo the "crime of 1873," but at best they saw in free silver only a step towards the goal they had in view. The *National Economist* severely criticized those advocates of free coinage who assumed to make it the one great reform proposition to the exclusion of all others. This journal argued that the increase in the currency to be expected from such a measure would be far too small to restore good times. The enactment of a free-coinage law was deemed desirable by the *National Economist* mainly because it would demonstrate the inadequacy of the silver panacea and would pave the way to "other and greater reforms."[41] Ignatius Donnelly reached the conclusion that it was "practical statesmanship to endorse free silver," although he looked forward to the day when both gold and silver might be discarded and an international legal tender paper money established; "but that," he said regretfully, "is a vast reformation which the

[39] Albert Shaw, "William V. Allen: Populist," *Review of Reviews*, 10:39.
[40] *Cincinnati Enquirer*, May 16, 1891.
[41] *National Economist*, 2:337; 6:386 (March 15, 1890; March 3, 1892).

world is not yet ready for; — the greater part of mankind have never yet heard of it." [42] A widely circulated Populist tract explained that the government had an enormous public debt, payable in gold and silver coin. These obligations had to be met, and free coinage would assist the government in meeting them. [43] Thus the old-fashioned Populists apologized for free silver more than they advocated it, and they regretted to see less discerning students of the money question attach an importance to the doctrine quite out of proportion to the benefits that could possibly be obtained from it.

But the influx into the Populist ranks of silverites of every sort and kind, if dangerous to the faith, had certain obvious advantages, and the temptation to ignore all academic differences that might impede the growth of the party was great, even among the very elect. If the men who called themselves Republicans or Democrats, but professed to believe in silver, could only be induced to come over to the Populists, the future of the new party would be assured. "I travel much and know the inner sentiment of a majority of the people of Iowa," wrote "Calamity" Weller, "and I declare unto you that if the money question was submitted at a non-partisan election . . . unlimited coinage would receive two votes to one anti-silver." [44] A better informed and less prejudiced observer declared, though doubtless with some exaggeration, that "every Populist, nine-tenths of the Democrats, and one-half of the Republicans are devoted to this heresy." [45]

As its popularity increased, the silver issue tended more and more to become the chief item in the Populist creed. The Populist national committee featured free silver in a proclamation urging members of the party to get together for great Fourth of July celebrations in 1893. Such notable

[42] *Representative*, August 2, 1893.
[43] Jory, *What is Populism?* 11.
[44] Weller to Senator Allen, October 13, 1893, in the Allen Papers.
[45] Tracy, in the *Forum*, 15:338.

orators of Populism as Donnelly and Weaver allowed free-silver arguments to absorb the greater part of their attention. Prominent Populists began to talk of removing from the Omaha platform such of its planks as might prove objectionable to silverites of any persuasion; and even so good a Populist as Senator Allen confessed that he could wait indefinitely on the matter of government ownership of railroads and might perhaps be satisfied with government control.[46] Governor Waite, frightened lest a separate silver party be formed, urged a national convention to revise the whole Populist creed. Chairman Taubeneck, replying to the Colorado governor for the Populist national committee, saw little but merit in Waite's plan. Something was needed, Taubeneck conceded, to unite more firmly the debtor South and the debtor West, and free silver seemed to be the thing.[47] Even the Farmers' Alliance was urged from within to abjure all impracticable and inconsequential demands in order to " come down to silver," and the annual session of the Southern Alliance, held at Topeka early in 1894, stood by the old declarations only after a bitter fight.[48] Moreover, converts to Populism who cared only for free silver rapidly altered the character of the party. The old-time Populist might cherish every line in the Omaha platform, but the free-silver Populist took seriously only the plank that he favored.[49]

The unmistakable tendency of silver men to concentrate under the Populist banner led old-party politicians to devise means for holding their own lines intact. " Bimetallism " in varied garbs was repeatedly offered as a substitute for free coinage. Some suggested putting more silver into the silver dollar in order to raise its value to a parity with the gold

[46] *People's Party Paper* (Atlanta), June 16, 1893; Shaw, in the *Review of Reviews*, 10:41; clippings in Donnelly Scrapbooks, Vol. 14.
[47] *People's Party Paper*, December 15, 22, 1893; January 5, 1894; February 23, 1894.
[48] *Kansas Weekly Capital and Farm Journal*, February 8, 15, 1894; *People's Party Paper*, February 16, 1894.
[49] McVey, *Populist Movement*, 174.

dollar, but this scheme met instant opposition from the silver miners, who wanted the price of silver to go up, and from the debtor farmers, who wanted the purchasing value of the dollar to go down.

"International bimetallism" was invoked, and hopes were raised that the consent of all nations might be obtained to some plan that would make possible a double monetary standard. In November, 1892, mainly through American influence, a conference on bimetallism was actually held in Brussels with twenty nations participating, but of all these nations the United States alone seemed much interested in the subject for which the conference was called. Henry Cabot Lodge advanced a plan of discriminatory tariff legislation to force England into an international agreement for the free coinage of silver — merely "a piece of Yankee ingenuity," so Senator Allen concluded. According to the senator from Nebraska, "this whole talk of 'international bimetallism' is a subterfuge by which the Republican party hopes to deceive the people in 1896." [50]

What would happen in 1896 might be forecast in a measure by the results of the elections of 1894. Would the old parties be able to evade the silver issue successfully and hold their ranks intact? Or would their evasions continue to play into the hands of the Populists with the result that what was now the third party would soon become the second or first in the nation?

[50] Shaw, in the *Review of Reviews*, 10:39. On the results of the Brussels Conference see the *Nation*, 55:424 (December 8, 1892). Commenting upon the platform adopted by the Democrats in 1896, the editor of the *Review of Reviews* observed: "Their assurance . . . that the free coinage of silver at the ratio with gold of 16 to 1 would mean bimetallism rather than silver monometallism, is an exercise of wondrous faith." *Review of Reviews*, 14:141 (August, 1896).

CHAPTER TWELVE

THE ELECTIONS OF 1894

BEGINNING in January, 1894, Cleveland resorted to his famous bond issues, borrowing gold to bolster up the depleted gold reserve. Because many people were still fearful that the government might not be able to maintain gold payments, the gold that Cleveland drew into the treasury was almost immediately withdrawn by the presentation of paper for redemption, and thus one bond issue followed another with disconcerting rapidity.

Early in 1895, when the gold reserve was down to about forty-one millions, Cleveland made a deal with J. P. Morgan and August Belmont to sell through the firms they represented some sixty-five million dollars' worth of four-per-cent gold bonds. The contract provided that half the desired gold should be obtained from outside the United States and that the brokers should use their influence to prevent any further drain upon the treasury reserve. Inasmuch as the bonds were delivered to the brokers at 104½ and sold by them at 112½, it was clear that the profits on the transaction must have been enormous. But whatever it cost the government in dollars and cents, the plan worked. Gold was kept in the treasury, and silver payments were avoided. Meantime, the unpopularity of the president in silver circles became little short of appalling. Tom Watson described the bond issues as "an outrage upon a liberty-loving people," and Senator

Allen argued convincingly that the president's actions were worse than outrageous, they were unconstitutional.[1]

Cleveland's handling of the serious industrial disorders that broke out during his administration also made him bitter enemies. The period of hard times that followed the panic of 1893 threw many laborers out of work and led to drastic cuts in the wages of those who held their jobs. Strikes were common, and in the most notable of these, the Pullman Strike, which occurred at Chicago in the summer of 1894, President Cleveland intervened with federal troops, ostensibly to maintain order but actually, so the laborers believed, to defeat the strike. The president's hostility to the rights of labor seemed the more evident in view of the fact that Governor John P. Altgeld of Illinois had refused to call for federal troops and afterwards vigorously criticized the president for sending them. During this strike, too, Eugene V. Debs, an official of the American Railway Union, was thrown into jail for ignoring an injunction that forbade the union to interfere with the running of trains.

Henry Demarest Lloyd saw in actions such as these evidence of a conspiracy between the government and the employing class — a conspiracy that he duly exposed in his famous tract *Wealth against Commonwealth.* As for the armies of unemployed men, they tended, encouraged by the preachings of "General" Jacob S. Coxey of Massillon, Ohio, to congregate in groups for marches upon Washington. Coxey's solution of the problem of unemployment was the familiar plan of public works, the necessary funds to be obtained by large issues of paper money. To urge this plan upon Congress his army actually marched upon Washington, only to be made ridiculous when its leaders on arrival were arrested for ignoring orders to keep off the grass of the Capitol grounds. Other "armies" under other "generals" were

---

[1] *People's Party Paper* (Atlanta). January 12, 1894; Shaw, in the *Review of Reviews*, 10:37; Arnett, *Populist Movement in Georgia*, 172-175.

much in evidence, but their plans for direct protests similarly miscarried.[2]

While old-party politicians and news vendors, Democrats and Republicans alike, were decrying the excesses of labor and making sport of the "industrial armies," the Populists were losing no opportunities to extend both their sympathy and their help. Already they had gone on record in protest against the use of "Pinkertons," especially as invoked during the Homestead Strike of 1892, and they now redoubled their efforts to make clear the identical interests of farmers and laborers. They saw nothing ridiculous about the march of Coxey's army. If "the bankers and usurer class, who at the present time so completely dominate our government," could appear before congressional committees, why not this "living petition" of workingmen?[3]

Senator Allen told his colleagues that instead of encouraging police preparations to receive the marching workingmen as enemies and invaders, the Senate should appoint a committee to wait upon Coxey and his followers and listen to their grievances.[4] Governor Lewelling of Kansas issued a circular in defense of the industrial armies in which he described the unemployed men not as tramps and vagabonds but as poor unfortunates, robbed and legislated out of their right to work. When a detachment of the "Commonweal Army" under "General" Bennett passed through Topeka in June, 1894, a convention of Kansas Populists, then in session, took up a collection amounting to one hundred and two dollars for the "laborers." Another detachment under "General" Kelley, which crossed Iowa, was cordially welcomed to Des Moines by General Weaver, who congratulated

[2] E. Benjamin Andrews, *The United States in Our Own Times*, 719–735; Donald R. McMurray, *Coxey's Army: A Study of the Industrial Army Movement of 1894;* W. R. Browne, *Altgeld of Illinois: A Record of his Life and Work,* chs. 12–15.

[3] B. O. Fowler to Allen, April 25, 1894, in the Allen Papers; *Representative* (St. Paul), August 1, 1894.

[4] Shaw, in the *Review of Reviews,* 10:37.

the men upon their "marching petition" and the objects
they sought to attain by it. Weaver also acted as "master
of ceremonies" when a delegation was sent to ask the governor's help in obtaining free transportation across the state
for the army.[5]

Their sympathy for the rights of labor, however, did not
lead the Populists to counsel or condone acts of violence.
For labor, as for agriculture, they insisted that a political
remedy must exist. "In this country," Ignatius Donnelly
told a Montana audience,

where the ballot box stands open, and the majority governs, if
the bulk of the people of the United States with the ballot box
in their hands are robbed of their liberties and reduced to serf-
dom, no one is to blame but themselves. . . . When the majority acting through the ballot box cannot preserve their liberties, then it will be time to talk about armed revolution.[6]

To another western audience he said, "We have no right to
speak of blood-shed when we have the power of the ballot."
And to another, "We do not believe that the path of reform
is through the torch and the rifle. We believe it is through
the ballot box."[7]

It was, indeed, one of the fondest of the Populist hopes
that the discontented laborers could be induced to throw in
their lot with the protesting farmers. Both classes suffered
apparently from the same plutocratic oppressors; both
classes must be represented in the new party if it were to
be in any true sense a "people's party." From the very first
the Alliance and Populist platforms championed boldly the
cause of labor, demanding all the popular labor reforms and
soliciting the cooperation of all the prominent labor organizations.

[5] Barr, in *Kansas and Kansans*, 2:1190; Haynes, *James Baird Weaver*, 353–
354; Donald R. McMurray, "The Industrial Armies and the Commonweal,"
Mississippi Valley Historical Review, 10:215–252.
[6] Clipping from the *Anaconda Standard*, September 6, 1893, in the Donnelly
Scrapbooks, Vol. 14.
[7] Clipping from *Denver Daily News*, February 24, 1894, in the Donnelly
Scrapbooks, Vol. 16; *Representative*, July 18, 1894.

For some reason labor remained singularly unimpressed. Perhaps the labor leaders realized that money inflation would mean higher prices as well as higher wages and that "more money" might leave the laborer relatively about where he had been before. They may also have seen that the interests of the farmer as a producer might frequently clash with the interests of the laborer as a consumer. Powderly and some other prominent Knights of Labor were in favor of joining forces with the Populists, but the Knights were a declining order, and what little help they were able to give the new party was of small consequence. As for the American Federation of Labor, which was growing rapidly in numbers, it was warned by Samuel Gompers himself to avoid all entanglements with "the employing farmer." Laborers in general, when too dissatisfied with the old parties to remain with them, went in for socialism rather than for Populism.[8]

As the campaign of 1894 approached, however, the Populists were full of hope. They felt confident that the prevailing hard times and unemployment would mean thousands of votes for the protesting third-party candidates. They were convinced that Cleveland's unpopularity, not only among the laboring classes but also among all those who favored free silver, would result in heavy Democratic desertions to the Populist standard. Democrats who agreed with Cleveland that the tariff legislation contained in the Wilson-Gorman Act of 1894 was a "piece of party perfidy" might also go in for Populism. And the record of the Harrison administration would serve as a warning to all reformers of what might be expected of Republicans.[9] Early in May, Donnelly's paper professed to have received

advices from all parts . . . that the hard times are making converts for us by the thousand, every day and hour. The princi-

---

[8] Stewart, in the *Indiana Magazine of History*, 15:73; Arnett, *Populist Movement in Georgia*, 125 note; John R. Commons and others, *History of Labour in the United States*, Vol. 2, ch. 13.

[9] *People's Party Paper*, July 6, 1894; Arnett, *Populist Movement in Georgia*, 176–178; Haynes, *Third Party Movements*, 279.

ples of the Omaha platform are becoming constantly more popular. There is no more talk about "calamity howlers." The people are all "howling" together now.[10]

About the same time Tom Watson's paper noted with sardonic glee that old-party statesmen were "suffering from the premonitory symptoms of the Belshazzar ague."[11] A few weeks later Taubeneck, still the Populist national chairman, claimed that in the preceding thirty days more recruits had been added to the third party than during any six months of its earlier history.[12]

In the West the confidence of the Populists was so strong that in making nominations for the year there was a general tendency to spurn fusion and "keep in the middle of the road." Taubeneck urged that separate Populist tickets should be nominated even if there was not a chance that they might win. His idea was that in this way the full Populist strength would be available for tabulation and the rapid growth of the party could be shown. Donnelly described fusion as a mere "sacrifice of opinions and beliefs for the sake of plunder" and proclaimed that "the middle of the road is the place for Populists."[13] In Kansas, where the experiment of fusion had been tried in 1892, an Anti-Fusion People's Party League was formed, which captured the state organization and dictated the Populist policy throughout the campaign.[14] When some of the silverites urged that united support be given a single silver candidate, regardless of his party, in each contest, Governor Waite described the plea as "one of the most deadly attacks that had ever been made" on Populism. His notion was to let the silverites come over to the Populists, and not to meet them half way.[15]

[10] *Representative*, May 2, 1894.
[11] *People's Party Paper*, May 11, 1894.
[12] *Ibid.*, June 1, 1894.
[13] *People's Party Paper*, June 1, 1894; *Representative*, June 13, September 12, 1894.
[14] Miller, "Populist Party in Kansas," 269.
[15] *People's Party Paper*, February 23, April 23, 1894.

Weaver, of Iowa, was decidedly less obstinate. "I am a middle-of-the-road man," he said, "but I don't propose to lie down across it so no one can get over me. Nothing grows in the middle of the road"; and presently he accepted a nomination for Congress at the hands of both Populists and Democrats.[16]

But in general Weaver's warning was disregarded, and the amount of fusion that took place in the West in 1894 was distinctly less than in 1892. In the several mountain states — Colorado, Nevada, Wyoming, Idaho — where close combinations had been effected two years before, fusion now turned out to be entirely impossible. For the most part separations took place by mutual consent. The Populists were unwilling to make any concessions of consequence to the Democrats, and the Democrats were tired of abjectly following the Populist lead. In Colorado, Waite and all his works were indignantly repudiated by the Democrats, and in Nevada the dominant Silver party, which had furnished the bulk of the Populist votes in 1892, went its own way, leaving the Populists high and dry. Nevada had an abundance of state tickets in 1894, Silver party, People's party, Republican party, and two varieties of Democrats.[17]

[16] Haynes, *James Baird Weaver*, 357–359. The phrase "keep in the middle of the road" was used as early as 1892 in a jingle published by the *Rocky Mountain News* and reprinted in Haynes, *James Baird Weaver*, 468–469:

> "Side tracks are rough, and they're hard to walk
> Keep in the middle of the road;
> Though we haven't got time to stop and talk
> We keep in the middle of the road.
> Turn your backs on the goldbug men,
> And yell for silver now and then;
> If you want to beat Grover, also Ben,
> Just stick to the middle of the road.
>
> "Don't answer the call of goldbug tools,
> But keep in the middle of the road;
> Prove that the West wasn't settled by fools,
> And keep in the middle of the road.
> They've woven their plots, and woven them ill,
> We want a Weaver who's got more skill,
> And mostly we want a Silver Bill,
> So we'll stay in the middle of the road."

[17] *Appletons' Annual Cyclopaedia, 1894*, 150, 513.

In the frontier states of the Middle West fusion was likewise at low ebb. Kansas, which, thanks to fusion, had been the banner Populist state in 1892, now had three separate state tickets in the field, and the Democrats were quite as hostile to the Populists as ever the Republicans had been.[18] In 1892 the Populist and Democratic organizations in Minnesota and South Dakota had kept apart; in 1894 they showed no less resistance to the idea of union.[19] In Nebraska, where there had been relatively little fusion in 1892, there was manifest some disposition for the two minority parties to get together. William Jennings Bryan, who had threatened in 1893 to go out and " serve his country and his God under some other name," in case the Democratic party was ever committed to the gold standard, tried hard to promote fusion, but the Populists, with a few important exceptions, showed little interest and nominated a full state ticket of their own, headed by Judge Silas A. Holcomb for governor. The Democrats indorsed Holcomb and two other names on the Populist ticket, but for this favor they apparently received nothing important in return. The Cleveland Democrats held a separate convention and nominated a " straight Democratic " ticket, which played no significant part in the campaign.[20] In North Dakota fusion between Democrats and Populists on the state ticket, except for the governorship, was accomplished, and the same candidate for Congress was indorsed by both parties.[21] Throughout the West there was less difficulty in securing fusion on congressional than on state tickets, probably because of the intense desire of the silverites to secure control of Congress. The election of state officials, who could actually accomplish nothing whatever in the way of currency reform, was from the standpoint of the mere silverite a matter of small consequence.

[18] *Ibid.*, 392–393.
[19] *Ibid.*, 490–491, 721.
[20] *Ibid.*, 506–507. See also Barnhart, " Farmers' Alliance and People's Party in Nebraska," 363–371.
[21] *Appletons' Annual Cyclopaedia, 1894,* 556.

In the South, however, the situation with regard to fusion was entirely different. There the cooperation between Republicans and Populists that had begun in 1892 was continued two years later with unabated zeal. If anything, the fact that the national government was now Democratic made fusion in the South easier and more natural in 1894 than in 1892 when the Republicans were still in control at Washington, for the two minority parties could now unite whole-heartedly in criticizing the record of a national administration. The success of fusion in the West during the preceding campaign had been due in no small part to the existence of a comparable situation. Western Democrats and western Populists could meet on common ground in 1892 when they discussed the shortcomings of the Harrison administration. This advantage the western advocates of fusion had lost by 1894, and further, thanks to their success in 1892, they had a record to defend. Southern fusionists, however, had not only a common point of attack, they had no embarrassing Waite or Lewelling in office to serve as an obstacle to the persistence of fusion. Moreover, both southern Populists and Republicans had a common and genuine local grievance. They knew that the Democrats had defrauded them shamefully two years before; so at least on the issues of an honest election and an honest count the two minority parties of the South were in complete accord.

The year 1892 was a presidential year and the year 1894 was not, but aside from this difference the two elections were very similar in the South. Tillmanism, still triumphant in South Carolina, held the great bulk of the votes in that state to their old-party allegiance. Otherwise thousands of South Carolina farmers would have gone over to Populism. Tillman's organization took up all the Populistic doctrines, free silver in particular, and Tillman himself, who aspired to a seat in the United States Senate, led in the denunciation of Cleveland. "I haven't got words," he told a Rock Hill audience, "to say what I think of that old bag of beef [President Cleveland]. If you send me to the Senate, I promise I won't be bulldozed by him." [22]

The Populists nominated a full state ticket in Georgia, and the Republicans made no nominations, leaving their supporters free to aid in the election of the third-party candidates. In most of the congressional districts of the state, also, it was a straight-out contest between the Populists, supported by the Republicans, and the Democrats. Watson, who sought once more to be returned to Congress from the tenth district, was the most spectacular campaigner, but W. H. Felton, who ran as a Populist in the "bloody seventh" also attracted much attention. There was practically no avoidable competition anywhere in the state between Populists and Republicans for seats in the state legislature.[23]

In Alabama the contest of 1892 between the supporters of Kolb on the one hand and the regular Democratic organization on the other was renewed. Kolb had back of him not only the entire Populist and Republican strength but also a faction of Democrats who called themselves "Jeffersonian Democrats" and sought to redeem the state from the clutches of the machine that had governed it for so long. The allies, mindful of their experience in 1892, made a free vote and an honest count their foremost demand. They urged also a contest law for state offices and, in the realm of national politics, silver coinage. Fusion extended also into the legislative and congressional campaigns, and the forces of the contenders appeared to be very evenly matched.[24]

In North Carolina the Republicans and the Populists were more completely in harmony than ever before. Many of the state officers held over, but a state treasurer and important judicial officers were to be chosen, besides congress-

[22] *Augusta Chronicle*, June 18, 1894, quoted in Simkins, *Tillman Movement*, 181 note. See also Haynes, *Third Party Movements*, 280.
[23] Arnett, *Populist Movement in Georgia*, 182–183.
[24] *Appletons' Annual Cyclopaedia, 1894*, 5.

men and members of the legislature. Because of the death of
Senator Vance, the legislature would have two vacancies in
the United States Senate to fill instead of the one that ordinarily would have occurred. A combination ticket of Populists and Republicans was arranged for the state and judicial
contests, and in the congressional and legislative districts
the two minority parties stood in each other's way as little
as possible.[25] Elsewhere in the South cooperation between
Republicans and Populists on minor offices was common, but
ordinarily it stopped short of agreement upon the same
candidates for state office.[26]

The Populist campaigns of 1894, North and South, repeated the tactics of 1890 and 1892. There were the usual
conventions and rallies, picnics and barbecues. Populist orators dwelt upon the hardness of the times and the great dis-
parity between the living conditions of the rich and the poor.
"We meet in the midst of turbulent times," Donnelly told
a Minnesota audience. "This government was founded by
plain men, not millionaires. But . . . we now have two parties arrayed against each other, Aristocracy against Com-
monality. Thirty thousand families own one-half the wealth
of the country, and they have no part in producing it. They
have stolen it from the labor and toil that has produced the
nation."[27] A cartoon, "Uncle Sam's Crown of Thorns,"
copyrighted in 1894 by the Vox Populi Company of St.
Louis, showed Uncle Sam wearing a hat of the usual shape,
but constructed of thorns. Each thorn was named after
some trust or some wicked anti-Populist doctrine such as
monometallism.[28] In the South the unfairness of the opposition in the preceding campaign came in for much discussion,
but throughout the nation the Populist campaigners found

[25] *Ibid.*, 552, 553; S. A. Delap, "The Populist Party in North Carolina,"
Historical Papers of the Trinity College Historical Society, 14:40–74.
[26] *Appletons' Annual Cyclopaedia, 1894*, under names of states.
[27] *Representative*, July 18, 1894.
[28] Reproduced in the *People's Party Paper*, June 8, 1894. Did Bryan find his
"crown of thorns" figure here?

Uncle Sam's "Crown of Thorns."

How long, O Lord! How long wilt Thou forget me? How long wilt Thou hide thy face from me? How long shall I take counsel in my soul having sorrow in my heart daily? How long shall my enemies be exalted over me? Consider and hear me, O Lord, my God; lighten mine eyes, lest I sleep the sleep of death.—Psalms, xiii, 1, 2, 3.

free silver their best talking point. Forgetful of their earlier
doctrines, their claims for silver passed all reasonable bounds.
It means work for the thousands who now tramp the streets
. . . not knowing where their next meal is coming from. It
means food and clothes for the thousands of hungry and ill-clad
women and children. . . . It means the restoration of confidence in the business world. It means the re-opening of closed
factories, the relighting of fires in darkened furnaces; it means
hope instead of despair; comfort in place of suffering; life instead
of death.[29]

As usual, however, the results of the election were by no
means all that the third-party leaders had hoped. Most of
the western states that had given Populist majorities in 1892
now returned to their former allegiance. All the silver states
that had voted for Weaver two years before were lost. In
Colorado and Idaho the Republican victory was complete;
in Nevada, the Silver party swept everything before it. In
Kansas the absence of Democratic support and the presence
of dissension among the Populists themselves gave an easy
and complete victory to the Republicans, who elected their
entire state ticket, a majority in the lower house of the legislature, and every congressman but one.[30] In North Dakota
and South Dakota the Republicans won complete victories;
in Minnesota they elected their entire state ticket, majorities in both houses of the legislature, and every congressman
but one; in Nebraska they lost the governorship and one
congressman to the fusionists but had everything else their
own way.[31] In Iowa even General Weaver went down in the
general slaughter.[32]

Although fusion was a feature of the campaign of 1894 in
the South to a far greater degree than in the West, the

[29] *Representative*, September 19, 1894.
[30] At the opening Populist rally " Mrs. Lease got up and attacked Mrs. Diggs.
Sister Diggs responded, and called her a liar, and the campaign was properly
launched." Barr, in *Kansas and Kansans*, 2:1191.
[31] *Appletons' Annual Cyclopaedia, 1894*, under names of states.
[32] Haynes, *James Baird Weaver*, 361.

southern Populists emerged from the election only a little more successful than their western brethren. In South Carolina Tillmanism swept everything before it as usual, but the Tillmanites were Democrats, however much they might applaud third-party views.[33] In Georgia the tactics of 1892 were again invoked to defeat the Populists. The colored voters were tempted by whatever bribes the opposing forces could muster, the standard price for votes, however, being only a dollar apiece. No doubt negro votes were purchased by both parties, but the Democrats were the better provided with funds, and furthermore they held control of the election machinery. Ballot boxes were stuffed, unfavorable returns were falsified, and whole precincts were thrown out on the flimsiest technicalities. "We had to do it!" one participant confessed years later. "Those d—— Populists would have ruined the country!"[34]

Nevertheless the Democrats were compelled to concede to Judge Hines, Populist candidate for governor, a total of 96,000 votes, only about 20,000 less than his successful opponent. Two years before, the Democratic majority had been about 80,000.[35] In Watson's district, the frauds that had been committed were so transparent and notorious that the successful Democratic candidate pledged himself to resign the fourth of March following the election in order to give the people another chance to choose between himself and Watson. He was as good as his word, and a special election was held, but Watson again lost both on the face of the returns and in a contest which he instituted before the House.[36] All the other Georgia seats in Congress went to Democrats, but the Populists made notable gains in the legislature, securing on joint ballot about one-fourth of the total number of votes. In general the Populists showed little

---

[33] Simkins, *Tillman Movement*, 183.
[34] Arnett, *Populist Movement in Georgia*, 183–184; *Appletons' Annual Cyclopaedia, 1894*, 313.
[35] *Ibid.*
[36] Brewton, *Thomas E. Watson*, 263–265.

strength in the cities and in the "black" counties but polled heavy votes in the rural districts where the whites predominated.[37]

In Alabama the same excitement over the election prevailed as in Georgia, and the same methods were employed. There the state election was held in August, and according to official returns the Democratic candidate for governor won by 110,830 votes to 83,309 for the fusionist candidate, Kolb. Kolb felt certain that he had actually received a majority of the votes cast but that a large percentage of them had been counted out. In the absence of any effective means of bringing a contest, however, his only recourse was to issue through his campaign committee an indignant address to the people, which urged the formation of "honest election leagues" to prevent such disasters in the future. The Democrats also elected the rest of their state ticket and a comfortable majority in both houses of the legislature. On the face of the election returns the Alabama delegation in Congress would consist of eight Democrats and one Populist. While generally admitting the existence of frauds in both elections, the Democrats claimed that the fusionists were by no means guiltless. Money from the North, they avowed, had been used freely in the vain attempt to wrest the state from Democratic control. As in Georgia, the urban population and the "black" counties furnished the bulk of the Democratic support, while the rural whites were strongly fusionist.[38]

In North Carolina, where the Republican party was far stronger than in most southern states and could contribute a heavy vote to fusionist candidates, the combination Populist and Republican state and judicial ticket won by a majority of about twenty thousand. Thirty-one counties were carried either by the Populists separately or by a Populist-

[37] Arnett, *Populist Movement in Georgia*, 184.
[38] *Appletons' Annual Cyclopaedia, 1894*, 5, 6; Manning, *Fadeout of Populism*, 46–48.

Republican coalition. The Populists won a majority over
all other parties in the state Senate, and the Populists and
Republicans together had a majority in the House. Once the
legislature was convened, a Republican was chosen to fill out
the unexpired term of Senator Vance, and a Populist, Marion
Butler, was chosen for the six-year term. In the newly
elected national House of Representatives North Carolina
was represented by three Populists, three Democrats, two
Republicans, and one Independent Protectionist, the latter
being elected over a Democrat by Populist and Republican
votes. In only three out of the nine districts had the minority parties failed to combine on a single candidate. The
results in North Carolina constituted the outstanding Populist victory of the year and pointed clearly to the means by
which in 1896 the fusionists might hope to obtain complete
control of the state.[39]

Following the election, many contests were brought from
southern congressional districts by Populists or Republicans
who claimed to have been counted out. Inasmuch as the
national House of Representatives was now Republican, it
was by no means unwilling to go behind the returns, and
many of the contestants were successful. From the first
South Carolina district a negro Republican took the place
of the Democrat whose election had been duly certified.
A North Carolina Democrat was unseated and his place
given to a Populist. A Virginia Democrat was forced to
yield to a Republican. In Alabama three contests were decided in favor of the contestees, and as many fusionists replaced Democrats; thus the total non-Democratic delegation
from the state was raised from four to nine. The Alabama
fusionists, on the ground that had a fair election been held
the majority of the legislature would have been fusionist,
not Democratic, also attempted to unseat a United States

[39] Delap, in *Historical Papers of the Trinity College Historical Society*,
14:56–57; *Appletons' Annual Cyclopaedia, 1894*, 553; *Official Congressional Directory*, 54 Congress, Session 2, p. 3.

DISTRIBUTION OF THE VOTE FOR POPULIST AND FUSION STATE TICKETS IN 1894

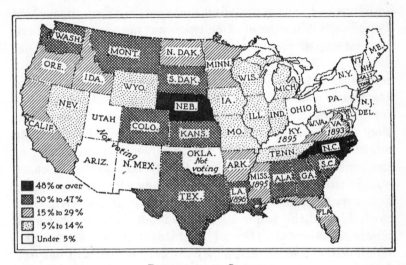

PERCENTAGES BY STATES

Alabama	47.64	Maine	4.91	Ohio	....
Arkansas	19.31	Maryland	....	Oregon	28.96
California	18.01	Massachusetts	2.70	Pennsylvania	2.04
Colorado	41.37	Michigan	7.20	Rhode Island	....
Connecticut	0.99	Minnesota	29.68	South Carolina	30.43
Delaware	....	Mississippi (1895)	26.99	South Dakota	35.47
Florida	20.68	Missouri	8.45	Tennessee	9.93
Georgia	44.46	Montana	30.94	Texas	36.13
Idaho	29.20	Nebraska	49.06	Vermont	....
Illinois	6.96	Nevada*	6.75	Virginia (1893)	28.60
Indiana	5.31	New Hampshire	....	Washington	39.04
Iowa	8.23	New Jersey	....	West Virginia	....
Kansas	38.96	North Carolina	53.78	Wisconsin	6.81
Kentucky (1895)	4.73	North Dakota	22.66	Wyoming	11.29
Louisiana (1896)	43.68	New York	0.60		

* A Silver party ticket won in Nevada.

senator who had been elected by the overwhelmingly Democratic legislature. But the Senate was unwilling to go this far.[40]

Actually the Populists had gained very little ground, if any, by the election of 1894. The men elected to the national

[40] Manning, *Fadeout of Populism*, 35–36, 47; *Congressional Directory*, 54 Congress, Session 2, pp. 3, 17, 18, 97–98, 118.

House or Senate by fusionist votes chose generally to affiliate with the Democrats or with the Republicans — only four senators and four representatives openly styled themselves Populists.[41] After the election there was not a state, southern or western, that could be cited as predominately Populist territory. The North Carolina legislature had a fusionist majority, but only one house was clearly Populist, and a Democrat was still governor of the state. The Nebraska governor was a Populist who owed his election to a combination with the Democrats, but all the other Nebraska state officers and the majority of the members in both houses of the legislature were Republicans. The Kansas state Senate was Populist, but this was due entirely to the presence of hold-over senators elected in 1892. The Republicans had everything their own way in the House and in the state offices. Furthermore, the silver states with one accord had repudiated Populism and all its works. Local conditions had had much to do with some of these results, but one factor that played a large part was the general mid-term desire to rebuke the national Democratic administration. Apparently the voters felt that the surest way to do this was to vote the Republican rather than the Populist ticket.[42]

The Populists, however, professed to believe that the election indicated a remarkable increase in their strength. They pointed to the greater number of Populist votes cast in the election — 1,471,590 in 1894 as against 1,041,028 in 1892, an increase of forty-two per cent. In most of the western states, including those in which the Populists had won in 1892 only to lose in 1894, the Populist vote actually bulked larger in the second election than in the first, and this without so much assistance from the Democrats. In some western states the increase in the Populist vote was little short of phenomenal. Minnesota, for example, cast 87,931 Populist

[41] Congressional Directory, 54 Congress, Session 2, pp. 189, 191.
[42] Haynes, Third Party Movements, 280–281.

votes in 1894 as against 29,313 in 1892. In California the
Populist vote was nearly double that of 1892, and the vigorous Populist denunciations of railway domination in California state government undoubtedly had much to do with
the success of the Democratic candidate for governor, who
urged that the Central and Union Pacific railroads should
be taken over and operated by the United States government. A Populist candidate, Adolph Sutro, builder of the
tunnel beneath the Comstock lode in Nevada, was elected
mayor of San Francisco in spite of all that the authorities of
the Southern Pacific Railway, the old-party political organizations, and the city newspapers could do.

Southern statistics, badly garbled as they were by the
Democratic returning boards, showed clearly that the Populists had made great gains in that region. Allowing for the
fact that many of the votes cast for Populist candidates were
actually cast by Republicans, the Populists could still claim
that with a free vote and an honest count they could have
won single-handed. Observing the results of the election
from the vantage point of Minnesota, Ignatius Donnelly declared that the Populist party was clearly the coming party.
The Democrats were "buried under the drift a thousand
feet deep." Ultimately, he argued, the Democratic party
would go the way the old Whig party once went, and only
two parties would be left, the Populists, representing the
views of the common people, and the Republicans, representing the views of the plutocrats.[43]

[43] *Representative*, November 7, 1894; McVey, *Populist Movement*, 198;
Haynes, *Third Party Movements*, 280–281.

CHAPTER THIRTEEN

THE TRIUMPH OF FUSION

FOLLOWING the elections of 1894 advocates of free silver redoubled their efforts, fixing their eyes now on the presidency in 1896 as their goal. Naturally their arguments eclipsed all else in most of the reform journals, but other newspapers also, the city dailies no less than the country weeklies, were forced by an irresistible public demand to give liberal space to discussions of the "money question." Reprints of the speeches made in Congress by silver senators and silver representatives were relentlessly distributed. Innumerable pamphlets, sold at popular prices or given away, flooded the country. Whole books were devoted to the subject.

Among these books no other could compare in popular appeal with *Coin's Financial School*, by W. H. Harvey, first published in 1894. Harvey was a Virginian by birth who, after practicing law for a time in Ohio, went further west and tried his hand with some success at ranching and silver prospecting. The panic of 1893, however, robbed him of his earnings, and he embarked upon a new career, this time as editor of a weekly journal in Chicago. But his journal failed, and he found success only when he had published his famous *Coin's Financial School*, which became the great silver classic and was sold by the hundreds of thousands of copies. "Coin," so the reader was led to believe, was a young financier who conducted a group of prominent business men through a series of classroom lectures on the money question. With simple language and apparently unanswerable logic he set forth the arguments for silver and confounded the reasoning of his opponents. No doubt most people who read the book understood the ruse, but to many "Professor Coin" was a very real person, who had actually conducted the school described and had actually routed in person some of the ablest gold-bug advocates in the country. For the author did not scruple to mention the names of Coin's pupils, and they were names everyone knew. As Dr. Haynes points out, Harvey's book could not have created the silver movement; rather the strength of the silver movement paved the way for the popularity of such a book.[1] But undoubtedly Coin's arguments brought thousands of wavering voters to the silver side. Extracts reprinted by newspapers extended indefinitely the range of Harvey's readers; even the gold-bug papers had to print what Coin had said if they were to refute his arguments, and they sometimes learned to their sorrow that what they said was forgotten while the words of Coin made a lasting impression.[2]

The success of *Coin's Financial School* naturally incited imitations, the most successful of which was probably Ignatius Donnelly's *The American People's Money*, published in 1895. This book is the record of an imaginary conversation between a banker and a farmer thrown together on a transcontinental railway journey. Much after the fashion of "Coin," but with little of his finesse, the farmer enlightens the banker on the money question. Donnelly, however, was less a silverite than a Populist. Careful reading of his book would have revealed that he favored "scientific paper money" ultimately as the ideal currency, and he generously

[1] Haynes, *Third Party Movements*, 295.
[2] *Review of Reviews*, 14:131-132 (August, 1896); Mark Sullivan, *The Turn of the Century, 1900-1904; Our Times, The United States, 1900-1925*, 1:175-180; Harry Thurston Peck, *Twenty Years of the Republic, 1885-1905*, 453-454.

included arguments in favor of a large variety of reforms that had nothing whatever to do with the money question. Donnelly's book did not compare in popularity with Harvey's, but it did reach a large audience, and whatever readers may have thought of the rest of it, the arguments for silver were well received.[3]

The silver propaganda was widely accepted in the Democratic party, particularly in the South. There the masses, long assured by reformers that the ills of the South were due mainly to a faulty currency, drifted steadily in the direction of silver. Some of the leaders held back, defending Cleveland and the gold-standard policy of his administration as good party regulars should.[4] Others, however, almost in a panic after the demonstration of Populist strength that the campaign of 1894 afforded, minced no words in their denunciations of the president's gold-bug tendencies. They feared that only by advocating free silver, the most popular of the third-party principles, could they hope to maintain their complete control of the South. Losses there might be from gold-bug defection, but these would be more than outweighed by additions from the ranks of the Populists.[5] And furthermore, by accepting free silver as a cardinal article of Democratic faith, the menace of Populism might be once and for all removed.

Western Democrats were not far behind southern Democrats in their readiness to respond to the pleas of the silverites. After the election of 1894 the American Bimetallic League continued its efforts, sponsoring conference after conference on the money question and sending a veritable army of lecturers over the country to proclaim silver views. Petitions poured in upon Congress to enact free-silver laws,

---

[3] Volume 17 of the Donnelly Scrapbooks contains many newspaper comments on the book.

[4] *Southern Mercury* (Dallas), February 23, 1893; Arnett, *Populist Movement in Georgia*, 185–188.

[5] Arnett, *op. cit.*, 185–189; *Southern Mercury*, May 16, 1895.

and free-silver bills galore made their appearance in response
to these demands. At Memphis, Tennessee, in June, 1895,
the silver men held a great convention primarily to commit
enough Democrats to the silver issue to insure silver control
of the Democratic party in the coming election. The convention was well attended, and while representatives of every
party were numbered among the delegates, the Democrats,
according to a local reporter, were in the majority by a ratio
of nine to one. Bryan came down from Nebraska to address
the conference, and southern delegates united with western
delegates to pledge themselves to work *within the party fold*
for the enactment of a free-silver law.[6]

If it was the policy of the silver propagandists to capture
the Democratic party for free silver, it was no less their
policy to wean the Populists away from their other doctrines
and to commit them to a program in which free silver should
be the only item of consequence. The long and variegated
Omaha platform, however, stood directly in the path of such
a policy. For this reason the demand for a modification of
the Populist creed became insistent. It was asserted that the
Omaha resolutions were adopted under great excitement
and on short notice, hence failed to represent the true position of the party. If a new convention must be held in order
to modify these demands, by all means hold it. The inference that the Omaha platform was hastily thrown together
and out of tune with Populist sentiment in 1892 was as far
as possible from the truth; but the demand for revision was
so strong that a convention, more or less official, was actually
held at St. Louis early in 1895 to suggest such changes in
the party creed as might seem desirable. The St. Louis convention, however, to the great disappointment of the silver-

[6] *Commercial Appeal-Avalanche* (Memphis), June 14, 15, 20, 1895; *Southern
Mercury*, June 16, 20, 1895; William Jennings and Mary B. Bryan, *The Memoirs
of William Jennings Bryan*, 102; Stewart, in the *Indiana Magazine of History*,
15:53.

ites, voted by a large majority to hold fast to the whole of
the Omaha platform.[7]

Some of the more prominent of the Populist leaders keenly
regretted the St. Louis action and hoped regardless of it to
reduce the number of the Populist demands. "Keep the
money question to the front," Taubeneck wrote to a third-
party state convention that met in the summer of 1895, "it
is the only living issue before the people. I hope your state
convention will build a platform making the 'money question' the great central idea, unencumbered with details or
side issues."[8]

Weaver spoke in similar vein. "I shall favor going before
the people in 1896," he said, "with the money question
alone, unencumbered with any other contention whatsoever.
Not on the silver issue alone, but distinctly favoring unre-
stricted coinage at the ratio of 16 to 1, and legal tender government paper, with neither bonds nor banks of issue."[9]
Out in Colorado he advised an audience that party lines
might well be swept aside, and that all Populists, Democrats,
and Republicans who believed in currency reform should
hold themselves in readiness to join hands.[10] This meant
fusion, exactly what the silverites desired. And it might
mean the shipwreck of the Populist party as such. Wise
after the event, a southern editor accused the advocates of
free coinage of "a deep-laid conspiracy to ruin our party and
destroy the reform movement." As the initial acts of this
conspiracy they "had secured control of our party machinery
and of nearly all those whom we had elected to positions of
honor and trust."[11]

But the conquest of Populism by the silverites was not

[7] *Southern Mercury*, January 3, February 21, March 7, November 14, 1894;
McVey, *Populist Movement*, 177–178.
[8] *Southern Mercury*, July 11, 1895.
[9] *Ibid.*, April 11, 1895.
[10] *Ibid.*, May 30, 1895.
[11] *Ibid.*, February 25, 1897.

accomplished without a struggle. Particularly did the leaders of Populist thought in the South inveigh against a policy the end and aim of which was obviously fusion with a free-silver Democracy. The Democratic party was not to be trusted. If it swallowed some of the People's party "fallacies" now, it was with a view to swallowing the People's party later on. Union with the Democrats on free silver meant in the long run unconditional surrender to the very enemy that southern Populism had been organized to fight, meant a humiliating return to the Bourbon fold.[12]

In the opinion of Southern Populist leaders the silver issue was not the only issue of the day. Orthodox Populists held that unlimited coinage "would not increase the money circulation of the country one dollar per capita"; hence the fight for more paper money as well as for free silver must be continued.[13] And what had become of the important issues of land and transportation, once so dear to the hearts of men like Weaver and Taubeneck, men who were now preaching union on a single issue?[14] These men professed to believe that a separate free-silver party might be formed, which would utterly destroy Populism, whereas anyone with half an eye could see that there was no such danger. The free-silver party was "like a Kentucky militia regiment, it consists of all colonels and no privates"; and its real reason for existence was to act as a bugaboo to frighten the People's party from the Omaha platform and into the Democratic camp.[15]

Taubeneck, with his arguments for a union of the silver forces on a single issue, was regarded as an active menace. He proposed to place the People's party at the "tail-end of the procession when it ought to have been in the lead." Let

[12] Arnett, *Populist Movement in Georgia*, 190–191.
[13] *Southern Mercury*, June 18, 1896.
[14] *Ibid.*, February 27, 1896.
[15] *Ibid.*, March 12, July 9, 1896.

him continue during the campaign of 1896 at the head of the
Populist organization and " there would be no People's party
at the close of the present campaign." [16]

O, come into my party, said the spider to the fly —
Then he sharpened up his pencil and winked the other eye.
The way into my party is across a single plank —
You can take it from your platform, the rest can go
to — blank.[17]

The elections of 1895 occurred while the debate over silver
and fusion was raging, but owing to the fact that most of
these elections were in safely Republican territory, they
failed to reflect fully what was going on. The insignificance
of Populism in the East was perhaps well enough attested
by the fact that in Massachusetts the Populist candidate
for governor polled less votes than the Prohibitionist.[18] In
Ohio, where the Populists nominated " General " Jacob S.
Coxey for governor on a platform apparently of his own
making, the third-party candidate emerged with approximately one-sixteenth of the total number of votes cast.[19]
In Iowa Weaver's best efforts to promote fusion with the
Democrats were unavailing. The administration forces retained control of the Democratic party, forcing the nomination of a gold man for governor and making inevitable a
Republican triumph.[20]

Elsewhere in the West the evidence pointed strongly
towards the speedy capture of the Democratic party by the
silverites. The free-silver Democrats in the Missouri legislature succeeded in inducing the state executive committee of
their party to call a convention to determine anew the attitude of the Missouri Democracy towards silver, and the
convention, with Congressman R. P. Bland as its chairman,

[16] Ibid., July 16, 1896. See also Haynes, Third Party Movements, 282–283;
Arnett, Populist Movement in Georgia, 189–191.
[17] Southern Mercury, April 25, 1895.
[18] Appletons' Annual Cyclopaedia, 1895, 466.
[19] Ibid., 624–625.
[20] Haynes, Third Party Movements, 356–357.

came out boldly for silver.[21] In Nebraska the free-silver element won complete control of the state Democratic convention, called to nominate a candidate for justice of the supreme court and two regents of the state university. When
a resolution commending Cleveland and the Democratic national platform of 1892 was presented to this convention,
there arose such a roar of protest " that order was not established for ten minutes, when the offending resolution was
tabled without comment." [22]

But, if the Nebraska election is to be taken as an index
of sentiment, the time was not yet ripe, even in the West, for
the complete fusion of Democrats and Populists on the silver
issue. The Populist nominee for the supreme court was Maxwell, who had been discarded by the Republicans two years
before on account of his independent views.[23] Perhaps this
nomination was intended to be a bid for support from all
who professed themselves to be opposed to partisanship on
the bench and resentful of judicial favoritism to corporations,
but the same delegates who voted to nominate Maxwell also
went on record as reaffirming their "party's principles,"
thereby suggesting their unwillingness to give up the whole
Omaha platform for the single issue of silver.

Nebraska Democrats were no more ready for fusion than
the Populists. They maintained their separate ticket and
refused to concur in the nomination of Maxwell, as they
might very well have done, especially since large numbers
of free-silver Democrats would be sure to vote for Maxwell
anyway. Moreover, the gold Democrats held a rump convention and nominated an administration ticket of their
own. Nebraska was thus favored with four sets of candidates
in 1895, Republican, Populist, free-silver Democrat, and
gold Democrat. In spite of the split opposition, the Republicans won by a comparatively small margin — eight or nine

[21] *Appletons' Annual Cyclopaedia, 1895*, 500.
[22] *Ibid.*, 522–523.
[23] Klotsche, in the *Nebraska Law Bulletin*, 6:451.

thousand votes. Had the free-silver Democrats given Maxwell their official indorsement, probably the Populists would have won. But even Bryan himself, who only the year before had seemed eager to promote fusion, now urged all Democrats to support their regular party nominee. District judges chosen at the same election numbered twenty Republicans, two Democrats, and six Populists or fusionists.[24]

The only southern state to hold an important election in 1895 was Mississippi. There the campaign was hot between the silverites and the " sound money " men, the advantage in numbers all lying with the former. At first the administration Democrats tried to urge that the silver question, being important in a national way only, should be excluded from consideration, but the fact that the legislature to be chosen must select a United States senator undermined that argument. Cleveland himself finally intervened in the campaign with a letter setting forth his views, and the office-holding administration Democrats with the support of many local newspapers put up a stiff fight. But the drift towards silver was unmistakable. The county conventions were overwhelmingly for silver, and finally the state convention also.

In giving its support to the Populistic doctrine of free coinage, however, the Democratic state convention declared itself on the question of the Populist party in no uncertain terms. " We firmly believe," it resolved,

that the Democratic party of Mississippi and of the Union affords the only security for the maintenance of white supremacy in the state, and that all political movements which tend to divide or weaken our party are in direct conflict with the true interests of the state and dangerous to the welfare and happiness of both races. We therefore deplore the movement now afoot for the organization of a third party, to be known as the People's party, and we conjure all good men and true patriots to stand by the Democratic party. [25]

[24] *Appletons' Annual Cyclopaedia, 1895,* 523; Barnhart, "Farmers' Alliance and People's Party in Nebraska," 381–385.
[25] *Appletons' Annual Cyclopaedia, 1895,* 497.

But the Populists nevertheless nominated candidates and adopted a thoroughly "Populistic" platform, which demanded "laws to guarantee fairness and honesty in elections." Whether because of the absence of such laws and a will to enforce them or merely because Democratic voters were in the majority, the Populists lost the election. The official vote for governor stood 48,873 to 17,466. There were no Republican candidates, and presumably the Republican voters supported the Populist ticket.[26]

As the year of the presidential election approached, about the only thing certain was that the money question would play a dominant role in the campaign. On this issue the Republicans would be certain either to stick by the gold standard or to trim with some kind of complicated statement on bimetallism. The Democrats would be split into two warring factions, gold Democrats, mostly from the East, who supported Cleveland, and silver Democrats from the West and South, who denounced him bitterly. Which faction would win in the national nominating convention was long a moot question, although predictions were common in the West and South that the silver men would be able to dictate both the platform and the candidate. The Populists were likewise split. The old-timers among them, at least those who had not become officeholders, were generally in favor of standing by every plank in the Omaha platform. The newer converts, aided and abetted by many of the third-party leaders who had achieved office, or now hoped to achieve office, favored yielding to the popular clamor for silver. On a strictly silver platform they could poll a greater number of votes than on an old-time Populist platform.

In these troubled waters the simon-pure silverites, led by the American Bimetallic League and supported by many local silver organizations, fished earnestly. They would split the Republican party if they could and capture the silver wing; they would commit the Democratic party to silver

[26] *Ibid.*, 498.

and a silver candidate; they would trim off the Populist "isms," holding the third party down to a "silver-only" policy; and, finally, they would secure the fusion of all the silver forces in the country to elect silver candidates all along the line from president to constable. "It may be," a wavering editor wrote, "that this is the way it has been decreed that humanity shall advance a step." [27]

The Populists considered carefully what strategy to adopt in making their nominations in the campaign of 1896. The southern antifusionist wing advocated a convention well in advance of the Republican and Democratic gatherings, perhaps as early as the twenty-second of February — a significant date in Populist history. "The sooner we get into line and march upon the enemy, the better will be our chances of scoring a victory," wrote a Texas editor.[28] But the Populist national committee, meeting in January, 1896, decided that the gold-bugs would dominate both the old-party conventions and that if the People's party met last it would then have "the easy task of gathering all the bolting silverites" together into the third-party camp.[29] So instead of February 22, July 22 was chosen as the date for the Populist convention to assemble.

This was not an unreasonable decision. The Republicans could not possibly come out for silver, and the administration Democrats, with the help of the president, might easily control the Democratic convention, regardless of the silver sentiment in West and South. B. O. Fowler of the *Arena*, one of the few prominent magazines to support the silver cause, wrote to Senator Allen about the time the Populist national committee met that the Democrats would make an effort to nominate Cleveland or Olney or, failing that, someone else "who will carry out Wall Street's plans." [30]

[27] *Broad-Axe* (St. Paul), July 2, 1896.
[28] *Southern Mercury*, October 10, 1895.
[29] H. D. Lloyd, "The Populists at St. Louis," *Review of Reviews*, 14:300; *St. Louis Globe-Democrat*, July 25, 1896.
[30] B. O. Fowler to Allen, January 24, 1896, in the Allen Papers.

Unfortunately all such Populist predictions were wrong, and those responsible for the late convention were never forgiven by the men who had wished the third-party candidates to take the field first — " a position to command respect, if not dictate terms, instead of a position demanding compromise." [31]

Well before the date set for the opening of the Republican convention, June 16, the Populist leaders received positive assurance that Teller and his silver friends would bolt. Taubeneck, therefore, and a number of other prominent Populists put in their appearance at St. Louis, determined to persuade the bolting Republicans to throw in their lot with the third party. But the Populists were not alone in their desire to obtain help from the silver Republicans. " The Democrats had a large and influential lobby here," Taubeneck wrote to Donnelly, who had found it impossible to be present, " moving heaven and earth to get the bolting Republicans to join the Democratic party and go to the Chicago Convention. Bryan was here the entire week. Bland also had a strong lobby on the ground. We got in touch with the bolting Republicans before the Convention opened, and agreed upon a policy " that permitted the bolting Republicans to maintain a provisional organization; planned fusion with the Populists in the western states on electors, congressmen, and state and local tickets; and proposed that the silver Republicans join the Populists in their July convention. " I think we have received the full benefit of the Republican bolt for our party in the future," Taubeneck concluded. " The Democrats are exceedingly sore; especially the silver wing of the party. They have the arrogance to claim that the bolting Republicans ought to join them and that the Populists ought to endorse their National ticket." [32]

But the definite stand taken by the regular Republicans in favor of the single gold standard worried the Populists.

[31] *Southern Mercury,* July 30, 1896.
[32] H. E. Taubeneck to Donnelly, June 10, 22, 1896, in the Donnelly Papers.

The logic of the situation called for a Democratic indorsement of free silver at Chicago, and the drift of Democratic
sentiment in that direction seemed unmistakable. Should
the gold wing win control at the Chicago convention, the
Populist course at St. Louis would be easy, but suppose the
Democrats should name a silver candidate upon a frankly
silver platform? What then could the Populists do in the
face of the nation-wide demand that the silver forces must
unite?

Taubeneck, foreseeing this situation and realizing that an
effort would be made by the Democrats "to swallow the
People's party," conferred with the Populist leaders on the
best course to pursue. Donnelly's advice, vouchsafed even
before the Republican nominations had been made, was to
induce the Democrats not to make nominations at Chicago
but to meet for that purpose with the Populists at St.
Louis.[33] This was a fantastic suggestion, for beyond a doubt
the Democrats would nominate their own candidates in their
own convention. In general, however, western Populists were
eager to get together with the Democrats on any possible
terms. The southern Populists were obdurate, insisting that
the third-party organization be maintained at all costs. Taubeneck told the southerners that he " would go as far as they
could to accomplish that end," [34] but he apparently decided
that the only way to do it was for the Populists to demand
that the Democrats nominate Teller. "We must take a firm
stand that we will not endorse the nominee of the Chicago
Convention if he should be a straight out Democrat," Taubeneck wrote to Donnelly. "They must take Teller or be
responsible for the division of the silver forces at the polls
next November." [35]

From the Populist standpoint, there was much to be said

[33] Indorsement on letter from Taubeneck to Donnelly, June 10, 1896, in the
Donnelly Papers.
[34] St. Louis Republic, July 22, 1896.
[35] Taubeneck to Donnelly, June 20, 1896, in the Donnelly Papers.

in support of the Teller candidacy. Teller's dramatic withdrawal from the Republican convention had advertised him widely and had won great applause from all silverites, regardless of party. He would be far more acceptable to the southern Populists, who had long been allied with the Republicans in local politics, than any Democrat could possibly be. His nomination, moreover, would insure the support of all silver Republicans, whose assistance might be sorely needed, for in case the Democrats and Populists should agree upon any reliable silver man, the gold Democrats would certainly go over to the Republicans. Moreover, with Teller as their candidate, the Democrats and the Populists would both be in the position of having nominated an outsider to head their respective tickets. Neither could look down upon the other. Both organizations could be maintained; separate platforms could be written. Bland, on the other hand, whom the silver Democrats seemed disposed to support, was a strict party man and was regarded by some of the Populists as " especially unsafe upon the questions of national banks and the government issue of all paper money." [36] " Suppose the Democrats give up their favorite sons Boies and Bland," wrote a Minnesota editor, " and the People's party forego the pleasure of nominating their great genius and splendid patriot Ignatius Donnelly, and all unite in the nomination of Henry M. Teller for president? " [37]

There was opposition to Teller, however. It was urged that he represented only the silver corporations, who owned him " soul, body and breeches " and required him to take the step he took at St. Louis. He was not a Populist, never had been, and probably never would be. He had no more interest in the vast majority of the Populistic reforms than if he had been a Democrat or a regular Republican instead of a bolter. Furthermore, his chances of securing either the

[36] W. L. Hand to Allen, June 27, 1896, in the Allen Papers.
[37] Broad-Axe, June 25, 1896.

Democratic nomination or the Populist nomination were not bright. When Taubeneck and some others came out openly for Teller, a Nebraska correspondent wrote to Senator Allen, "Sorry to see those fellows of yours indorsing Teller — *won't do at all.* Tail can't wag the dog this time." [38] Robert Schilling wrote an even more emphatic disapproval to Donnelly: "I am raising hell about the Teller address. They wanted me to sign it, but I refused point blank. These fellows run off after strange gods at every opportunity. Wonder if the Gresham blunder at Omaha was not enough?" [39] Nevertheless, when Taubeneck called a conference to meet at Chicago on the eve of the Democratic convention "to see what can be done towards getting the silver democrats to unite with us or in some way cooperate so that we can all vote for one electoral ticket in the next campaign," the weight of opinion still seemed to be that Teller's name at the head of both tickets would be the best solution of the problem.[40]

The Chicago convention came over to silver, as had been expected, but Teller was not its nominee, even though Governor Altgeld of Illinois, "the Hanna of the Democratic convention," [41] seemed disposed at the outset to swing the delegates in that direction. Passing by both Teller and Bland, the convention took Bryan of Nebraska, an out-and-out Democrat, but one who had long flirted with Populistic doctrines. For vice president, however, it nominated Arthur M. Sewall of Maine, a well-to-do bank president, railway director, and shipbuilder, who had nothing whatever in common with the Populists except his belief in free silver.[42]

To say that Populism seethed and boiled after the Chicago

[38] T. B. Carr to Allen, June 26, 1896, in the Allen Papers.
[39] Robert Schilling to Donnelly, June 22, 1896, in the Donnelly Papers.
[40] Taubeneck to Donnelly, July 5, 1896, in the Donnelly Papers; *Southern Mercury,* July 23, 1896; *Review of Reviews,* 14:137–138 (August 1896).
[41] Haynes, *Third Party Movements,* 288.
[42] *Review of Reviews,* 14:140 (August, 1896). Much light is thrown on Bryan's early connections with the Populists in a master's thesis by Jesse E. Boell entitled "William Jennings Bryan before 1896," submitted to the University of Nebraska in 1929.

KILLING THE GOOSE THAT LAID THE GOLDEN EGG

[Drawn for the New York *Herald;* reproduced in the *Review of Reviews,*
August, 1896.]

results became known is to put it mildly. Populist sentiment
in the West was overwhelmingly in favor of the nomination
of Bryan by the St. Louis convention. Westerners knew that,
Democrat though he professed to be, Bryan was in fact the
product of Populism. "We put him to school," Donnelly remarked later on, "and he wound up by stealing the school-
books." [43] True, the Chicago platform did not go the whole
length of Populism, but it did indorse free silver, "without
waiting for the aid or consent of any other nation"; it condemned the bond-selling policy of the Cleveland administration; it demanded as an offset to the Supreme Court decision
invalidating the income-tax legislation of 1894 an amendment to the Constitution authorizing Congress to levy such
a tax; it urged stricter control of the railway systems by the
federal government; it denounced "government by injunction"; and it branded Cleveland's recent military intervention in the Pullman Strike at Chicago as "a crime against
free institutions." [44] The Chicago platform so far as it went
was satisfactory, and there were many to argue that the
Populists would do well to take a satisfactory candidate
running on a partially satisfactory platform with some
chance of being elected rather than to name a candidate of
their own on a platform more to their liking, with the full
knowledge that he had not the slightest chance to win.

The plea, duly fostered by the Democrats, that there
should be only one silver leader in the campaign was enormously effective in the West. Populists were reminded that
they had always professed to believe that principle should
be placed ahead of personal or party advantage. Here was
a chance to prove that they had meant what they said. Let
the reform forces be divided and the gold bugs would continue in control of the national government. Let all who
stood for silver unite, and at least this one reform stood a

[43] *Representative* (St. Paul), September 28, 1898.
[44] *Review of Reviews*, 14:141 (August, 1896).

good chance of success. "I care not for party names," said Jerry Simpson, "it is the substance we are after, and we have it in William J. Bryan." So also thought Weaver and Allen and a host of minor lights,[45] some of whom had an eye on the loaves and fishes. "Should we be able to endorse Bryan at St. Louis," one enthusiastic Nebraskan pointed out, "and continue the combination for our state offices, . . . we can make a clean sweep."[46] Practical politicians pointed out, too, that the bulk of the Populist vote in the West would go to Bryan, whether he received the Populist nomination or not. Donnelly was told by some of his Minnesota friends that nine-tenths of the Minnesota Populists would vote for Bryan under any circumstances.[47]

Others, however, feared the results of fusion and sounded a note of warning. Peffer of Kansas was afraid of it and came around to Bryan most reluctantly. "There are only two points of which I feel tenacious," he said as the Populists gathered at St. Louis. "The first is to maintain our party organization; the second is to combine the silver vote of the country."[48] A few preferred the frank abandonment of all third-party organization to fusion. "I am in favor of meeting the new democratic party half way," wrote one Nebraska Populist. "If they will give up the name 'Democratic' we should be willing to give up the name 'People's Party.' In union there is *strength*. Let us all unite as *one* party with *one banner, one name* and *no fusion*. A union is not fusion."[49] A fair number of northern Populists — intransigents such as Donnelly, Coxey, and Weller — protested against the drift towards Bryan. Weller felt that Bryan's

[45] Arnett, *Populist Movement in Georgia*, 197; Howard S. Taylor and others, *The Battle of 1900*, 508.

[46] J. G. Kruse to Allen, July 15, 1896, in the Allen Papers; *St. Louis Globe-Democrat*, July 23, 1896.

[47] Telegram of W. H. Smallwood and others to Donnelly, July 21, 1896, in the Donnelly Papers; *St. Louis Republic*, July 21, 1896.

[48] *St. Louis Republic*, July 19, 1896.

[49] M. A. Courtright to Allen, July 20, 1896, in the Allen Papers.

nomination at St. Louis would be "construed as a surrender of the People's party movement and the destruction of their present magnificent organization."[50] According to Donnelly, "the Democratic party has now moved up and taken possession of the ground we occupied four years ago. We are glad to see it. This result of the prowess of education is encouraging, but we do not propose to abandon the post of teacher and turn it over to our slow and stupid scholar."[51]

As for the southern Populists, they were almost a unit in their opposition to Bryan. This was not due to any great dislike for the man himself or for his principles. Rather they agreed with Weller that an indorsement of Bryan by the Populists would come dangerously near to ringing the death knell of the new party. The Democratic platform, they argued, was a deliberate plagiarism of Populist doctrines, designed not to be carried out but only to win third-party men back to their old allegiance. What had fusion done to the Greenback party a few years before? "Don't be deceived," one Populist editor cautioned his readers. "Experience should teach us to 'fear the Greeks even when bearing gifts.'"[52] Nor could southern Populists forget the rancor that their local campaigns against strongly intrenched Democratic machines had aroused. Men who had turned Populist had been ostracized socially, had been discriminated against in business, had suffered personal insults and even physical injuries. How could they now unite with the enemy? "For God's sake don't indorse Bryan," a Texas Populist wrote to one of the St. Louis delegates. "Our people are firm, confident and enthusiastic; don't betray their trust. Don't try to force us back into the Democratic party; we won't go."[53]

The atmosphere at St. Louis, while the delegates gathered, was heavily charged. Rumors were rife that if the con-

[50] *St. Louis Republic*, July 20, 1896.
[51] *St. Louis Globe-Democrat*, July 20, 1896.
[52] *Southern Mercury*, July 16, 1896.
[53] *St. Louis Republic*, July 20, 1896; Taylor and others, *Battle of 1900*, 508.

vention indorsed Bryan the mid-roaders would bolt and that
if it failed to indorse Bryan those who favored fusion would
bolt. The Democrats were on hand, led by Senator Jones of
Arkansas, to make sure that the Chicago ticket was indorsed.
Hanna's "paid agents" were likewise on hand, or at least
were supposed to be, in order to carry the convention against
fusion.[54]

Taubeneck was in a most perplexed state of mind. Following the Chicago convention he had been disposed to favor
Bryan, but after communicating with Populist leaders all
over the country, he had reached the conclusion that not
over sixty per cent of the Populist vote could be delivered
to the Chicago ticket and that a separate ticket must therefore be constructed. Presently, however, he reversed himself,
or, as the mid-roaders contended, he "got in out of the wet,"
fearing that otherwise he would lose his job in case Bryan
were nominated.

Plans of compromise were suggested, the most common
being for the Populists to name a separate national ticket
but to combine with the Democrats on fusion electoral tickets. After the election all the silver electors could support
the candidate of the stronger faction for president and the
candidate of the weaker faction for vice president.[55] Senator
Jones announced for the Democrats that this would never
do; but the number of mid-road delegates appeared to be
appallingly large, and clearly some concessions would have
to be made to them.

Fortunately for the fusionists there was no obvious mid-
road candidate for the presidential nomination. Donnelly
of Minnesota and Vandervoort of Nebraska were willing
enough, but they lacked followers. Debs was suggested, but
he was a Socialist, not a Populist. Allen would have suited

[54] St. Louis Republic, July 19, 21, 1896; St. Louis Globe-Democrat, July 20,
1896.
[55] St. Louis Globe-Democrat, July 19, 1896; St. Louis Republic, July 19, 21,
1896.

many, but he was known to be ardently for Bryan. Mimms of Tennessee, Towne of Minnesota, Davis of Texas, and many others were mentioned, but none aroused enthusiasm. At the end of the first day's session irreconcilables from twenty-one states met together and agreed to support S. F. Norton of Illinois for president and Frank Burkett of Mississippi for vice president; but they had very little reason to prefer these candidates over any of the others. The fusionists, by way of contrast, had only one candidate, and they knew exactly what they wanted.[56]

When the time came for the selection of a temporary chairman of the St. Louis convention, the national committee put forward Senator Marion Butler of North Carolina, as a man who might draw support from both factions. Butler shared with other southerners the view that the Populist organization must be maintained at all costs, but he had come around to the view that the Populists, by nominating Bryan outright instead of merely indorsing him, might retain their separate identity. The Bryanites would have preferred Weaver or Field as temporary chairman, and the mid-roaders talked of Donnelly, O. D. Jones of Missouri, and others; but finally all opposition to Butler died down, and he was selected by acclamation. As keynoter the young North Carolinian — he was still in his early thirties — talked too long and wearied his audience. But his protest against making the Populist party an annex to either of the old parties was well received. "Let us find the truth in the middle way," he counseled.[57]

The compromise on the temporary chairmanship merely

[56] G. A. Luikart to Allen, June 23, 1896, and Paul Vandervoort to Allen, March 11, 1896, in the Allen Papers; St. Louis Globe-Democrat, July 23, 1896; Arnett, Populist Movement in Georgia, 199–200; clippings in the Donnelly Scrapbooks, 18:83, 109, 113. A convention of the "National Silver Party of America," which met in St. Louis while the Populist convention was in progress and indorsed the nomination of Bryan and Sewall, was almost entirely ignored by the Populists.

[57] St. Louis Globe-Democrat, July 19, 22, 23, 1896; St. Louis Republic, July 22, 23, 1896.

postponed the inevitable struggle for supremacy, which came on the matter of permanent organization. After the temporary organization was completed the convention adjourned until eight o'clock in the evening to give the committee on permanent organization time to formulate its report. The mid-road element, still confident that it could control the convention, planned a great demonstration for the evening session, but on reassembling they found that the lights were off. Candles were lighted, but the demonstration fizzled out in the gloom, and Butler adjourned the convention to meet next day at ten o'clock. Twenty-five minutes after Butler had declared the convention adjourned, however, the lights once again burned brightly, and the mid-roaders, remembering that a similar maneuver was reputed to have defeated Blaine for the Republican nomination in 1876, asserted vociferously that they had been the victims of foul play. "The fusion gang manipulated the convention like an expert machine," one irate editor reported. "They even had a string to the electric light switch board, and when darkness was thought to be needed to confound the middle-of-the-roaders, the electric lights were turned off." [58]

Whatever the situation might have been had the lights not gone out, next morning the fusionists were clearly in the majority. The convention call had awarded to each state a delegate for every senator and representative it had in Congress and an additional delegate for each two thousand votes or majority fraction thereof cast for Populist candidates in 1892, 1894, or 1895, the highest vote cast controlling. This gave undue strength to the states where fusion had been undertaken at one time or another and tended to favor the West at the expense of the South.[59] Furthermore, the Bryan propaganda was exceedingly effective, and many who came as mid-roaders presently voted with the fusionists. The ma-

[58] *Southern Mercury*, July 30, August 6, 1896; *St. Louis Republic*, July 23, 1896; *St. Louis Globe-Democrat*, July 23, 1896.
[59] *Southern Mercury*, February 27, 1896.

jority report of the committee on permanent organization named Senator William V. Allen of Nebraska — a strong advocate of fusion — for the permanent chairmanship, and the minority report named James E. Campion of Maine, a somewhat obscure mid-roader. The vote on adoption of the majority report stood 758 to 564 and fairly stated the ratio of strength between the two contending factions. In taking the chair Allen called attention to the banners that read, "Keep in the Middle of the Road," urging the Populists to go even further and "occupy the whole road."[60] Perhaps his figure of speech was far-fetched, but everyone knew that he meant to advocate a union of the silver forces.

The election of Allen as permanent chairman indicated for a certainty the nomination of Bryan, and the contest between the mid-roaders and the fusionists at once shifted to Sewall. It might be argued that Bryan was a Populist in all but name, but by no stretch of the imagination could such a thing be said of Sewall. There was no sounder Populistic doctrine than that the national banks should be abolished, and Sewall was a national banker. Government ownership, or at the very least rigid government control, of the railroads had always been advocated by the Populists, and Sewall was a railway director. Add to these disqualifications such others as that Sewall was the head of one trust, the partial owner of others, an easterner, and a man of wealth, and the reasons why many Populists struck at registering any approval of him whatever seem abundantly clear. Many third-party men who were strongly for Bryan balked at Sewall. "If we elect Bryan," said one of these, "and he should be killed by some assassin by order of the money power, we would want a different man from Sewall."[61]

Nevertheless, the more influential fusionists such as Wea-

[60] *Ibid.*, July 30, 1896; *St. Louis Republic*, July 24, 1896; *St. Louis Globe-Democrat*, July 24, 1896.
[61] M. A. Courtright to Allen, July 20, 1896, in the Allen Papers; Arnett, *Populist Movement in Georgia*, 196–198; Bryan, *Memoirs*, 117.

ver and Allen advocated nominating Sewall as well as Bryan. In case this were done the Democrats were known to be willing to agree on fusion electoral tickets and Senator Jones, chairman of the Democratic national committee, promised also to give the Populists and the silver Republicans representation on a union executive committee to carry on the campaign. Moreover, the same candidate for president and different candidates for vice president would complicate the situation immeasurably. Senator Jones held that the Populists had no choice but to name both Democratic candidates or neither; and the fusionists among the Populists, mostly of this opinion also, planned to nominate Bryan at once, then adjourn until they could talk the delegates into taking Sewall.[62]

But the danger of a split in the party should Sewall receive the Populist nomination loomed larger as the convention wore on. Butler and others urged the Democrats to drop Sewall and permit the nomination of a Populist as the fusion candidate for vice president, but Jones refused point-blank to consider such a proposition. Then someone suggested that the convention reverse the usual order and name the vice-presidential candidate first. The idea spread like wildfire, and presently a minority report from the committee on rules, advocating this procedure, brought the question directly before the convention.

Naturally enough all the anti-Bryan people supported this modification of the rules. They knew that Sewall would be far easier to defeat than Bryan, and they hoped that with Sewall off the ticket they could yet persuade the convention to nominate a mid-road candidate for president. Others felt that, even if Bryan were to win the nomination for president, the Populists were entitled to their own man for vice president. With a Populist candidate actually named, probably the Democrats could yet be persuaded to withdraw Sewall. Furthermore, with at least a vice-presidential candidate of their own the Populists could preserve their own organization and would be in less danger of suffering complete assimilation by the Democrats. That the man chosen should be a southern Populist was generally conceded. When the matter was brought to a vote, the mid-road element scored a victory. There was much changing of sides, but on every count the minority report was adopted by a majority of about a hundred.[63]

This striking mid-road victory took the breath of the Democratic leaders at St. Louis. Senator Jones telegraphed hurriedly to Bryan, "Populists nominate Vice-President first. If not Sewall, what shall we do? I favor your declination in that case. Answer quick." To which Bryan promptly replied: "I entirely agree with you. Withdraw my name if Sewall is not nominated."[64] And by way of emphasis the Democratic nominee also wired to Chairman Allen, whom he knew well, "I shall not be candidate before the Populist convention unless Sewall is nominated."[65] Bryan's wire to Jones, which was made public before the vote on vice president was taken, had all the appearance of an ultimatum, and on the strength of this statement Senator Jones gave out the news that if Sewall were not named, Bryan would decline even if nominated. But Senator Allen, as chairman of the Populist convention, refused to take the threat of Bryan's declination seriously. Allen kept to himself the information he had just received, and he declined to permit an announcement of Bryan's attitude to come before the convention, which meantime had launched upon the business

[63] St. Louis Republic, July 21–25, 1896; St. Louis Globe-Democrat, July 21–25, 1896; Southern Mercury, July 30, 1896; Stewart, in the Indiana Magazine of History, 15:55–56.
[64] St. Louis Republic, July 25, 1896; St. Louis Globe-Democrat, July 25, 1896; Arnett, Populist Movement in Georgia, 200.
[65] This telegram, W. J. Bryan to Allen, dated July 24, 1896, is in the Allen Papers.

of making nominating speeches and was in the midst of a tremendous oratorical orgy.[66]

The nominating speeches put before the convention the names of Arthur Sewall of Maine, Harry Skinner of North Carolina, Frank Burkett of Mississippi, A. L. Mimms of Tennessee, Mann Page of Virginia, and Thomas E. Watson of Georgia. The contest that ensued was in a sense three-cornered. Sewall received the support of the extreme fusionists. Burkett, in particular, and probably all the other nominees except Watson were put forward by the extreme mid-roaders. Watson was the candidate of the compromisers, who, as Donnelly said in a speech seconding Watson's nomination, were "willing to swallow Democracy gilded with the genius of a Bryan" but were unable to "stomach plutocracy in the body of Sewall."[67]

The Watson candidacy was carefully planned. At the end of the first day's session a group of compromisers, foreseeing that if the Bryanites and the mid-roaders could not be brought together, the convention would be hopelessly split, wired Watson of Georgia — as extreme a mid-road Populist as ever breathed or wrote — asking him if he would consent to run on a ticket with Bryan. Watson consented, believing undoubtedly that the Democrats would withdraw Sewall and that the names of Bryan and Watson would head both tickets. Most of the compromisers at St. Louis believed the same thing, and many of them insisted that Chairman Jones of the national Democratic committee had promised them as much.[68]

When the vote was taken it became clear that Watson and the policy with which his candidacy was identified had a large majority in the convention. At sixteen minutes to one on the morning of July 25 his nomination was an-

[66] *St. Louis Globe-Democrat*, July 25, 1896; *St. Louis Republic*, July 25, 1896.
[67] *Ibid.*
[68] Arnett, *Populist Movement in Georgia*, 199–200; Brewton, *Life of Thomas E. Watson*, 267–270.

nounced. Most of the Populists were satisfied, but the Democrats sulked. "Wall Street bankers and McKinley managers wild with delight over convention's action yesterday," one New Yorker wired to Allen. "They felt crushed at prospect silver forces being combined. Today they bet ten to one on McKinley and gold." [69]

A long and typically Populistic platform, featuring as usual finance, transportation, and land, had already been adopted,[70] so that all that now remained for the convention to do was to nominate Bryan and adjourn. The one serious obstacle in the way of this program was Bryan's declaration that he would not be a candidate in case Sewall were not nominated. Allen wired Bryan asking that his telegram of declination be withdrawn, but apparently received no answer; [71] so Bryan's name went before the convention along with that of S. F. Norton, on whom the mid-roaders had united for a last stand. While the vote was being taken, however, word was passed about that Allen had just received and had refused to communicate to the convention a telegram announcing that Bryan would not run if nominated. One delegate asked Allen point-blank if this were true. Allen promptly replied that it was not.[72] Meantime Governor Stone of Missouri asked permission to read to the convention a message that he had received from Bryan, but Allen refused to give Stone the floor, explaining later that the Populist convention could have nothing to do with a purely Democratic negotiation. When the vote was finally counted it showed 1,042 for Bryan to 321 for Norton, with 12 votes scattered. Doubtless as a means of insurance against what

[69] W. B. Jaxson to Allen, July 25, 1896, in the Allen Papers; *Appletons' Annual Cyclopaedia, 1896*, 768.

[70] Edward Stanwood, *A History of the Presidency*, 1:551–554.

[71] *St. Louis Republic*, July 25, 1896.

[72] Years later Allen declared that the telegram that he was accused of putting in his pocket without opening or reading to the convention "was from an old gentleman of Resaca, Georgia . . . and not from Mr. Bryan at all." *Illinois State Register*, July 6, 1904.

Bryan might do, Allen asked and received for the national committee permission to exercise plenary power after the adjournment of the convention. To some observers, however, it appeared that this right of the committee to do all things that the convention might do if it were in session was obtained in order to make it possible later on to "roll" Watson in favor of Sewall.[73]

Apparently there were very few Populists who were entirely satisfied with the outcome at St. Louis. Western Populists were happy at the nomination of Bryan and extremely appreciative of the part Chairman Allen had played in bringing about that result, but they regretted that Sewall's name had been left off the ticket and were deeply concerned as to how they could manage with two vice-presidential candidates.[74] Southern Populists doubted the wisdom of the Bryan nomination and denounced Allen, whom they blamed for bringing it about, in unmeasured terms. "Never did Reed of Maine exercise more autocratic power over the lower house than did Allen in his capacity as presiding officer," one southern editor complained.[75] With Watson the southerners were satisfied, however, and to make way for the fiery Georgian they demanded vociferously that Sewall either resign his place on the ticket or be "taken down." A common opinion was that "Mr. Bryan must take his choice of a running mate. If he is a friend of the people, as he claims to be, he must turn from the millionaire banker and railroad magnate to the poor man, whose heart beats in sympathy with the common people." [76]

Northern mid-roaders found little comfort in anything that the St. Louis convention had done. Some of them held

[73] St. Louis Republic, July 26, 1896; St. Louis Globe-Democrat, July 26, 1896.

[74] Theodore W. Ivory to Allen, July 27, 1896; Matt Kean to Allen, August 3, 1896; R. F. Pettigrew to Allen, August 1, 1896; all in the Allen Papers.

[75] Southern Mercury, August 6, 1896; see also E. C. Baldwin to Allen, December 13, 1896, in the Allen Papers.

[76] Southern Mercury, August 6, 13, 1896; Arnett, Populist Movement in Georgia, 203–206.

a meeting immediately after the St. Louis convention had adjourned and passed a resolution that Bryan's unsuccessful rival for the Populist nomination, S. F. Norton of Chicago, should be declared the regular nominee in case Bryan failed to accept within thirty days. They were powerless, however, to enforce their resolution and could only lament the unhappy fate that had overtaken their party. "Our convention never should have been postponed until after those of the old parties with the object of catching the crumbs that might fall from their tables," a correspondent wrote to "Calamity" Weller. Donnelly probably felt worse than most of the mid-roaders, for when they had talked at St. Louis of making him their candidate, even the Minnesota delegation had gone back on him. "Our soul is weary of this whole business," he mourned. "We shall retire to our library. . . . In the domain of literature we have a realm of our own . . . and in it we will bestow the remaining years of our life." [77]

Under the circumstances it was not easy for the Populists to organize for the campaign. The St. Louis convention thoughtfully left to the national committee, rather than to the presidential nominee, the task of choosing its own chairman, and the committee replaced Taubeneck with Butler of North Carolina.[78] At once the problem of notifying the successful nominees presented itself. How was Bryan to accept the Populist nomination after having accepted the Democratic nomination? On certain items the two platforms were contradictory. To stand on both of them the versatility of the Nebraskan would be sorely taxed, for he would have to favor redeemable as well as irredeemable paper money and government control of the railroads no less than government ownership.[79] Furthermore, how could Bryan, after all he had said, accept the Populist nomination when to do so im-

[77] G. W. Everts to Weller, July 26, 1896, in the Weller Papers; *St. Louis Globe-Democrat*, July 26, 1896; *Representative*, July 29, 1896.

[78] *St. Louis Globe-Democrat*, July 25, 1896; *Southern Mercury*, August 27, 1896.

[79] *Southern Mercury*, July 30, 1896.

plied his approval of Watson for vice president? Bryan himself passed the word along to Senator Allen, who as chairman of the Populist convention would naturally head the notification committee, that perhaps the notification might best be sent by mail. Further, since the Populists had nominated him after his nomination on the Chicago platform, they could not ask him to go beyond that platform or to abandon Mr. Sewall.[80] Ultimately Butler announced that the People's party custom was not to notify its nominees and that no notifications would be made. Bryan therefore never accepted or rejected the Populist nomination. Watson unofficially accepted his nomination many times and refused to get out of Sewall's way, although often besought to do so. The Democratic idea of fusion, he insisted, was "that we play Jonah while they play whale."[81] Small wonder that a correspondent urged Allen to "try and shut off Tom Watson. His egotism and injudicious talk is hurting us."[82] But Watson talked on.

Under the leadership of Butler the Populist campaign was conducted in much closer harmony with the Democrats than the attitude of the St. Louis convention would seem to have justified. Fusion tickets of electors were arranged in twenty-eight states, the number of places allotted to Democrats and Populists following as closely as possible the ratio of votes polled by the two parties in the preceding election. Generally the Democrats got the larger number of electors, only seventy-eight Populist names occurring on the fusion tickets to one hundred and ninety-eight Democratic names.

In the South, where the only fusion that the Populists had ever known had been fusion with the Republicans, " Popocrat " tickets were hard to obtain, but five southern, or near-southern states — Missouri, Arkansas, Louisiana, Kentucky, and North Carolina — succeeded in effecting such a com-

---

[80] J. A. Edgerton to Allen, August 28, 1896, in the Allen Papers.

[81] *Southern Mercury*, August 27, 1896; Arnett, *Populist Movement in Georgia*, 204.

[82] L. E. Lincoln to Allen, September 13, 1896, in the Allen Papers.

bination. In the others the Democrats would give no quarter to the Populists, and even in Watson's own state, Georgia, the Populist ticket was finally withdrawn. Watson's insistence on remaining in the campaign was severely criticized, particularly in view of the fact that, even if Bryan should have been elected, there would probably have been no majority for any one of the three candidates for vice president, and the Republican Senate would surely have chosen Hobart in preference to Sewall.[83]

In those western states where local elections were held in 1896, fusion on state and congressional tickets followed national fusion as a matter of course. This was true not only in the states of the Middle Border, Kansas, Nebraska, the Dakotas, and Minnesota, but also in some of the farther western states, such as Montana, Washington, and Idaho. In Colorado the unhappy experience of the state under Waite militated against any combination of Populists and Democrats. The best the Populists could do was to secure the help of the National Silver party; while the bulk of the Republicans, as silver Republicans, joined forces with the Democrats. In the southern states fusion between the Populists and the Democrats on anything more than the national ticket was altogether too much to expect, except, perhaps, in South Carolina, where the Tillmanites held their lines intact. In Missouri, which may hardly be considered a southern state, the Populist candidate for governor withdrew, and his place on the ticket was not filled. Presumably this was to further the chances of Democratic success. In Arkansas the Populists ran candidates of their own for state offices but made no nominations for Congress. In Tennessee efforts to achieve fusion between the Democrats and the Populists were made, but they were unavailing.[84]

In a number of the southern states the Populists accom-

[83] *Appletons' Annual Cyclopaedia, 1896,* 770; Stanwood, *History of the Presidency,* 1:564–565; Arnett, *Populist Movement in Georgia,* 203.
[84] *Appletons' Annual Cyclopaedia, 1896,* under names of states mentioned.

plished "what a circus poster would advertise as a great dual feat — they fused with both the Democrats and Republicans. They voted the national ticket with the Democrats and the state ticket with the Republicans."[85] This was not so difficult in Lousiana in view of the fact that the state election came early in the year and was over before Bryan was nominated. A number of Louisiana sugar planters who were out of sorts with the Democratic party because of the tariff law of 1894 joined with the Populists and the Republicans to nominate a full state ticket, which the regular Democrats were able to defeat only by use of the most drastic methods. In some of the parishes there was riot and bloodshed on election day, and the fusionists claimed that they were deliberately counted out.[86] In Alabama the state executive committee of the People's party had met early in the year to consider a program of fusion in the state with the Republicans. "There are some kickers in our ranks against it," one who favored it wrote to Senator Allen. "*Without it we are helpless.*"[87] A strong minority of Alabama Republicans also opposed fusion, but the majority agreed to it, and a complete fusion ticket was finally named for state offices, although in most of the congressional districts each party went its own way.[88] In Georgia no Republican state ticket was nominated. The Republican executive committee advised the Republican voters that they were at liberty to choose as individuals between the Populist and the Democratic state tickets; but the chairman of this committee later issued an unofficial circular urging all Republicans to support the Populist ticket.[89] A somewhat similar arrangement existed in Texas, and in North Carolina the Populists and Republicans supported the same candidates for all state offices except those of gov-

[85] *Nashville American*, July 8, 1897.
[86] *Appletons' Annual Cyclopaedia, 1896*, 424.
[87] W. S. Reese to Allen, January 17, 1896, in the Allen Papers.
[88] *Appletons' Annual Cyclopaedia, 1896*, 11; *Tribune Almanac, 1897*, 228.
[89] *Appletons' Annual Cyclopaedia, 1896*, 313.

ernor and lieutenant governor.[90] In spite of this apparent
disagreement with regard to the leading places on the state
ticket, fusion in North Carolina actually went much further
than elsewhere. In nearly all of the congressional districts,
fusionist candidates received the support of both parties,
and in most of the counties combination tickets for membership in the state legislature were arranged. Fusion seems
ordinarily not to have extended, however, to candidacies for
county offices.[91]

In general, these various fusion measures resulted from a
popular demand and were well received by the rank and file
of third-party adherents. Men who had fought against one
another for years rejoiced to "lay aside their petty differences and unite for the cause of humanity."[92] Bryan was
pictured as less a Democrat than a Populist and his nomination as evidence that, at least along national lines, the Democratic party was coming around to the tenets of Populism.
Reform journals vied with orthodox Democratic sheets in
their devotion to the "silver-tongued orator of the Platte"
and the doctrines for which he stood. "I feel this year we
are squarely aligned," one of Allen's correspondents rejoiced,
"the mass against the class, the honest yeomanry of the
land against the pampered owners of wealth, grown arrogant
by sanction of law. I hope and trust the time of our emancipation has come."[93]

There were nevertheless some doubting Thomases who
questioned the sincerity of the Democrats and feared that
the chief result of the campaign would be the destruction of
the Populist party.[94] Especially was this true in the South,
where the unscrupulous methods that the Democrats had
long used against their adversaries were not easily forgotten.

[90] *Ibid.*, 537, 733.
[91] Delap, *Populist Party in North Carolina*, 60–64.
[92] *Broad-Axe*, August 20, 1896.
[93] Matt Kean to Allen, August 3, 1896, in the Allen Papers.
[94] Stewart, in the *Indiana Magazine of History*, 15:56–61.

"No consent on the part of the People's Party voters of Texas to support Mr. Bryan," ran one Southern pronouncement, "can be construed to mean an endorsement of the corrupt gang in Texas." Southern Populists were also much offended by the unwillingness of the Democrats to accept the Populist candidate for vice president. "If Bryan does want our support," the same editor remarked, "he ought to be willing to adopt the policy which will most certainly assure him of that support, and that policy is to accept as a running mate a southern Populist, named by the unanimous vote of the Populist national convention."[95]

On one thing the Populists scored a complete victory. Their crusading ardor was communicated in full to the Democrats, who, under Bryan's leadership, treated the country to the most spectacular campaign it had ever seen. Bryan himself traveled over thirteen thousand miles, visited two-thirds of the states in the Union, and made during a period of fourteen weeks no less than four hundred speeches. Lesser lights, Democrats and Populists alike, imitated him to the best of their ability. Newspapers teemed with news of the campaign, and editorial writers could think of little else to write about. The whole campaign was marked by a tensely earnest study of the money question. According to Professor Woodburn of Indiana University there had not been such a tremendous interest in any political issue on the part of the masses since the Civil War. "On the street, in the schoolhouse meeting, in the debating club, wherever several are gathered together, the money question has been seriously discussed. Voters have sought anxiously to know the truth. Men who have never thought about the money question before have given it their earnest attention." Professor Macy of Grinnell noted this same "intense and unusual interest in the debate. Men and women sit for hours listening to a

[95] *Southern Mercury*, August 6, 1896.

presentation of facts and statistics. . . . Wherever men meet in shop or by the way they engage in financial discussion." [96]

It was the tenacious insistence of the "Demopop" campaigners that forced the issue along this line. At the beginning of the canvass the Republicans tried other tactics. Ridicule was a standard weapon, one that William Allen White tried in his famous editorial, "What's the matter with Kansas?"

We all know; yet here we are at it again. We have an old moss-back Jacksonian who snorts and howls because there is a bath-tub in the State House. We are running that old jay for Governor. We have another shabby, wild-eyed, rattle-brained fanatic who has said openly in a dozen speeches that "the rights of the user are paramount to the rights of the owner." We are running him for Chief Justice, so that capital will come tumbling over itself to get into the State. We have raked the ash-heap of failure in the State and found an old human hoop-skirt who has failed as a business man, who has failed as an editor, who has failed as a preacher, and we are going to run him for Congressman-at-large. He will help the looks of the Kansas delegation at Washington. Then we have discovered a kid without a law practice and have decided to run him for Attorney-General. Then for fear some hint that the State had become respectable might percolate through the civilized portions of the nation, we have decided to send three or four harpies out lecturing, telling the people that Kansas is raising hell and letting corn go to weeds. [97]

But the Populists had been called names before, and at least for the moment the Democrats were more interested in principle than in persons. The Republicans also tried to draw off the opposition into a discussion of the tariff, but all such efforts proved fruitless. The only thing left to do was to meet argument with argument, and this the Republicans finally did. Such men as Roosevelt and Hanna toured the

---

[96] *Review of Reviews*, 14:524–526 (November, 1896).

[97] Peck, *Twenty Years of the Republic*, 524, quoting *Emporia Gazette*, August 15, 1896.

Populist areas and reasoned earnestly with audiences that had already studied the financial question and understood what the orators were talking about. Tons of literature were distributed, much of which was carefully read and pondered. The fallacies of the silver arguments were painstakingly exposed, and sound money advocates vied with "Professor Coin" in the aptness of their illustrations and the lucidity of their statements. Probably also there was much intimidation of voters. Certainly the silver forces believed that the employes of "railroads and other great corporations and industrial establishments" were told that, if Bryan won, their wages would be cut and their jobs might be forfeited. Rumors were rife, too, of the unlimited financial resources that the Republicans, under Hanna's leadership, would have at their disposal on election day. But the fusionists themselves, with at least one vested interest — the silver miners — on their side, did not lack entirely the sinews of political war.[98]

The election results were a great disappointment to the fusion forces, who had confidently expected to win. Bryan's vote in the South and the West was magnificent, but even so he lost five states west of the Mississippi River and four states south of Mason and Dixon's Line.[99] Could he have won all or most of these, as optimists among his supporters had hoped, he would have won the election without a single electoral vote from the region east of the Mississippi and north of the Ohio. The Bryan and Watson ticket polled only a little over two hundred thousand popular votes, but the Populist strength was not fairly indicated by this figure, since many Populists voted for Bryan and Sewall. Watson finally received twenty-seven votes in the electoral college,

---

[98] *Farm, Stock and Home* (Minneapolis), 13:2 (November 15, 1896); *Review of Reviews*, 14:522–526 (November, 1896); Herbert D. Croly, *Marcus Alonzo Hanna; His Life and Work*, 335–345; Arnett, *Populist Movement in Georgia*, 211. A remarkable collection of pamphlets on the money question, in part personally collected by E. Benjamin Andrews, is in the library of the University of Nebraska.

[99] *Review of Reviews*, 14:644 (December, 1896).

Can the American producer, already heavily weighed down, stand the additional burden of the Permanent Gold Standard?

A POPULAR SILVER POSTER

[From the *Review of Reviews*, November, 1896.]

although some regularly chosen Populist electors deserted him for Sewall.[100] In Congress the Republicans won a comfortable majority in both houses over all other parties. The Populists, however, showed an increase. Some twenty-five members of the new House of Representatives classified themselves as Populists, silverites, or fusionists, and in the next *Congressional Directory* six senators were designated as Populists.[101]

In the West, wherever fusion tickets for state office had been agreed upon, the odds were strongly in favor of the combination. In Kansas, Nebraska, South Dakota, Montana, Idaho, and Washington the fusionists swept everything before them, usually electing a majority in both houses of the state legislature, as well as all the state executive officers. In Colorado, however, the Democratic–Silver Republican ticket emerged far in the lead, and in Minnesota and North Dakota the triumph of the Republicans was complete.[102]

In the South the attempts of the Populists to combine with the Republicans on state offices were more successful than might have been expected under the circumstances. In Georgia, Alabama, and Texas the Populists, with the help of the Republicans, were able, according to the election returns, to poll over forty per cent of the votes cast. But for the frauds, which were as apparent as in the two preceding campaigns, the Populists were confident that they would have won. Tillmanism carried the day in South Carolina as usual; and the combination of Republicans and Populists won a sweeping victory in North Carolina.

The election of 1896 in North Carolina constitutes one of the few outstanding episodes in the history of southern Populism. The legislature of 1895, which had a Republican-Populist majority in both houses, had passed a drastic elec-

[100] Stanwood, *History of the Presidency*, 1:561–568.
[101] *Congressional Directory*, 55 Congress, Extraordinary Session, 161, 163.
[102] *Tribune Almanac, 1897*, 228 ff.

tion law to prevent corruption at the polls. Under the terms of this act, which a Democratic governor had signed, each party was guaranteed a representation on all registration and election boards, and heavy punishments were prescribed for both vote-buying and vote-selling. The law worked reasonably well, and for perhaps the first time since the emergence of the "solid South" non-Democratic voters in a strictly southern state had a chance to show approximately their full strength. The Republican candidates for governor and lieutenant governor were elected, and also the full fusionist slate for the minor state offices. Three Republicans and four Populists out of a total delegation of nine were chosen to Congress, and a Populist-Republican majority was returned in both houses of the state legislature.[103]

So far as the number of Populists elected to office in 1896 was concerned, the third party had done very well, but there was nevertheless abundant reason for discouragement. Populists had been urged to drop everything else for a joint campaign with the Democrats to secure the free and unlimited coinage of silver. They had been assured that, if only they would do this, victory would crown their cause. They had made the sacrifice, had subordinated their party to principle, as requested; but the victory was lacking. Thousands of Populists, moreover, judging by the popular vote, had deliberately gone over to the Democrats, and thanks to the welding influence of a heated campaign against a common enemy, they bade fair to stay where they were. Also, the third-party machinery had suffered from too much Democratic tinkering and would be difficult to repair. Optimists there were to shout that the reform forces were not downhearted, that "just four years after the defeat at Bull Run, General Lee signed the treaty of peace at Appomattox,"[104] but the pessi-

---

[103] Delap, *Populist Party in North Carolina*, 57–58. See also under names of states in *Appletons' Annual Cyclopaedia, 1896*, and in the *Tribune Almanac, 1897*.

[104] W. J. Reinke to Allen, November 7, 1896, in the Allen Papers.

mists were much more in evidence. They blamed Butler for his campaign policy of working hand-in-glove with the Democrats. They blamed Bryan and the Democrats for their heartless disregard of Watson. They blamed themselves for ever consenting to an unholy alliance with the enemy. And they conceded freely that the Populist party as a great and independent organization was a thing of the past.

"SOME OF THE 'ANARCHISTS' WHO RAISE OUR WHEAT"

[Drawn for the New York *Journal*; reproduced in the *Review of Reviews*, September, 1896.]

CHAPTER FOURTEEN

"KEEP IN THE MIDDLE OF THE ROAD"

THE news of Bryan's defeat was followed promptly by an agitation on the part of the mid-road Populists against any future indulgence in fusion. The leadership of this anti-fusion movement was speedily assumed by the National Reform Press Association, which was composed of the more radical editors from all over the country. Many of these editors had supported Bryan, but for the most part they had done so with great reluctance. It had not been easy for a reform editor to retract views that he had long heatedly proclaimed, and it had been even more difficult for him to contemplate with serenity the drift of Populist subscribers into the devouring jaws of an only slightly regenerated Democracy.

Paul Vandervoort, the editor of a radical sheet published at Omaha, Nebraska, was president of the National Reform Press Association, and when he announced that the " supreme duty of the hour " was to find a way out of fusion and back to the original Populist faith and practice, he represented well the sentiment of the organization over which he presided.[1] On the twenty-second of February the National Reform Press Association held a meeting at Memphis, Tennessee, in which the chief subject of discussion was how to salvage the greatest possible number of Populists from the

[1] *Southern Mercury* (Dallas), February 25, 1897.

380

holocaust of fusion. The Association finally voted that as a first step the Populist national committee should be asked to arrange for a delegate convention to be held not later than July 4, 1897. At this convention the future policy of the party should be decided.[2]

Senator Butler, chairman of the Populist national committee, felt no particular enthusiasm for this plan. His course during the campaign had been caustically criticized by the mid-road element, which was now demanding insistently that his place as chairman be taken from him. Accordingly, he made no effort to call a meeting of the national committee, as requested by the editors. He did, however, get together the entire Populist delegation in Congress for a conference on the subject. These individuals, elected in the main by fusion arrangements, were apparently in complete accord with Butler. They feared that a conference on the state of the party would result in much friction between the mid-road and the fusionist elements, and they strongly advised against it.[3] The reform editors had apparently foreseen some such result as this, for they had empowered a committee to act in case Butler would not; and presently this committee, in session at Girard, Kansas, on April 15, 1897, issued an address to the Populists, containing a call for a national conference of the party to assemble at Nashville, Tennessee, on July 4, 1897.[4]

Hundreds of delegates and observers appeared at Nashville on the appointed date, most of them, however, hailing from southern states. It was at once evident that the sentiment of the conference would be strongly against fusion in the future — one delegate described the meeting as " a suit for divorce " from the Democracy. According to Donnelly, who took a prominent part in the proceedings, the Demo-

[2] *Ibid.*, March 11, 1897.
[3] *Ibid.*, March 25, 1897.
[4] *Ibid.*, April 22, 1897; January 20, 1898.

cratic party might be "fit to assist in handling a single temporary side issue like free silver, but the welfare of mankind
and the interests of all the ages demand the continued existence of the People's Party." Free silver, however, seemed to
be almost as much discredited among the delegates as was
fusion. Someone remarked that it had "lost its stinger" and
was no longer the universal panacea that would end the
country's woes. It was even suggested that the subject might
appropriately be eliminated entirely from the Populist platform. More important, however, was the task of organizing
to prevent adoption of a fusionist policy in any subsequent
campaign. Finally a committee was appointed to be formed
of three representatives from each state having delegates
present and to be known as the National Organization Committee. By this action the Nashville conference really created a separate mid-road Populist organization, charged
with the duty of watching the regular organization and cooperating with it if possible but prepared to go its own way
should necessity demand. An address, obviously from Donnelly's pen, was also issued, proclaiming that "the people's
party was born to live and not to die." [5]

On call of its own executive committee the National Organization Committee held two sessions at St. Louis, one in
November, 1897, and one in January, 1898. The call for the
January meeting had also invited the attendance of the
national committee of the Populist party, but Butler chose
to ignore the invitation, although some of the regular committeemen apparently went to St. Louis unofficially of their
own accord. The mid-road committee, claiming that the will
of the people had been too much ignored by Populist leaders,
now proposed a referendum, to be voted on by township
and county conventions of Populists in May and by state
conventions in June.

The questions referred were, first, should a national con-

vention of the party be held? And, second, if such a convention were to be held, should it be on July 4, 1898, May 26,
1899, or February 22, 1900? The state conventions were requested to elect delegates to the national convention on the
usual basis; and the national committee of the Populist party
was requested to convene in June, 1898, in order to carry out
the mandate of the voters. The national committee was also
asked to devise machinery by which in the future important
questions of party policy might be submitted on referendum
direct to the people.[6]

The threat of independent mid-road action could be ignored during the year 1897, since that was not an election
year, but the approach of the mid-term elections of 1898
forced the regulars to take cognizance of what was going on.
It was quite impossible to tabulate the results of the so-called
referendum submitted by the National Organization Committee; but plenty of evidence was at hand to show that the
insurgent mid-roaders had strong support. The *Southern
Mercury*, for example, advocated a convention on the Fourth
of July, 1898, and asked, "Shall we wait until it is too late
to reform our shattered ranks before moving into action, or
shall we not?" Should the regular committee fail to call such
a convention, the National Organization Committee was
urged to issue the call on its own authority.[7]

Butler finally decided to call a meeting of his committee,
ostensibly to hear reports, recommend policies in the congressional campaign, organize speakers and writers, and so
on, but actually to see if some plan of action could be devised that would satisfy the National Organization Committee. Ultimately both committees were called to meet in
Omaha, Nebraska, on the fifteenth of June. This city was
an attractive meeting place in the summer of 1898 because
it was the scene of the Trans-Mississippi Exposition, then in

[6] *Southern Mercury*, November 25, 1897; January 20, 1898.
[7] *Ibid.*, January 27, 1898; February 10, 1898.

progress. The National Reform Press Association and the Nebraska Farmers' Alliance both found occasion to call meetings at Omaha for the middle of June. So the place was full of Populists when the two rival national committees came together.[8]

It was a stormy meeting. The Populist national committee contained within its membership many mid-road sympathizers who were in hearty accord with the plans of the National Organization Committee. The contest, therefore, took place within the national committee, with the National Organization Committee, in theory at least, merely observing what was going on and holding itself in readiness to act in case the regulars failed to yield to the mid-road demands. In fact, however, the National Organization Committee used its best efforts to secure control of the national committee.

The fight began in the credentials committee, which worked a day before it could make a report. Only twenty-seven national committeemen were actually on hand, but all the rest seemed to be represented by proxies, some evidently by more than one. The report that the credentials committee at length presented was favorable to the fusionist element and was adopted by a vote of sixty-four to thirty-seven; hence the regular organization seemed to be safely in the hands of Butler and his fusionist friends. There was a strong disposition on their part, however, to make concessions, and it was finally agreed to appoint another committee of six, three mid-roaders and three fusionists, to confer on the existing differences and to report back a plan of compromise. The deliberations of this committee led to the famous "Omaha Contract." According to the terms of this contract, it was agreed that the national chairman should not lend his influence or aid in national, state, county, or municipal elections embodying fusion; and, further, that a Populist national convention should be called, not in 1898

[8] *Omaha World-Herald,* June 13–15, 1898.

but early in 1900, at least thirty days before the dates commonly set for the holding of old-party conventions.[9]

Undoubtedly the framers of the " Omaha Contract " supposed that they had entirely satisfied the National Organization Committee, but some of the mid-road leaders refused to be bound by what had been done and promptly issued from Omaha the long-threatened call for an early convention to restate the Populist creed and to name candidates for the election of 1900. September 4, 1898, was the date set for the meeting of this convention, and Cincinnati was designated as the place. The score or so of mid-roaders who signed this call were mainly from the South and alleged as an excuse for their action that the referendum vote, asked for by the St. Louis meeting of the National Organization Committee, required it. Just how they or anyone else could know the result of this vote was not revealed. As for Butler and his committee, these officers, according to the revolters, had forfeited " all rightful claim to the leadership of the people's party " by " selling out to the Democratic party in the campaign of 1896." [10]

The convention that presently assembled at Cincinnati in response to this call represented only the extreme left wing of Populism. Each state and territory was authorized to send two delegates at large, and an additional delegate for every two thousand votes cast for Populist candidates at any election since 1890; [11] but at the opening session a reporter from the *Cincinnati Enquirer* counted only seventy-two men and four women present, while only twelve out of the forty-five states had sent delegates. Except for Ignatius Donnelly and " General " Jacob S. Coxey, there were few prominent Populists in attendance. Most of the delegates were well along in years — men who had in their earlier days been Grangers or

[9] *Omaha World-Herald*, June 15–18, 1898; *Southern Mercury*, June 23, 30, 1898.
[10] *Omaha World-Herald*, June 18, 1898.
[11] *Ibid.*

Greenbackers and were now blessed with an "astonishing assortment of whiskers."[12] By wholesale resort to proxies the slim attendance was somewhat disguised, but even so the totals in the votes taken seemed meager.

The convention adopted a long-winded review of Populist history and principles written by Donnelly in the florid rhetoric that had overtaken him in his old age. Free silver was abandoned in favor of "legal tender paper money," and the initiative and referendum were given pronounced emphasis. In the future, it was determined, the people should rule the People's party. All delegate conventions for making platforms and nominations should give way to direct vote by members of the party at primary elections. In similar fashion candidates for office and members of party committees might be recalled. The system of law-making in state and nation should also be drastically revised. All laws should be submitted to a vote of the people before adoption and should go upon the statute books only in case they found favor with a majority of the voters. On the question of immediate or deferred presidential nominations for the election of 1900, the convention broke squarely in two. About half the delegates went ahead and nominated Wharton Barker of Philadelphia for president and Ignatius Donnelly of Minnesota for vice president. The other half left the hall and later issued a protest against what the rest of the convention had done.[13]

One of the arguments that had been used in favor of a Populist national convention for 1898 was that such a gathering would stimulate interest in the state and congressional elections of that year and would perhaps help in laying plans for the campaign.[14] It could hardly be said that the Cincinnati convention had served any such purposes. Rather,

---

[12] *Cincinnati Enquirer*, September 6, 1898.
[13] *Ibid.*, September 6, 7, 1898; *Representative* (St. Paul), September 14, 1898; *Southern Mercury*, December 15, 1898.
[14] *Southern Mercury*, February 10, 1898.

it had called striking attention to the division in Populist
ranks that had appeared first at St. Louis in 1896 and with
each succeeding month had become more and more apparent. Populists were at war among themselves, calling each
other names, wildly accusing former friends and colleagues
of betraying the cause of reform. "For the last year everything seems to have gone wrong with the organization," an
Arizona Populist wrote to Senator Allen, "and we have become almost discouraged and sick at heart, and if there is
a rift in the clouds that seem to have settled down about us
that is visible from any point of view we want to know it." [15]
Undoubtedly the heated dissensions that had followed in the
wake of the election of 1896 had done much to insure this
obvious decline in Populist fortunes. "That the People's
Party is passing must be evident to all observers," former
Senator Peffer confessed. "*Why* it is going, and *where*, are
obviously questions of present public concern." [16]

Far more significant than the mere presence of dissension
within the ranks of the reformers, however, was the unmistakable fact that prosperity was returning. At first Populists were inclined to scoff at the assertions of Republican
editors that this was so. "All this newspaper talk about a
revival of business in the East, particularly in Massachusetts, is a lie made out of whole cloth," one eastern workingman wrote to Allen. "As a matter of fact business is worse
to-day here in Boston than it was before election. . . ." [17]
"Business is 'booming' in this section?" the same correspondent asked ironically.

From every town comes the news of closed factories, and continued cuts in wages. Why, my boss who was so in earnest for
McKinley's election, because it meant return of prosperity, has
given or rather will give each employee a Christmas present in

[15] J. Q. White to Allen, July 8, 1897, in the Allen Papers.
[16] Article, "People's Party," *Harpers' Encyclopaedia of United States History,*
Vol. 7.
[17] E. C. Baldwin to Allen, December 17, 1896, in the Allen Papers.

the shape of a notice that on January 1 the shop will close for good.[18]

Ignatius Donnelly wrote sarcastically in his paper about the inauguration of President McKinley:

To-morrow is the 4th of March.

McKinley will move into the White House. The reign of Confidence will begin. The Advance Agent of Prosperity will advance. Every man and woman will have plenty to eat and drink and wear. Hunger will depart forever from the land. Wheat will be worth a dollar a bushel, oats fifty cents, barley seventy-five cents. The workingman will have steady employment at big wages. Prosperity will

" Make the widow's heart to sing
Though the tear was in her eye."

There shall be no more mortgages nor taxes nor sheriff's executions. And the Populists will hang their burning cheeks with shame to think that they opposed the great and good McKinley and the greater and gooder Mark Hanna.[19]

Nevertheless prosperity was returning by the year 1898, not only in the industrial East but noticeably also in the West. Crops were better, prices were higher. Even in the South a new prosperity was in sight, depending in part upon the rapid expansion of manufacturing into that region.

Moreover, the argument that gold was too scarce to admit of being used as a monetary standard — perhaps the most potent of all the Populist doctrines — was ceasing to carry conviction. For beyond a doubt gold was becoming more plentiful. It was the irony of fate that Bryan should have staged his frenzied campaign against the crucifixion of mankind upon a cross of gold at the very time when forces were at work to undo all his best arguments. For even as Bryan fought, the production of gold, which for nearly a generation had failed to keep pace with the increased demands of business, was mounting phenomenally. What had happened to

[18] E. C. Baldwin to Allen, December 13, 1896, in the Allen Papers.
[19] *Representative*, March 3, 1897.

silver a little earlier was now happening to gold. The cyanide process, introduced in 1890, made possible the use of lean ores, previously wasted, and the extraction from the richer ores of more gold than was obtainable by the old methods. New gold fields were also opened up, in Australia, South Africa, and presently in the Klondike. The inflation of the currency, so long demanded by every variety of Populist, was taking place, but in a way that could hardly have been foreseen. Instead of the unlimited coinage of silver or the issuance of large quantities of paper money, it was the addition of enormous quantities of gold to the currency that was bringing about the expansion of the circulating medium, so long desired. The following figures on the world's annual production of gold go far towards explaining the decline of Populism:

WORLD'S ANNUAL PRODUCTION OF GOLD, 1890–1899 [20]

1890	$118,848,700	1895	$198,763,600
1891	130,650,000	1896	202,251,600
1892	146,651,500	1897	236,073,700
1893	157,494,800	1898	286,879,700
1894	181,175,600	1899	306,724,100

It was true, too, that the outbreak of the war with Spain did much to turn the public mind from the demands for reform that the Populist platforms were accustomed to recite. At least one reform editor was conscious of the danger that the war might be "used to divert the popular mind from unjust matters at home which are pressing for solution." [21] But in general the humanitarian pleas that were so effective in arousing both Democrats and Republicans against Spanish tyranny in Cuba were at least equally effective with the Populists; temporarily the reformers were more concerned

[20] *Statistical Abstract of the United States*, 1922, p. 709. See also Sullivan, *Our Times*, 1:298.
[21] *Southern Mercury*, May 5, 1898.

about the sufferings of the downtrodden Cubans than about the sufferings of American farmers and workingmen. Early in 1896 Senator Allen spoke vigorously in the Senate on the duty of maintaining inviolate the principles stated in the Monroe Doctrine;[22] and when William Jennings Bryan entered military service, he was applauded for his patriotism by Populists no less than by Democrats. The voting of bond issues to aid in financing the war drew fire from the Populists, who would have preferred issues of treasury notes,[23] but for the most part the traditional principles of Populism were forgotten in the excitement engendered by the war. Populists early saw and denounced the dangers lurking in a policy of "imperialism"; but when it came time to ratify the Paris treaty, Populist senators showed no greater disposition than Democratic senators to block the administration's program. Senator Allen had no apologies to make for his favorable vote on the treaty and resented deeply the inference that it was cast for ratification as a result of the persuasive arguments of Colonel Bryan.[24]

Perhaps the policy of fusion had less to do with the decline of Populism than the mid-roaders avowed, but fusion undoubtedly did lose many adherents for the new party. In national affairs the Democrats were generally advertised, during and after the campaign of 1896, as having gone squarely over to Populist principles, and many Populists, assuming that this was the case, stood ready to forget the fiction of a separate party of their own. Especially was this true of those Populists who came into the party primarily because of its stand on silver. In the West, moreover, the Democratic party usually accepted whatever demands the Populists cared to make on local issues, so that in this section the difference between Democrats and Populists closely ap-

---

[22] *Congressional Record*, 54 Congress, Session 1, pp. 25, 134, 254.

[23] *Southern Mercury*, June 9, 1898. See also A. E. Sheldon to Allen, April 12, 1898, in the Allen Papers.

[24] G. G. Vest to Allen, April 3, 1904, in the Allen Papers.

proached the vanishing point. In North Dakota the Populists were even ready by 1898 to give up their separate name and to call themselves, along with all other fusionists, Independent Democrats.[25] Mid-roaders declared, not without reason, that Populist managers in the West and in the nation at large were treating the third party as if it were doomed to death and as if in the meantime its main purpose was to serve as a mere adjunct to the Democracy.[26]

The situation was different in the South, but it was no less unfavorable to the perseverance of Populism. There the unholy alliance with the Republicans drove many third-party adherents back to the Democratic fold; while others, especially the more radical leaders, who could never hope to recover their standing in Democratic circles, drifted over to the Republicans. Furthermore, the absurdity of fusing with the Democrats in one section of the country and with the Republicans in another could not fail to break down Populist morale. Obviously an enduring party could not be built upon such shifting sands.[27]

The breakdown of Populism in the South is associated closely with the results of the fusion experiment in North Carolina. There the Republicans and Populists had won in the election of 1896 not only the legislature and the state offices but many local and county offices as well. In some of the eastern counties, where the bulk of the Republican voters were negroes, this meant negro officeholders in large numbers.[28] Whole counties were said to be " negroized," and the horrors of Reconstruction, if not exactly repeated, were at least feared. A white Republican postmaster at Wilmington, North Carolina, warned his party that it was permitting too many negroes to hold office. With thirty-six colored magistrates in the county, a colored register of deeds, and

---

[25] *Appletons' Annual Cyclopaedia, 1898*, 703.
[26] *Representative*, January 13, 1897; *Southern Mercury*, January 27, 1898; February 17, 1898; *Farm, Stock and Home*, July 1, 1898.
[27] *Southern Mercury*, February 17, 1898.
[28] Delap, *Populist Party in North Carolina*, 57–58, 60–64, 71.

some colored appointees of the national administration, he felt certain that to avoid a conflict between the races the local offices must be promptly turned over to the whites. A Raleigh correspondent wrote to the *Atlanta Constitution* that the negroes, with no less than a thousand members of their race holding office in North Carolina, were preparing to make the state a place of refuge for their kind from all over the South; and this item, copied by most of the North Carolina papers as a startling revelation of the truth, aroused intense excitement throughout the state.[29]

Even the Populists were affected. If the existence of a third party was to mean in practice that the negroes were to have the upper hand, many reformers were ready to abandon their third-party tickets. Democrats announced boldly, as the elections of 1898 approached, that they were determined to overthrow negro rule, "peaceably if possible, but by revolution if they must." Populists were urged to return to a Democratic party that was now, so it was alleged, thoroughly renovated and in entire harmony with Populist principles. As an organization, however, the Populist party in North Carolina continued its cooperation with the Republicans, and in the election of 1898 the Democrats still had to face a powerful and well-united opposition.[30]

But once the issue of white supremacy was successfully raised, the triumph of the Democrats was easy. In vain did the Populist leaders maintain that the Democratic leopard had not changed its spots; that all its protestations of reform were insincere. In vain did the third-party orators defend fusion as the only practical means of obtaining results and saving the party. "If we go in the middle of the road," said one of them, "which might be the best course for us to pursue if we could pursue it and live as an organization, the canvass will be made against our candidates by the Democratic party that 'you have no chance for election. Our

[29] *Appletons' Annual Cyclopaedia, 1898,* 509–511.
[30] Delap, *Populist Party in North Carolina,* 69–70.

candidates will be elected or the Republicans will be. You are simply not in it.' "[31]

The election returns showed a complete Democratic victory. Seven Democratic congressmen were returned; two Republicans; and no Populists. The state Senate in 1896 had contained forty-three fusionists and seven Democrats; the state Senate elected in 1898 contained seven fusionists and forty-three Democrats. The state House of Representatives elected in 1896 had contained ninety fusionists and thirty Democrats; the House elected in 1898 contained twenty-six fusionists and ninety-four Democrats. Local offices likewise usually fell to Democrats, and in Wilmington, following the election, a riotous mob destroyed a negro newspaper office, killed a dozen negroes, deposed the Republican city officials, and installed white Democrats in their places.[32]

With events like these occurring in North Carolina, Populists found it almost impossible to hold their ranks intact anywhere else in the South. In Georgia and Alabama, where Populism with Republican assistance had almost triumphed in preceding elections, the Populist tickets in 1898 polled only about thirty per cent of the total vote cast for state officers, and in the congressional elections they scarcely figured at all. In such states as Florida, Mississippi, and Louisiana the Populist party had practically ceased to exist. In South Carolina, where there had never been a Populist organization worthy of the name, Populist principles continued to triumph under the leadership of Tillman, but an anti-negro election riot showed that the race question there, as in North Carolina, aroused more feeling than any reforms that Populism had to offer. Only in Texas did the southern third-party men make a showing that could compare favorably with the results of preceding years.[33]

[31] *Ibid.*, 72–73.
[32] *Appletons' Annual Cyclopaedia, 1898*, 512–513.
[33] See under names of states in *Appletons' Annual Cyclopaedia, 1898*, and in the *Tribune Almanac, 1899*.

Out in the West there was no embarrassing race issue to add to the difficulties of the Populists, but there were difficulties enough, as the Populist leaders themselves knew well. Could the fusion forces that had won the state elections in Kansas, Nebraska, and elsewhere justify the trust that had been given them? Could Democrats and Populists work harmoniously together in victory as in defeat? Past performances were not reassuring. According to Breidenthal of Kansas, who wrote out his doubts shortly after the election of 1896 in a letter to Senator Allen of Nebraska, much hard work would be required " to hold together the forces that contributed to our success," and only by " exceedingly good management " could defeat in 1898 be prevented.[34]

Little more can be said for the fusion state governments of this period, however, than that they were undistinguished. Many of the local reforms demanded by the Populists had already been enacted into law, frequently enough with Republican as well as Democratic assistance.[35] The silver crusade, culminating in Bryan's recent race for the presidency, had focused attention upon a national panacea that had effectively eclipsed in importance the various local demands, once deemed so significant. For the most part fusionist governors and fusionist legislatures merely marked time or played politics.[36]

In the elections of 1898 fusion tickets appeared again in Kansas, Nebraska, Minnesota, the Dakotas, and even out in Oregon and California. Minor mid-road defections occurred almost everywhere in the West, and in some of the mountain states the silver combinations that were effected in 1896 could not be revived in 1898. But the fusionists won few victories. In Kansas, where Populism had once ruled su-

---

[34] John Breidenthal to Allen, December 5, 1896, in the Allen Papers. Another letter to Allen, from W. F. Bryant, December 6, 1897, describes a similar situation in Nebraska.
[35] For example, the Newberry Railroad Act in Nebraska. See Nebraska, *House Journal, 1893*, 707; *Senate Journal, 1893*, 957.
[36] *Appletons' Annual Cyclopaedia, 1897*, under names of western states.

preme, the Republicans filled all the state offices, secured control of the state legislature, and chose every congressman but one. In North Dakota, Oregon, and California, not to mention states where fusion had broken down, the Republicans won victories equally pronounced. In South Dakota and Minnesota fusion candidates for governor were elected, but the Republicans got everything else.[37]

Only in Nebraska could the fusionists claim that they had come measurably close to holding their own. There the extraordinary efforts of the Republicans to discredit Bryanism in Bryan's own state had roused the "Demopops" to a highly effective counteractivity. As early as 1897 the battle for supremacy was on. That year a "three in one" convention of silver Republicans, silver Democrats, and Populists had shrewdly named a Bryan Democrat for the supreme court and had won.[38] In 1898 both Bryan and fusion were strong in the state, and the fusionists scored another victory. Their entire state ticket was elected, and they captured four out of the six Nebraska seats in Congress. The great disappointment in this election for the Populists, however, was the loss of the legislature, which on joint ballot turned out to be safely Republican. This disappointment was due less to an abiding interest in local reforms than to the fact that Senator Allen's term was expiring, and a Republican would surely be chosen to replace him. But luck was with the Populists for once. The Republican chosen by the legislature died without ever having had an opportunity to take his seat in the Senate, and the fusionist governor promptly sent Allen back to Washington, where he continued to represent Nebraska until 1901.[39]

[37] *Ibid.* See also the *Tribune Almanac, 1899,* under the names of the states mentioned.

[38] Benton Maret to Allen, August 16, 1897; W. F. Bryant to Allen, September 6, 1897; S. F. Maxwell to Allen, September 4, 1897; all in the Allen Papers. See also the *Tribune Almanac, 1898,* 279.

[39] *Appletons' Annual Cyclopaedia, 1898,* 474–475; Morton and Watkins, *Illustrated History of Nebraska,* 3:494.

The well-nigh universal disasters that overtook the Populists in 1898 gave a kind of morbid satisfaction to the uncompromising mid-road faction. Their own showing was decidedly less consequential than that of the fusionist wing of the party, but fusion was blamed for every unfortunate result. Donnelly urged that the strength of the mid-road ticket in Minnesota "was immensely reduced by our disgusted people going squarely over to the Republican party. They sought revenge on the Democrats who had invaded our ranks, bought up our leaders, and forced their loathsome nuptials on our unhappy people." [40] The *Southern Mercury* blamed fusion, which it had always opposed, for the return of thousands of southern Populists to the Democratic party and thousands of northern Populists to the Republican party.[41] Most mid-roaders believed that only by a frank disavowal of fusion and a definite return to pure Populist principles could the one-time glory of Populism be restored.

This had been the stand of the Cincinnati convention of September, 1898, which had nominated Barker and Donnelly. But the trouble with the action taken at Cincinnati was that it presupposed a willingness, not shared by all the mid-roaders, to cut loose from the regular organization. Many antifusionists believed that they might yet capture the machinery of the party and dictate the future Populist policy. At Kansas City, in May, 1899, a meeting of the National Organization Committee, held jointly with a meeting of the National Reform Press Association, urged mid-roaders to resume control of their party if they could, and to secede from it only in case they must. Delegates pledged to independent action should be sent to the regular Populist nominating convention in 1900, and contesting delegations whenever the fusionists won the regular delegations. Should the convention presume to nominate a Democrat, it would then

[40] *Representative*, November 16, 1898.
[41] *Southern Mercury*, December 17, 1898.

be time enough to bolt. In the meantime the work of organizing precinct Populist clubs, through which to obtain a referendum on every important party policy, should be continued.[42]

Mid-road members of the regular national committee also attempted to forestall the adoption of a fusionist policy in 1900. Several of them met at Memphis, Tennessee, in December, 1899, to urge on Chairman Butler the necessity of calling an early meeting of the national committee. February 12 was named as a desirable date for the meeting and Chicago a desirable place. But Butler, in spite of the fact that probably more than half the members of the national committee signed memorials requesting the Chicago meeting, took his own poll of the committee and called the meeting for Lincoln, Nebraska, a week later than had been requested.[43]

The choice of Lincoln as a meeting place was obviously determined upon in order to put the mid-roaders at a disadvantage. Lincoln was the home of Bryan, and Nebraska was the strongest fusion state in the union. Meetings of Populist national committees were certain to be attended in large part by local citizens serving as substitutes for distant members, the rule being that only one proxy should be voted by one person. The number of Lincoln mid-roaders who could be on hand to vote proxies was known to be limited. Both sides, however, busily collected proxies, and the mid-roaders claimed that there were plenty of Populists in Bryan's home town who were "ready to oppose the 'prostitution of the party.'" Jo A. Parker, head of the National Organization Committee, led the mid-road forces, appearing in Lincoln with very few followers but with proxies enough to run the antifusion vote up to more than fifty. The fusionists had more national committeemen in actual attendance but ap-

[42] *Ibid.*, May 25, June 1, 1899; January 11, 18, 1900.
[43] *Nebraska State Journal*, February 21, 1900; *Southern Mercury*, February 22, 1900.

parently fewer proxies, and the chances for a battle royal
seemed good. But Chairman Butler was unwilling to run
the risk of defeat. On calling the meeting to order he
promptly announced a temporary roll made up chiefly of
fusionists, then appointed a strongly fusionist credentials
committee, and arbitrarily adjourned the session.[44]

Claiming that Butler's adjournment of the meeting was
illegal, the mid-roaders continued in session with one of their
own men as chairman, adjourning presently in strictly legal
fashion to meet again in the evening when a compromise
plan for the settlement of contests could be suggested. The
plan adopted by the mid-roaders failed, however, to meet the
approval of the fusionists. Their credentials committee,
dominated by Senator Allen and General Weaver, seated a
fusionist majority, and the convention voted to exclude all
committeemen or proxy-holders who had violated the
"Omaha Contract" by participating in the Barker-Donnelly
convention at Cincinnati. "We have thrown them over the
transom," said Senator Allen.

Thereafter the mid-road members and the fusionists met
separately, each claiming to be the regular Populist national
committee. The mid-roaders voted to "gather the clans into
a mighty conclave on the ninth of May at the birthplace of
our party [Cincinnati], and kindle anew the fires of liberty
in our ranks which have been dimmed by the faithlessness of
our chosen leaders. . . ."[45] The fusionists, mainly concerned that the Populist nomination should go again to
Bryan, finally agreed to meet on the date selected by the
mid-roaders, but at Sioux Falls, South Dakota, instead of
Cincinnati. Many of the fusionists would have preferred to
meet at the same time and place as the Democrats, but the
majority held that it would be poor policy to go back so

[44] *Nebraska State Journal*, February 20–22, 1900; *Southern Mercury*, March 8,
1900.
[45] *Nebraska State Journal*, February 21, 1900.

openly on the "Omaha Contract." Moreover, there was
nothing to prevent the Populists from nominating Bryan
before the Democrats got around to it. "The Democrats
stole our platform in 1896," someone remarked, "why
haven't we just as good a right to steal their candidate in
1900?" [46]

The two conventions that met in response to these calls
attested eloquently the bankruptcy of Populism. Sioux Falls
made ready for a mighty host, providing for the use of the
convention a tent with seats for eighty-four hundred. But
the crowds failed to appear. Not over four hundred delegates, including alternates, were in attendance, and not over
fifteen hundred strangers all told were in the city while the
convention met. The nomination of Bryan was a cut-and-
dried affair and was made by acclamation. Extreme fusionists, such as Senator Allen, favored making no nomination
whatever for vice president, leaving this for later agreement
with the Democrats. But Butler and the southerners again
demanded a separate Populist nominee, with the result that
the convention chose Charles A. Towne of Minnesota, a
silver Republican, for second place on the ticket. The Democrats, it was hoped, would nominate both Bryan and Towne.
The utter unselfishness of the fusionists at Sioux Falls in
nominating a Democrat for president and a silver Republican for vice president provoked both pity and amusement
on the part of outsiders.[47]

The mid-roaders at Cincinnati had little more to boast
of. They were reported to have a few more delegates in attendance, but probably this report emanated from the Republican press, which was doing its best to discredit the proceedings at Sioux Falls. And they were far from harmonious.
Many, perhaps a majority, of the delegates would have pre-

[46] Ibid.
[47] Daily Argus Leader (Sioux Falls, South Dakota), May 9–11, 1900; National Watchman (Washington, D. C.), July 12, 1900. The two Populist platforms are printed in Stanwood, History of the Presidency, 1:39–43.

ferred to nominate former Congressman M. W. Howard of Alabama, but they were bluffed off by the Barker men, who declared they would bolt if their candidate were turned down. And so the ticket that had been nominated in 1898, Wharton Barker and Ignatius Donnelly, was reconstituted. Donnelly was a Populist, tried and true, and perhaps, as he claimed, the father of the People's party. But he had worn himself out in the advocacy of lost causes and was known to be on the brink of the grave. Barker was likewise an old man; he was an easterner of some private means, and he had been prominent in Populist circles only since the election of 1896. Viewed from any standpoint, the ticket was ridiculous.[48]

The campaign and election made it clear even to the few remaining Populists that the day of their party was done. The fusionists were humiliated at the hands of the Democrats, who nominated Bryan, as had been expected, but ignored Towne. "It was entirely unnecessary," ran one comment, "that a great organization like the Democratic party, embracing among its membership so many distinguished statesmen, should go outside of its ranks for a candidate, no matter how great or how deserving the candidate might be."[49] Towne withdrew, and the fusion Populists swallowed the Democratic nominee with what grace they could muster. As for the mid-roaders, they considered themselves fortunate whenever they were able to place their ticket on an official ballot. On election day neither faction won any victories worthy of mention — national, state, or local. The once powerful and united Populist party had now dwindled down to two pitiful remnants, one group bound by the gift of a few score offices to do the bidding of the Democrats, the other steadfast in principle to the last, but looking backward to a glory that had gone.

[48] *National Watchman*, July 17, 1900; *Cincinnati Enquirer*, May 10, 11, 1900.
[49] *National Watchman*, July 12, 1900.

" THE SILVER WAVE HAS RECEDED AND LEFT THE POPOCRATIC FISH
HIGH AND DRY "

[Drawn for *Judge;* reproduced in the *Review of Reviews*, November, 1896.]

"We refuse to get into the grave where the Greenback party lies, fused to death," said Donnelly in 1898; and even after the election of 1900 a few of the faithful held on to their hopes. At Kansas City, in September, 1901, a conference of all the reform parties — fusion Populists, mid-road Populists, single taxers, liberal socialists, anybody who would come — decided to launch a new party to be known as the Allied party.[50] This small but motley group called a convention to meet at Louisville, Kentucky, in April, 1902. The Louisville convention, composed, as Senator Allen remarked, of the "political odds and ends of the United States," actually met and after a day of bitter wrangling announced the birth of the "Allied People's party." The mid-road organization of 1900, however, was continued with only slight amendment and was authorized to make the convention call in 1904.[51] Since it was generally believed that the next Democratic candidate for the presidency would be a conservative, the reformers counted on winning the Bryan Populists, and perhaps even Bryan himself, to their cause.

For the most part their hopes were dashed. When the Populist convention to make presidential nominations was held in Springfield, Illinois, on July 5 and 6, 1904, only about two hundred delegates attended. Some of them did, indeed, represent the fusionist wing of the party, which was now officially welcomed back home, but Bryan was not with them, nor were many who had followed him. Thomas E. Watson of Georgia was nominated for president and T. H. Tibbles of Nebraska for vice president,[52] but in the campaign that followed they cut as little figure as the Prohibitionist candidates. Four years later at St. Louis another small group of old-timers renominated Watson for the presidency, with Samuel W. Williams of Indiana as second on the ticket, and

[50] *Representative*, June 15, 1898; *Kansas City Star*, September 18–19, 1901.
[51] *Louisville Courier Journal*, April 2–4, 1902.
[52] *Illinois State Register*, July 5–6, 1904. The platform is printed in Stanwood, *History of the Presidency*, 2:115–117.

confessed among themselves that their party was dead. Grief over the passing of Populism as an organization, however, was somewhat tempered by the firm conviction that Populist principles could never die. "Populism," said one delegate, "is written across the face of the Oklahoma constitution"; and he went on to observe, even more significantly, that "Roosevelt's messages read like the preamble to the Populist platform."[53]

[53] *St. Louis Globe-Democrat*, April 3, 1908. The platform is printed in Stanwood, *History of the Presidency*, 2:159-161.

CHAPTER FIFTEEN

THE POPULIST CONTRIBUTION

EARLY in 1890, when the People's party was yet in the embryo stage, a farmer editor from the West set forth the doctrine that "the cranks always win." As he saw it,

The cranks are those who do not accept the existing order of things, and propose to change them. The existing order of things is always accepted by the majority, therefore the cranks are always in the minority. They are always progressive thinkers and always in advance of their time, and they always win. Called fanatics and fools at first, they are sometimes persecuted and abused. But their reforms are generally righteous, and time, reason and argument bring men to their side. Abused and ridiculed, then tolerated, then respectfully given a hearing, then supported. This has been the gauntlet that all great reforms and reformers have run, from Galileo to John Brown.[1]

The writer of this editorial may have overstated his case, but a backward glance at the history of Populism shows that many of the reforms that the Populists demanded, while despised and rejected for a season, won triumphantly in the end. The party itself did not survive, nor did many of its leaders, although the number of contemporary politicians whose escutcheons should bear the bend sinister of Populism is larger than might be supposed; but Populistic doctrines showed an amazing vitality.

[1] *Farmers' Alliance* (Lincoln), February 15, 1890. This chapter follows in the main an article on "The Persistence of Populism," *Minnesota History*, 12:3–20 (March, 1931).

In formulating their principles the Populists reasoned that the ordinary, honest, willing American worker, be he farmer or be he laborer, might expect in this land of opportunity not only the chance to work but also, as the rightful reward of his labor, a fair degree of prosperity. When, in the later eighties and in the "heart-breaking nineties," hundreds of thousands — perhaps millions — of men found themselves either without work to do or, having work, unable to pay their just debts and make a living, the Populists held that there must be "wrong and crime and fraud somewhere." What was more natural than to fix the blame for this situation upon the manufacturers, the railroads, the money-lenders, the middlemen — plutocrats all, whose "colossal fortunes, unprecedented in the history of mankind," grew ever greater while the multitudes came to know the meaning of want. Work was denied when work might well have been given, and "the fruits of the toil of millions were boldly stolen." [2]

And the remedy? In an earlier age the hard-pressed farmers and laborers might have fled to free farms in the seemingly limitless lands of the West, but now the era of free lands had passed. Where, then, might they look for help? Where, if not to the government, which alone had the power to bring the mighty oppressors of the people to bay? So to the government the Populists turned. From it they asked laws to insure a full redress of grievances. As Dr. Turner puts it, "the defences of the pioneer democrat began to shift from free land to legislation, from the ideal of individualism to the ideal of social control through regulation by law." [3] Unfortunately, however, the agencies of government had been permitted to fall into the hands of the plutocrats. Hence, if the necessary corrective legislation were to be obtained, the people must first win control of their government.

[2] Donnelly's preamble to the St. Louis and Omaha platforms stated not unfairly the Populist protest. See Appendixes E and F.
[3] Turner, *Frontier in American History*, 277.

The Populist philosophy thus boiled down finally to two fundamental propositions; one, that the government must restrain the selfish tendencies of those who profited at the expense of the poor and needy; the other, that the people, not the plutocrats, must control the government.

In their efforts to remove all restrictions on the power of the people to rule, the Populists accepted as their own a wide range of reforms. They believed, and on this they had frequently enough the evidence of their own eyes, that corruption existed at the ballot box and that a fair count was often denied. They fell in line, therefore, with great enthusiasm when agitators, who were not necessarily Populists, sought to popularize the Australian ballot and such other measures as were calculated to insure a true expression of the will of the people.[4] Believing as they did that the voice of the people was the voice of God, they sought to eliminate indirect elections, especially the election of United States senators by state legislatures and of the president and the vice president by an electoral college. Fully aware of the habits of party bosses in manipulating nominating conventions, the Populists veered more and more in the direction of direct primary elections, urging in some of their later platforms that nominations even for president and vice president should be made by direct vote. Woman suffrage was a delicate question, for it was closely identified with the politically hazardous matter of temperance legislation, but, after all, the idea of votes for women was so clearly in harmony with the Populist doctrine of popular rule that it could not logically be denied a place among genuinely Populistic reforms. Direct legislation through the initiative and referendum and through the easy amendment of state constitutions naturally appealed strongly to the Populists — the more so as they

---

[4] At St. Louis in December, 1889, the Northern Alliance demanded the Australian system of voting. See Appendix A. Thereafter nearly every Alliance or Populist platform gave the subject favorable mention.

saw legislatures fail repeatedly to enact reform laws to which
a majority of their members had been definitely pledged. "A
majority of the people," said the Sioux Falls convention,
"can never be corruptly influenced." [5] The recall of faithless
officials, even judges, also attracted favorable attention from
the makers of later Populist platforms.

To list these demands is to cite the chief political innovations made in the United States during recent times. The
Australian system of voting, improved registration laws, and
other devices for insuring "a free ballot and a fair count"
have long since swept the country. Woman suffrage has won
an unqualified victory. The election of United States senators by direct vote of the people received the approval of
far more than two-thirds of the national House of Representatives as early as 1898; it was further foreshadowed by
the adoption, beginning in 1904, of senatorial primaries in a
number of states, the results of which were to be regarded
as morally binding upon the legislatures concerned; and it
became a fact in 1913 with the ratification of the seventeenth
amendment to the constitution.

The direct election of president and vice president was a
reform hard to reconcile with state control of election machinery and state definition of the right to vote. Hence this
reform never made headway; but the danger of one presidential candidate receiving a majority of the popular vote and
another a majority of the electoral vote, as was the case in
the Cleveland-Harrison contest of 1888, seems definitely to
have passed. Recent elections may not prove that the popular voice always speaks intelligently; but they do seem to
show that it speaks decisively.

In the widespread use of the primary election for the making of party nominations, the Populist principle of popular
rule has scored perhaps its most telling victory. Tillman

[5] See Stanwood, *History of the Presidency*, 2:39–42, for the Sioux Falls platform.

urged this reform in South Carolina at a very early date, but on obtaining control of the Democratic political machine of his state, he hesitated to give up the power that the convention system placed in his hands. At length, however, in 1896 he allowed the reform to go through.[6] Wisconsin, spurred on by the La Follette forces, adopted the direct primary plan of nominations in 1903, and thereafter the other states of the Union, with remarkably few exceptions, fell into line. Presidential preference primaries, through which it was hoped that the direct voice of the people could be heard in the making of nominations for president and vice president, were also adopted by a number of states, beginning with Oregon in 1910.

Direct legislation by the people became almost an obsession with the Populists, especially the middle-of-the-road faction, in whose platforms it tended to overshadow nearly every other issue; and it is perhaps significant that the initiative and referendum were first adopted by South Dakota, a state in which the Populist party had shown great strength, as close on the heels of the Populist movement as 1898. Other states soon followed the South Dakota lead, and particularly in Oregon the experiment of popular legislation was given a thorough trial.[7] New constitutions and numerous amendments to old constitutions tended also to introduce much popularly made law, the idea that legislation in a constitution is improper and unwise receiving perhaps its most shattering blow when an Oklahoma convention wrote for that state a constitution of fifty thousand words. The recall of elected officials has been applied chiefly in municipal affairs, but some states also permit its use for state officers and a few allow even judges, traditionally held to be immune from popular reactions, to be subjected to recall. Thus many of the favorite ideas of the Populists, ideas that had once

---

[6] Simkins, *Tillman Movement*, 239–243.

[7] Ellis P. Oberholtzer, *The Referendum in America together with Some Chapters on the Initiative and the Recall.*

been "abused and ridiculed," were presently "respectfully given a hearing, then supported." [8]

Quite apart from these changes in the American form of government, the Populist propaganda in favor of independent voting did much to undermine the intense party loyalties that had followed in the wake of the Civil War. The time had been when for the Republican voter "to doubt Grant was as bad as to doubt Christ," [9] when the man who scratched his party ticket was regarded as little, if any, better than the traitor to his country. The Alliance in its day had sought earnestly to wean the partisan voter over to independence. It had urged its members to "favor and assist to office such candidates only as are thoroughly identified with our principles and who will insist on such legislation as shall make them effective." [10] And in this regard the Alliance, as some of its leaders boasted, had been a "great educator of the people." The Populist party had to go even further, for its growth depended almost wholly upon its ability to bring voters to a complete renunciation of old party loyalties. Since at one time or another well over a million men cast their ballots for Populist tickets, the loosening of party ties that thus set in was of formidable proportions.

Indeed, the man who became a Populist learned his lesson almost too well. When confronted, as many Populist voters thought themselves to be in 1896, with a choice between loyalty to party and loyalty to principle, the third-party adherent generally tended to stand on principle. Thereafter, as Populism faded out, the men who once had sworn undying devotion to the Omaha platform, were compelled again to transfer their allegiance. Many Republicans became Democrats via the Populist route; many Democrats became Re-

[8] For satisfactory general discussions of these reforms see Charles A. Beard, *American Government and Politics*, 4th ed., ch. 24; Charles A. and Mary R. Beard, *The Rise of American Civilization*, Vol. 2, ch. 27; and David S. Muzzey, *The United States of America*, Vol. 2, ch. 7.

[9] Barr, in *Kansas and Kansans*, 2:1194.

[10] Appendix A.

publicans. Most of the Populists probably returned to the
parties from which they had withdrawn, but party ties, once
broken, were not so strong as they had been before. The
rapid passing of voters from one party to another and the
wholesale scratching of ballots, so characteristic of voting
today, are distinctly reminiscent of Populism; as are also the
nonpartisan ballots by which judges, city commissioners, and
other officers are now frequently chosen, wholly without regard to their party affiliations.

In the South the Populist demands for popular government produced a peculiar situation. To a very great extent
the southern Populists were recruited from the rural classes
that had hitherto been politically inarticulate. Through the
Populist party the " wool hat boys " from the country sought
to obtain the weight in southern politics that their numbers
warranted but that the Bourbon dynasties had ever denied
them. In the struggle that ensued, both sides made every
possible use of the negro vote, and the bugaboo of negro
domination was once again raised. Indeed, the experience of
North Carolina under a combination government of Populists and Republicans furnished concrete evidence of what
might happen should the political power of the negro be restored. Under the circumstances, therefore, there seemed to
be nothing else for the white Populists to do but to return
to their former allegiance until the menace of the negro voter
could be removed.

With the Democratic party again supreme, the problem of
negro voting was attacked with right good will. Indeed, as
early as 1890 the state of Mississippi, stimulated no doubt
by the agitation over the Force Bill, adopted a constitution
that fixed as a prerequisite for voting a two years' residence
in the state and a one year's residence in the district or town.
This provision, together with a poll tax that had to be paid
far in advance of the dates set for elections, diminished
appreciably the number of negro voters, among whom indi-

gence was common and the migratory propensity well developed. To complete the work of disfranchisement an amendment was added to the Mississippi constitution in 1892 that called for a modified literary test that could be administered in such a way as to permit illiterate whites to vote, while discriminating against illiterate, or even literate, blacks. The Tillmanites in South Carolina found legal means to exclude the negro voter in 1895; Louisiana introduced her famous "grandfather clause" in 1898; North Carolina adopted residence, poll-tax, and educational qualifications in 1900; Alabama followed in 1901; and in their own good time the other southern states in which negro voters had constituted a serious problem did the same thing. Some reverses were experienced in the courts, but the net result of this epidemic of anti-negro suffrage legislation was to eliminate for the time being all danger that negro voters might play an important part in southern politics.[11]

With this problem out of the way, or at least in the process of solution, it became possible for the rural whites of the South to resume the struggle for a voice in public affairs that they had begun in the days of the Alliance and had continued under the banner of Populism. They did not again form a third party, but they did contend freely in the Democratic primaries against the respectable and conservative descendants of the Bourbons. The Tillman machine in South Carolina continued for years to function smoothly as the agency through which the poorer classes sought to dominate the government of that state. It regularly sent Tillman to the United States Senate, where after his death his spirit lived on in the person of Cole Blease.[12] In Georgia the struggle for supremacy between the two factions of the Democratic party was a chronic condition, with now one side and

[11] Paul Lewison, "The Negro in the White Class and Party Struggle," *Southwestern Political and Social Science Quarterly*, 8:358–382. For an excellent brief statement see Holland Thompson, *The New South*, ch. 3.

[12] Simkins, *Tillman Movement*, ch. 10.

now the other in control. Former Populists, converted by the lapse of time into regular organization Democrats, won high offices and instituted many of the reforms for which they had formerly been derided. Even Tom Watson rose from his political deathbed to show amazing strength in a race for Congress in 1918 and to win an astounding victory two years later when he sought a seat in the United States Senate.[13]

For better or for worse, the political careers of such southern politicians as James K. Vardaman and Theodore G. Bilbo of Mississippi, the Honorable "Jeff." Davis of Arkansas, and Huey P. Long of Louisiana demonstrate conclusively the fact that the lower classes in the South can, and sometimes do, place men of their own kind and choosing in high office. In these later days rural whites, who fought during Populist times with only such support as they could obtain from Republican sources, have sometimes been able to count as allies the mill operatives and their sympathizers in the factory districts; and southern primary elections are now apt to be as exciting as the regular elections are tame. Populism may have had something to do with the withdrawal of political power from the southern negro, but it also paved the way for the political emancipation of the lower class of southern whites.

The control of the government by the people was to the thoughtful Populist merely a means to an end. The next step was to use the power of the government to check the iniquities of the plutocrats. When the Populists at Omaha were baffled by the insistence of the temperance forces, they pointed out that before this or any other such reform could be accomplished they must "ask all men to first help us to determine whether we are to have a republic to administer." [14] The inference is clear. Once permit the people really to rule, once insure that the men in office would not or could

---

[13] Arnett, *Populist Movement in Georgia*, 220–226; Brewton, *Thomas E. Watson*, chs. 44, 45.

[14] Appendix F.

not betray the popular will, and such regulative measures as would right the wrongs from which the people suffered would quickly follow. The Populist believed implicitly in the ability of the people to frame and enforce the measures necessary to redeem themselves from the various sorts of oppression that were being visited upon them. They catalogued in their platform the evils from which society suffered and suggested the specific remedies by which these evils were to be overcome.

Much unfair criticism has been leveled at the Populists because of the attitude they took towards the allied subjects of banking and currency. To judge from the contemporary anti-Populist diatribes and from many subsequent criticisms of the Populist financial program, one would think that in such matters the third-party economists were little better than raving maniacs. As a matter of fact, the old-school Populists could think about as straight as their opponents. Their newspapers were well edited, and the arguments therein presented usually held together. Populist literature, moreover, was widely and carefully read by the ordinary third-party voters, particularly by the western farmers, whose periods of enforced leisure gave them ample opportunity for reading and reflection. Old-party debaters did not tackle their Populist antagonists lightly, for as frequently as not the bewhiskered rustic, turned orator, could present in support of his arguments an array of carefully sorted information that left his better-groomed opponent in a daze. The appearance of the somewhat irrelevant silver issue considerably confused Populist thinking, but even so many of the old-timers kept their heads and put silver in its proper place.

The Populists observed with entire accuracy that the currency of the United States was both inadequate and inelastic. They criticized correctly the part played by the national banking system in currency matters as irresponsible and susceptible of manipulation in the interest of the creditor class. They demanded a stabilized dollar, and they believed that it could be obtained if a national currency " safe, sound, and flexible " should be issued direct to the people by the government itself in such quantities as the reasonable demands of business should dictate. Silver and gold might be issued as well as paper, but the value of the dollar should come from the fiat of government and not from the "intrinsic worth " of the metal.

It is interesting to note that since the time when Populists were condemned as lunatics for holding such views legislation has been adopted that, while by no means going the full length of an irredeemable paper currency, does seek to accomplish precisely the ends that the Populists had in mind. Populist and free-silver agitation forced economists to study the money question as they had never studied it before and ultimately led them to propose remedies that could run the gauntlet of public opinion and of Congress. The Aldrich-Vreeland Act of 1908 authorized an emergency currency of several hundred million dollars, to be lent to banks on approved securities in times of financial disturbance. A National Monetary Commission, created at the same time, reported after four years' intensive study in favor of a return to the Hamiltonian system of a central Bank of the United States. Instead Congress in 1914, under Wilson's leadership, adopted the federal reserve system. The Federal Reserve Act did not, indeed, destroy the national banks and avoid the intervention of bankers in all monetary matters, but it did make possible an adequate and elastic national currency, varying in accordance with the needs of the country, and it placed supreme control of the nation's banking and credit resources in the hands of a federal reserve board, appointed not by the bankers but by the president of the United States with the consent of the Senate. The Populist diagnosis was accepted, and the Populist prescription was not wholly ignored.[15]

[15] E. W. Kemmerer, *The A B C of the Federal Reserve System*, 5th ed.; H. Parker Willis, *The Federal Reserve System, Legislation, Organization, and Operation*.

Probably no item in the Populist creed received more thorough castigation at the hands of contemporaries than the demand for subtreasuries, or government warehouses for the private storage of grain; but the subtreasury idea was not all bad, and perhaps the Populists would have done well had they pursued it further than they did. The need that the subtreasury was designed to meet was very real. Lack of credit forced the farmer to sell his produce at the time of harvest, when the price was lowest. A cash loan on his crop that would enable him to hold it until prices should rise was all that he asked. Prices might thus be stabilized; profits honestly earned by the farmers would no longer fall to the speculators. That the men who brought forward the subtreasury as a plan for obtaining short-term rural credits also loaded it with an unworkable plan for obtaining a flexible currency was unfortunate; but the fundamental principle of the bill has by no means been discredited. Indeed, the Warehouse Act of 1916 went far towards accomplishing the very thing the Populists demanded. Under it the United States Department of Agriculture was permitted to license warehousemen and authorize them to receive, weigh, and grade farm products, for which they might issue warehouse receipts as collateral. Thus the owner might borrow the money he needed — not, however, from the government of the United States.[16]

In addition to the credits that the subtreasury would provide, Populist platforms usually urged also that the national government lend money on farm lands directly at a low rate of interest. This demand, which received an infinite amount of condemnation and derision at the time, has since been treated with much deference. If the government does not now print paper money to lend the farmer, with his land as security, it nevertheless does stand back of an elaborate system of banks through which he may obtain the credit he

[16] *Federal Statutes Annotated, Supplement*, 1918, pp. 1057–1065. See also Wiest, *Agricultural Organization in the United States*, 467–469.

needs. Under the terms of the Federal Reserve Act national banks may lend money on farm mortgages — a privilege they did not enjoy in Populist times — and agricultural paper running as long as six months may be rediscounted by the federal reserve banks. From the farm loan banks, created by an act of 1916, the farmers may borrow for long periods sums not exceeding fifty per cent of the value of their land and twenty per cent of the value of their permanent improvements. Finally, through still another series of banks, the federal intermediate credit banks, established by an act of 1923, loans are made available to carry the farmer from one season to the next or a little longer, should occasion demand; the intermediate banks were authorized to rediscount agricultural and live-stock paper for periods of from six months to three years. Thus the government has created a comprehensive system of rural credits through which the farmer may obtain either short-term loans, loans of intermediate duration, or long-term loans, as his needs require, with a minimum of difficulty and at minimum interest rates.[17]

It would be idle to indulge in a *post hoc* argument in an attempt to prove that all these developments were due to Populism; but the intensive study of agricultural problems that led ultimately to these measures did begin with the efforts of sound economists to answer the arguments of the Populists. And it is evident that in the end the economists conceded nearly every point for which the Populists had contended.

More recent attempts to solve the agricultural problem, while assuming, as readily as even a Populist could have asked, the responsibility of the government in the matter, have progressed beyond the old Populist panacea of easy credit. Agricultural economists now have their attention

[17] H. Parker Willis and William H. Steiner, *Federal Reserve Banking Practice*, chs. 10–14; W. S. Holt, *The Federal Farm Loan Bureau, Its History, Activities, and Organization;* Herbert Myrick, *The Federal Farm Loan System;* A. C. Wiprud, *The Federal Farm Loan System in Operation.*

fixed upon the surplus as the root of the difficulty. In industry, production can be curtailed to meet the demands of any given time, and a glutted market with the attendant decline in prices can be in a measure forestalled. But in agriculture, where each farmer is a law unto himself and where crop yields must inevitably vary greatly from year to year, control of production is well-nigh impossible and a surplus may easily become chronic. Suggestions for relief therefore looked increasingly towards the disposal of this surplus to the greatest advantage.[18]

The various McNary-Haugen bills that have come before Congress in recent years proposed to create a federal board through which the margin above domestic needs in years of plenty should be purchased and held, or disposed of abroad at whatever price it would bring. Through an "equalization fee" the losses sustained by "dumping" the surplus in this fashion were to be charged back upon the producers benefited. Although this proposition was agreeable to a majority of both houses of Congress, it met opposition from two successive presidents, Coolidge and Hoover, and was finally set aside for another scheme, less "socialistic." In 1929 Congress passed and the president signed a law for the creation of an appointive federal farm board, one of whose duties it is, among others, to encourage the organization of cooperative societies through which the farmers may themselves deal with the problem of the surplus. In case of necessity, however, the board may take the lead in the formation of stabilization corporations, which under its strict supervision may buy up such seasonal or temporary surpluses as threaten to break the market and hold them for higher prices. A huge revolving fund, appropriated by Congress, is made available for this purpose, loans from this fund being obtainable by the stabilization corporations at low interest rates. There is

[18] The Agricultural Crisis and Its Causes. Report of the Joint Commission of Agricultural Inquiry, 1921. *House Report* No. 408, 67 Congress, Session 1.

much about this thoroughly respectable and conservative law that recalls the agrarian demands of the nineties. Indeed, the measure goes further in the direction of government recognition of and aid to the principle of agricultural cooperation than even the most erratic Alliancemen could have dared to hope. Perhaps it will prove to be the "better plan" that the farmers called for in vain when the subtreasury was the best idea they could present.[19]

To the middle western Populist the railway problem was as important as any other — perhaps the most important of all. Early Alliance platforms favored drastic governmental control of the various means of communication as the best possible remedy for the ills from which the people suffered, and the first Populist platform to be written called for government ownership and operation only in case "the most rigid, honest, and just national control and supervision" should fail to "remove the abuses now existing."[20] Thereafter the Populists usually demanded government ownership, although it is clear enough from their state and local platforms and from the votes and actions of Populist officeholders that, pending the day when ownership should become a fact, regulation by state and nation must be made ever more effective.

Possibly government ownership is no nearer today than in Populist times, but the first objective of the Populists, "the most rigid, honest, and just national control," is as nearly an accomplished fact as carefully drawn legislation and highly efficient administration can make it. Populist misgivings about governmental control arose from the knowledge that the Interstate Commerce Act of 1887, as well as most regulatory state legislation, was wholly ineffectual during the nineties; but beginning with the Elkins Act of

[19] E. R. A. Seligman, *Economics of Farm Relief;* R. W. Kelsey, *Farm Relief and Its Antecedents.*

[20] Appendix D.

1903, which struck at the practice of granting rebates, a long series of really workable laws found their way into the statute books. The Hepburn Act of 1906, the Mann-Elkins Act of 1910, and the Transportation Act of 1920, not to mention lesser laws, placed the Interstate Commerce Commission upon a high pinnacle of power. State laws, keeping abreast of the national program, supplemented national control with state control; and through one or the other agency most of the specific grievances of which the Populists had complained were removed.[21] The arbitrary fixing of rates by the carriers, a commonplace in Populist times, is virtually unknown today. If discriminations still exist between persons or places, the Interstate Commerce Commission is apt to be as much to blame as the railroads. Free passes, so numerous in Populist times as to occasion the remark that the only people who did not have passes were those who could not afford to pay their own fare, have virtually ceased to be issued except to railway employes. Railway control of state governments, even in the old Granger states, where in earlier days party bosses took their orders directly from railway officials, has long since become a thing of the past. The railroads still may have an influence in politics, but the railroads do not rule. Governmental control of telephones, telegraphs, and pipe lines, together with such later developments as the radio and the transmission of electric power, is accepted today as a matter of course, the issues being merely to what extent control should go and through what agencies it should be accomplished.

For the trust problem, as distinguished from the railroad problem, the Populists had no very definite solution. They agreed, however, that the power of government, state and national, should be used in such a way as to prevent "indi-

[21] William Z. Ripley, *Railroads; Rates and Regulation;* Homer B. Vanderblue and Kenneth F. Burgess, *Railroads. Rates — Service — Management;* David Philip Locklin, *Railroad Regulation since 1920.*

viduals or corporations fastening themselves, like vampires, on the people and sucking their substance." [22] Antitrust laws received the earnest approval of Alliancemen and Populists and were often initiated by them. The failure of such laws to secure results was laid mainly at the door of the courts, and when Theodore Roosevelt in 1904 succeeded in securing an order from the United States Supreme Court dissolving the Northern Securities Company, it was hailed as a great victory for Populist principles. Many other incidental victories were won. Postal savings banks "for the safe deposit of the earnings of the people" encroached upon the special privileges of the bankers. An amendment to the national constitution in 1913, authorizing income taxes, recalled a contrary decision of the Supreme Court, which the Populists in their day had cited as the best evidence of the control of the government by the trusts; and income and inheritance taxes have ever since been levied. The reform of state and local taxation so as to exact a greater proportion of the taxes from the trusts and those who profit from them has also been freely undertaken. Labor demands, such as the right of labor to organize, the eight-hour day, limitation of injunctions in labor disputes, and restrictions on immigration were strongly championed by the Populists as fit measures for curbing the power of the trusts and were presently treated with great consideration. The Clayton Antitrust Act and the Federal Trade Commission Act, passed during the Wilson régime, were the products of long experience with the trust problem. The manner in which these laws have been enforced, however, would seem to indicate that the destruction of the trusts, a common demand in Populist times, is no longer regarded as feasible and that by government control the interests of the people can best be conserved.[23]

On the land question the Populist demands distinctly

[22] See the Cincinnati platform in the *American*, 27:167 (September 10, 1898).
[23] Eliot Jones, *The Trust Problem in the United States;* Henry R. Seager and Charles A. Gulick, *Trust and Corporation Problems;* Myron W. Watkins, *Industrial Combinations and Public Policy.*

foreshadowed conservation. "The land," according to the Omaha declaration, "including all the natural resources of wealth, is the heritage of all the people and should not be monopolized for speculative purposes." Land and resources already given away were of course difficult to get back, and the passing of the era of free lands could not be repealed by law, but President Roosevelt soon began to secure results in the way of the reclamation and irrigation of arid western lands, the enlargement and protection of the national forests, the improvement of internal waterways, and the withdrawal from entry of lands bearing mineral wealth such as coal, oil, and phosphates. At regular intervals, since 1908, the governors of the states have met together in conference to discuss the conservation problem, and this once dangerous Populist doctrine has now won all but universal acceptance.[24]

It would thus appear that much of the Populist program has found favor in the eyes of later generations. Populist plans for altering the machinery of government have, with but few exceptions, been carried into effect. Referring to these belated victories of the Populists, William Allen White, the man who had once asked, "What's the matter with Kansas?" wrote recently, "They abolished the established order completely and ushered in a new order."[25] Mrs. Mary E. Lease looked back proudly in 1914 on her political career:

In these later years I have seen, with gratification, that my work in the good old Populist days was not in vain. The Progressive party has adopted our platform, clause by clause, plank by plank. Note the list of reforms which we advocated which are coming into reality. Direct election of senators is assured. Public utilities are gradually being removed from the hands of the few and placed under the control of the people who use them. Woman suffrage is now almost a national issue. . . . The seed we sowed out in Kansas did not fall on barren ground.[26]

[24] Theodore Roosevelt, *Autobiography*, ch. 11; Charles R. Van Hise, *The Conservation of Natural Resources in the United States.*
[25] White, in *Scribners' Magazine*, 79:564.
[26] *Kansas City Star*, March 29, 1931.

Thanks to this triumph of Populist principles, one may almost say that, in so far as political devices can insure it, the people now rule. Political dishonesty has not altogether disappeared and the people may yet be betrayed by the men they elect to office, but on the whole the acts of government have come to reflect fairly clearly the will of the people. Efforts to assert this newly won power in such a way as to crush the economic supremacy of the predatory few have also been numerous and not wholly unsuccessful. The gigantic corporations of today, dwarfing into insignificance the trusts of yesterday, are, in spite of their size, far more circumspect in their conduct than their predecessors. If in the last analysis " big business " controls, it is because it has public opinion on its side and not merely the party bosses.

To radicals of today, however, the Populist panaceas, based as they were upon an essentially individualistic philosophy and designed merely to insure for every man his right to " get ahead " in the world, seem totally inadequate. These latter-day extremists point to the perennial reappearance of such problems as farm relief, unemployment, unfair taxation, and law evasion as evidence that the Populist type of reform is futile, that something more drastic is required. Nor is their contention without point. It is reasonable to suppose that progressivism itself must progress; that the programs that would provide solutions for the problems of one generation might fall far short of meeting the needs of a succeeding generation. Perhaps one may not agree with the view of some present-day radicals that only a revolution will suffice and that the very attempt to make existing institutions more tolerable is treason to any real progress, since by so doing the day of revolution is postponed; but one must recognize that when the old Populist panaceas can receive the enthusiastic support of Hooverian Republicans and Alsmithian Democrats these once startling demands are no longer radical at all. One is reminded of the dilemma that

Alice of Wonderland encountered when she went through the looking-glass into the garden of live flowers. On and on she ran with the Red Queen, but however fast they went they never seemed to pass anything.

"Well, in our country," said Alice, still panting a little, "you'd generally get to somewhere else — if you ran very fast for a long time as we've been doing."

"A slow sort of country!" said the Queen, "Now here, you see, it takes all the running you can do to keep in the same place. If you want to get somewhere else, you must run twice as fast as that!"

APPENDIXES

APPENDIX A

ST. LOUIS DEMANDS, DECEMBER, 1889

SOUTHERN ALLIANCE AND KNIGHTS OF LABOR [1]

Agreement made this day by and between the undersigned committee representing the National Farmers' Alliance and Industrial Union on the one part, and the undersigned committee representing the Knights of Labor on the other part, Witnesseth: The undersigned committee representing the Knights of Labor having read the demands of the National Farmers' Alliance and Industrial Union which are embodied in this agreement hereby endorse the same on behalf of the Knights of Labor, and for the purpose of giving practical effect to the demands herein set forth, the legislative committees of both organizations will act in concert before Congress for the purpose of securing the enactment of laws in harmony with the demands mutually agreed.

And it is further agreed, in order to carry out these objects, we will support for office only such men as can be depended upon to enact these principles in statute law uninfluenced by party caucus.

The demands hereinbefore referred to are as follows:

1. That we demand the abolition of national banks and the substitution of legal tender treasury notes in lieu of national bank notes, issued in sufficient volume to do the business of the country on a cash system; regulating the amount needed, on a per capita basis as the business interests of the country expand; and that all money issued by the Government shall be legal tender in payment of all debts, both public and private.

2. That we demand that Congress shall pass such laws as shall effectually prevent the dealing in futures of all agricultural and mechanical productions, preserving a stringent system of

[1] From the *National Economist*, 2:214–215 (December 21, 1889). See also the *St. Louis Republic*, December 7, 1889.

427

procedure in trials as shall secure the prompt conviction, and imposing such penalties as shall secure the most perfect compliance with the law.

3. That we demand the free and unlimited coinage of silver.

4. That we demand the passage of laws prohibiting the alien ownership of land, and that Congress take early steps to devise some plan to obtain all lands now owned by aliens and foreign syndicates; and that all lands now held by railroad and other corporations in excess of such as is actually used and needed by them, be reclaimed by the Government and held for actual settlers only.

5. Believing in the doctrine of " equal rights to all, and special privileges to none," we demand that taxation, National or State, shall not be used to build up one interest or class at the expense of another.

We believe that the money of the country should be kept as much as possible in the hands of the people, and hence we demand that all revenues, National, State, or county, shall be limited to the necessary expenses of the Government economically and honestly administered.

6. That Congress issue a sufficient amount of fractional paper currency to facilitate exchange through the medium of the United States mail.

7. We demand that the means of communication and transportation shall be owned by and operated in the interest of the people as is the United States postal system.

PLATFORM OF THE NORTHERN ALLIANCE [2]

Whereas, the farmers of the United States are most in number of any order of citizens, and with the other productive classes have freely given of their blood to found and maintain the nation; therefore be it

Resolved, that the public land, the heritage of the people, be reserved for actual settlers only, and that measures be taken to prevent aliens from acquiring titles to lands in the United States

[2] From the *Political Science Quarterly*, 6:293–294 (January, 1891). See also the *St. Louis Republic*, December 6, 7, 1889.

and Territories, and that the law be rigidly enforced against all railroad corporations which have not complied with the terms of their contract, by which they have received large grants of land.

2. We demand the abolition of the national banking system and that the government issue full legal tender money direct to the people in sufficient volume for the requirements of business.

3. We favor the payment of the public debt as rapidly as possible, and we earnestly protest against maintaining any bonds in existence as the basis for the issue of money.

4. We favor a graded income tax, and we also favor a tax on real-estate mortgages.

5. We demand economy and retrenchment as far as is consistent with the interests of the people in every department of the government, and we will look with special disfavor upon any increase of the official salaries of our representatives or government employees.

6. We favor such a revision and reduction of the tariff that the taxes may rest as lightly as possible upon productive labor and that its burdens may be upon the luxuries and in a manner that will prevent the accumulation of a United States Treasury surplus.

7. The stability of our government depends upon the moral, manual and intellectual training of the young, and we believe in so amending our public school system that the education of our children may inculcate the essential dignity necessary to be a practical help to them in after life.

8. Our railroads should be owned and managed by the government, and be run in the interest of the people upon an actual cash basis.

9. That the government take steps to secure the payment of the debt of the Union and Central Pacific railroads and their branches by foreclosure and sale, and any attempt to extend the time again for the payment of the same beyond its present limit will meet with our most emphatic condemnation.

10. We are in favor of the early completion of a ship canal connecting the great lakes with the Gulf of Mexico, and a deep

water harbor on the southern coast in view of opening trade relations with the Central and South American states, and we are in favor of national aid to a judicious system of experiments to determine the practicability of irrigation.

11. We sympathize with the just demands of labor of every grade and recognize that many of the evils from which the farming community suffers oppress universal labor, and that therefore producers should unite in a demand for the reform of unjust systems and the repeal of laws that bear unequally upon the people.

12. We favor the Australian system, or some similar system of voting, and ask the enactment of laws regulating the nomination of candidates for public office.

13. We are in favor of the diversification of our productive resources.

14. We will favor and assist to office such candidates only as are thoroughly identified with our principles and who will insist on such legislation as shall make them effective.

APPENDIX B

THE OCALA DEMANDS, DECEMBER, 1890[1]

1. a. We demand the abolition of national banks.

b. We demand that the government shall establish sub-treasuries or depositories in the several states, which shall loan money direct to the people at a low rate of interest, not to exceed two per cent per annum, on non-perishable farm products, and also upon real estate, with proper limitations upon the quantity of land and amount of money.

c. We demand that the amount of the circulating medium be speedily increased to not less than $50 per capita.

2. We demand that Congress shall pass such laws as will effectually prevent the dealing in futures of all agricultural and

[1] From the *Proceedings of the Supreme Council of the National Farmers' Alliance and Industrial Union,* 1890, pp. 32–33.

mechanical productions; providing a stringent system of procedure in trials that will secure the prompt conviction, and imposing such penalties as shall secure the most perfect compliance with the law.

3. We condemn the silver bill recently passed by Congress, and demand in lieu thereof the free and unlimited coinage of silver.

4. We demand the passage of laws prohibiting alien ownership of land, and that Congress take prompt action to devise some plan to obtain all lands now owned by aliens and foreign syndicates; and that all lands now held by railroads and other corporations in excess of such as is actually used and needed by them be reclaimed by the government and held for actual settlers only.

5. Believing in the doctrine of equal rights to all and special privileges to none, we demand—

a. That our national legislation shall be so framed in the future as not to build up one industry at the expense of another.

b. We further demand a removal of the existing heavy tariff tax from the necessities of life, that the poor of our land must have.

c. We further demand a just and equitable system of graduated tax on incomes.

d. We believe that the money of the county should be kept as much as possible in the hands of the people, and hence we demand that all national and state revenues shall be limited to the necessary expenses of the government economically and honestly administered.

6. We demand the most rigid, honest and just state and national government control and supervision of the means of public communication and transportation, and if this control and supervision does not remove the abuse now existing, we demand the government ownership of such means of communication and transportation.

7. We demand that the Congress of the United States submit an amendment to the Constitution providing for the election of United States Senators by direct vote of the people of each state.

APPENDIX C

ABSTRACT OF THE OMAHA RESOLUTIONS, JANUARY, 1891[1]

That we most emphatically declare against the present system of government as manipulated by the Congress of the United States and the members of the legislatures of the several states; therefore,

We declare in favor of holding a convention on February 22, 1892, to fix a date and place for the holding of a convention to nominate candidates for the office of President and Vice-President.

We declare that in the convention to be held on February 22, 1892, that representation shall be one delegate from each State in the Union.

That we favor the abolition of national banks, and that the surplus funds be loaned to individuals upon land security at a low rate of interest.

That we demand the foreclosure of mortgages that the government holds on railroads.

That the President and Vice-President of the United States should be elected by popular vote, instead of by an electoral college.

That the Alliance shall take no part as partisans in a political struggle by affiliating with Republicans or Democrats.

That we favor the free and unlimited coinage of silver.

That the volume of currency be increased to $50 per capita.

That all paper money be placed on an equality with gold.

That we as land-owners pledge ourselves to demand that the Government allow us to borrow money from the United States at the same rate of interest as do the banks.

That Senators of the United States shall be elected by vote of the people.

[1] From the *National Economist*, 4:333 (February 7, 1891). See also the *Omaha Daily Bee*, January 28, 1891.

APPENDIX D

CINCINNATI PLATFORM, MAY, 1891[1]

1. That in view of the great social, industrial and economical revolution now dawning on the civilized world and the new and living issues confronting the American people, we believe that the time has arrived for a crystalization of the political reform forces of our country and the formation of what should be known as the People's party of the United States of America.

2. That we most heartily endorse the demands of the platforms as adopted at St. Louis, Missouri, in 1889; Ocala, Florida, in 1890, and Omaha, Nebraska, in 1891, by industrial organizations there represented, summarized as follows:

a. The right to make and issue money is a sovereign power to be maintained by the people for the common benefit. Hence we demand the abolition of national banks as banks of issue, and as a substitute for national bank notes we demand that legal tender Treasury notes be issued in sufficient volume to transact the business of the country on a cash basis without damage or especial advantage to any class or calling, such notes to be legal tender in payment of all debts, public and private, and such notes, when demanded by the people, shall be loaned to them at not more than 2 per cent per annum upon non-perishable products, as indicated in the sub-treasury plan, and also upon real estate, with proper limitation upon the quantity of land and amount of money.

b. We demand the free and unlimited coinage of silver.

c. We demand the passage of laws prohibiting alien ownership of land, and that Congress take prompt action to devise some plan to obtain all lands now owned by alien and foreign syndicates, and that all land held by railroads and other corporations in excess of such as is actually used and needed by them be reclaimed by the government, and held for actual settlers only.

[1] From the *National Economist*, 5:162 (May 30, 1891). See also the *Farmers' Alliance*, May 28, 1891.

d. Believing the doctrine of equal rights for all and special privileges to none, we demand that taxation, national, state, or municipal, shall not be used to build up one interest or class at the expense of another.

e. We demand that all revenue — national, state, or county — shall be limited to the necessary expenses of the government, economically and honestly administered.

f. We demand a just and equitable system of graduated tax on income.

g. We demand the most rigid, honest, and just national control and supervision of the means of public communication and transportation, and if this control and supervision does not remove the abuses now existing we demand the government ownership of such means of communication and transportation.

h. We demand the election of President, vice-President and United States Senators by a direct vote of the people.

3. That we urge the united action of all progressive organizations in attending the conference called for February 22, 1892, by six of the leading reform organizations.

4. That a national central committee be appointed by this conference to be composed of a chairman, to be elected by this body, and of three members from each state represented, to be named by each state delegation.

5. That this central committee shall represent this body, attend the national conference on February 22, 1892, and, if possible, unite with that and all other reform organizations there assembled. If no satisfactory arrangement can be effected, this committee shall call a national convention not later than June 1, 1892, for the purpose of nominating candidates for President and vice-President.

6. That the members of the central committee for each state where there is no independent political organization, conduct an active system of political agitation in their respective states.

Resolved,[2] That the question of universal suffrage be recommended to the favorable consideration of the various states and territories.

[2] The following resolutions were regarded as the sense of the convention, but not as an integral part of the platform.

Resolved, That while the party in power in 1879 pledged the faith of the nation to pay a debt in coin that had been contracted on a depreciated currency, thus adding nearly $1,000,000,000 to the burdens of the people, which meant gold for the bondholders and depreciated currency for the soldier, and holding that the men who imperiled their lives to save the life of a nation should have been paid in money as good as that paid to the bondholders — we demand the issue of legal tender and treasury notes in sufficient amount to make the pay of the soldiers equal to par with coin, or such other legislation as shall do equal and exact justice to the Union soldiers of this country.

Resolved, That as eight hours constitute a legal day's work for government employees in mechanical departments, we believe this principle should be further extended so as to apply to all corporations employing labor in the different states of the Union.

Resolved, That this conference condemns in unmeasured terms the action of the directors of the World's Columbian Exposition on May 19, in refusing the minimum rate of wages asked for by the labor organizations of Chicago.

Resolved, That the Attorney General of the United States should make immediate provision to submit the act of March 2, 1889, providing for the opening of Oklahoma to homestead settlement, to the United States Supreme Court, so that the expensive and dilatory litigation now pending there be ended.

APPENDIX E

ST. LOUIS PLATFORM, FEBRUARY, 1892[1]

Preamble

This, the first great labor conference of the United States, and of the world, representing all divisions of urban and rural organized industry, assembled in national congress, invoking upon its action the blessing and protection of Almighty God, puts

[1] From the *National Economist*, 6:396 (March 5, 1892). See also the *Farmers' Alliance*, March 3, 1892.

forth, to and for the producers of the nation, this declaration of union and independence.

The conditions which surround us best justify our co-operation.

We meet in the midst of a nation brought to the verge of moral, political and material ruin. Corruption dominates the ballot box, the legislatures, the Congress, and touches even the ermine of the bench. The people are demoralized. Many of the States have been compelled to isolate the voters at the polling places in order to prevent universal intimidation or bribery. The newspapers are subsidized or muzzled; public opinion silenced; business prostrated, our homes covered with mortgages, labor impoverished, and the land concentrating in the hands of capitalists. The urban workmen are denied the right of organization for self-protection; imported pauperized labor beats down their wages; a hireling standing army, unrecognized by our laws, is established to shoot them down, and they are rapidly degenerating to European conditions. The fruits of the toil of millions are boldly stolen to build up colossal fortunes, unprecedented in the history of the world, while their possessors despise the republic and endanger liberty. From the same prolific womb of governmental injustice we breed two great classes — paupers and millionaires. The national power to create money is appropriated to enrich bondholders; silver, which has been accepted as coin since the dawn of history, has been demonetized to add to the purchasing power of gold by decreasing the value of all forms of property as well as human labor; and the supply of currency is purposely abridged to fatten usurers, bankrupt enterprise and enslave industry. A vast conspiracy against mankind has been organized on two continents and is taking possession of the world. If not met and overthrown at once it forbodes terrible social convulsions, the destruction of civilization, or the establishment of an absolute despotism.

In this crisis of human affairs the intelligent working people and producers of the United States have come together in the name of peace, order and society, to defend liberty, prosperity and justice.

We declare our union and independence. We assert one purpose to support the political organization which represents our principles.

We charge that the controlling influences dominating the old political parties have allowed the existing dreadful conditions to develop without serious effort to restrain or prevent them. They have agreed together to ignore in the coming campaign every issue but one. They propose to drown the outcries of a plundered people with the uproar of a sham battle over the tariff, so that corporations, national banks, rings, trusts, " watered stocks," the demonetization of silver, and the oppressions of usurers, may all be lost sight of. They propose to sacrifice our homes and children upon the altar of Mammon; to destroy the hopes of the multitude in order to secure corruption funds from the great lords of plunder. We assert that a political organization, representing the political principles herein stated, is necessary to redress the grievances of which we complain.

Assembled on the anniversary of the birth of the illustrious man who led the first great revolution on this continent against oppression, filled with the sentiments which actuated that grand generation, we seek to restore the government of the republic to the hands of the " plain people," with whom it originated. Our doors are open to all points of the compass. We ask all men to join with us and help us.

In order to restrain the extortions of aggregated capital, to drive the money changers out of the temple, " to form a [more] perfect union, establish justice, insure domestic tranquility, provide for the common defense, promote the general welfare and secure the blessings of liberty for ourselves and our posterity," we do ordain and establish the following platform of principles:

First — We declare the union of the labor forces of the United States this day accomplished permanent and perpetual. May its spirit enter into all hearts for the salvation of the republic and the uplifting of mankind.

Second — Wealth belongs to him who creates it. Every dollar taken from industry without an equivalent is robbery. If any one will not work, neither shall he eat. The interests of rural and urban labor are the same, their enemies are identical.

Platform [2]

First — We demand a national currency, safe, sound and flexible, issued by the general government only, a full legal tender for all debts, public and private; and that without the use of banking corporations a just, equitable and efficient means of distribution direct to the people at a tax not to exceed 2 per cent be provided as set forth in the sub-treasury plan of the Farmers' Alliance, or some better system; also, by payments in discharge of its obligations for public improvements.

a. We demand free and unlimited coinage of silver.

b. We demand that the amount of circulating medium be speedily increased to not less than $50 per capita.

c. We demand a graduated income tax.

d. We believe that the money of the country should be kept as much as possible in the hands of the people, and hence we demand all national and state revenue shall be limited to the necessary expenses of the government, economically and honestly administered.

e. We demand that postal savings banks be established by the government for the safe deposit of the earnings of the people and to facilitate exchange.

Second — The land, including all the natural resources of wealth, is the heritage of all the people and should not be monopolized for speculative purposes, and alien ownership of land should be prohibited. All land now held by railroads and other corporations in excess of their actual needs, and all lands now owned by aliens, should be reclaimed by the government and held for actual settlers only.

Third — Transportation being a means of exchange and a public necessity, the government should own and operate the railroads in the interests of the people.

[2] This platform, as originally reported by the committee, had no numbers in it, and the division into three planks was made after the adjournment of the convention, probably in the office of the *National Economist*. The three resolutions at the end of the platform were held not to be a part of the platform proper but merely an expression of the sentiments of the convention. The St. Louis papers, however, printed the platform and resolutions together, numbering fourteen separate paragraphs. Hence the charge, frequently made, that the convention adopted two platforms, one to be published in the South, and another with flattering promises for the union veterans, to be published in the North. *National Economist*, 6:402; 7:200 (March 19, June 11, 1892).

a. The telegraph and telephone, like the postoffice system, being a necessity for transmission of news, should be owned and operated by the government in the interest of the people.

RESOLUTIONS

Resolved, That the question of female suffrage be referred to the legislatures of the different States for favorable consideration.

Resolved, That the government should issue legal tender notes and pay the Union soldier the difference between the price of the depreciated money in which he was paid and gold.

Resolved, That we hail this conference as the consummation of a perfect union of hearts and hands of all sections of our common country. The men who wore the gray and the men who wore the blue meet here to extinguish the last smoldering embers of civil war in the tears of joy of a united and happy people, and we agree to carry the stars and stripes forward to the highest point of national greatness.

APPENDIX F

OMAHA PLATFORM, JULY, 1892[1]

Assembled upon the 116th anniversary of the Declaration of Independence, the People's Party of America, in their first national convention, invoking upon their action the blessing of Almighty God, puts forth, in the name and on behalf of the people of this country, the following preamble and declaration of principles: —

The conditions which surround us best justify our cooperation: we meet in the midst of a nation brought to the verge of moral, political, and material ruin. Corruption dominates the ballot-box, the legislatures, the Congress, and touches even the ermine of the bench. The people are demoralized; most of the States have been compelled to isolate the voters at the polling-places to prevent universal intimidation or bribery. The newspapers are largely subsidized or muzzled; public opinion si-

[1] From Stanwood, *History of the Presidency*, 1:509–513. See also the *National Economist*, 7:257–258 (July 9, 1892), and the *Tribune Almanac, 1892*, 38–40.

lenced; business prostrated; our homes covered with mortgages; labor impoverished; and the land concentrating in the hands of the capitalists. The urban workmen are denied the right of organization for self-protection; imported pauperized labor beats down their wages; a hireling standing army, unrecognized by our laws, is established to shoot them down, and they are rapidly degenerating into European conditions. The fruits of the toil of millions are boldly stolen to build up colossal fortunes for a few, unprecedented in the history of mankind; and the possessors of these, in turn, despise the republic and endanger liberty. From the same prolific womb of governmental injustice we breed the two great classes — tramps and millionaires.

The national power to create money is appropriated to enrich bondholders; a vast public debt, payable in legal tender currency, has been funded into gold-bearing bonds, thereby adding millions to the burdens of the people. Silver, which has been accepted as coin since the dawn of history, has been demonetized to add to the purchasing power of gold by decreasing the value of all forms of property as well as human labor; and the supply of currency is purposely abridged to fatten usurers, bankrupt enterprise, and enslave industry. A vast conspiracy against mankind has been organized on two continents, and it is rapidly taking possession of the world. If not met and overthrown at once, it forebodes terrible social convulsions, the destruction of civilization, or the establishment of an absolute despotism.

We have witnessed for more than a quarter of a century the struggles of the two great political parties for power and plunder, while grievous wrongs have been inflicted upon the suffering people. We charge that the controlling influences dominating both these parties have permitted the existing dreadful conditions to develop without serious effort to prevent or restrain them. Neither do they now promise us any substantial reform. They have agreed together to ignore in the coming campaign every issue but one. They propose to drown the outcries of a plundered people with the uproar of a sham battle over the tariff, so that capitalists, corporations, national banks, rings, trusts, watered stock, the demonetization of silver, and the oppressions of the usurers may all be lost sight of. They propose to sacrifice our homes, lives and children on the altar of mammon; to destroy the multitude in order to secure corruption funds from the millionaires.

Assembled on the anniversary of the birthday of the nation, and filled with the spirit of the grand general and chieftain who established our independence, we seek to restore the government of the Republic to the hands of "the plain people," with whose class it originated. We assert our purposes to be identical with the purposes of the National Constitution, "to form a more perfect union and establish justice, insure domestic tranquillity, provide for the common defence, promote the general welfare, and secure the blessings of liberty for ourselves and our posterity." We declare that this republic can only endure as a free government while built upon the love of the whole people for each other and for the nation; that it cannot be pinned together by bayonets; that the civil war is over, and that every passion and resentment which grew out of it must die with it; and that we must be in fact, as we are in name, one united brotherhood of freemen.

Our country finds itself confronted by conditions for which there is no precedent in the history of the world; our annual agricultural productions amount to billions of dollars in value, which must, within a few weeks or months, be exchanged for billions of dollars of commodities consumed in their production; the existing currency supply is wholly inadequate to make this exchange; the results are falling prices, the formation of combines and rings, the impoverishment of the producing class. We pledge ourselves, if given power, we will labor to correct these evils by wise and reasonable legislation, in accordance with the terms of our platform. We believe that the powers of government — in other words, of the people — should be expanded (as in the case of the postal service) as rapidly and as far as the good sense of an intelligent people and the teachings of experience shall justify, to the end that oppression, injustice, and poverty shall eventually cease in the land.

While our sympathies as a party of reform are naturally upon the side of every proposition which will tend to make men intelligent, virtuous, and temperate, we nevertheless regard these questions — important as they are — as secondary to the great issues now pressing for solution, and upon which not only our individual prosperity but the very existence of free institutions depends; and we ask all men to first help us to determine whether we are to have a republic to administer before we differ as to the conditions upon which it is to be administered; believing that the forces of reform this day organized will never cease to move forward until every wrong is remedied, and equal rights and equal privileges securely established for all the men and women of this country.

We declare, therefore,—

First. That the union of the labor forces of the United States this day consummated shall be permanent and perpetual; may its spirit enter all hearts for the salvation of the republic and the uplifting of mankind!

Second. Wealth belongs to him who creates it, and every dollar taken from industry without an equivalent is robbery. "If any will not work, neither shall he eat." The interests of rural and civic labor are the same; their enemies are identical.

Third. We believe that the time has come when the railroad corporations will either own the people or the people must own the railroads; and, should the government enter upon the work of owning and managing all railroads, we should favor an amendment to the Constitution by which all persons engaged in the government service shall be placed under a civil service regulation of the most rigid character, so as to prevent the increase of the power of the national administration by the use of such additional government employees.

First, *Money.* We demand a national currency, safe, sound, and flexible, issued by the general government only, a full legal tender for all debts, public and private, and that, without the use of banking corporations, a just, equitable, and efficient means of distribution direct to the people, at a tax not to exceed two per cent per annum, to be provided as set forth in the sub-treasury plan of the Farmers' Alliance, or a better system; also, by payments in discharge of its obligations for public improvements.

(a) We demand free and unlimited coinage of silver and gold at the present legal ratio of sixteen to one.

(b) We demand that the amount of circulating medium be speedily increased to not less than fifty dollars per capita.

(c) We demand a graduated income tax.

(d) We believe that the money of the country should be kept as much as possible in the hands of the people, and hence we demand that all state and national revenues shall be limited to the necessary expenses of the government economically and honestly administered.

(e) We demand that postal savings banks be established by the government for the safe deposit of the earnings of the people and to facilitate exchange.

Second, *Transportation.* Transportation being a means of exchange and a public necessity, the government should own and operate the railroads in the interest of the people.

(a) The telegraph and telephone, like the post-office system, being a necessity for the transmission of news, should be owned and operated by the government in the interest of the people.

Third, *Land.* The land, including all the natural sources of wealth, is the heritage of the people, and should not be monopolized for speculative purposes, and alien ownership of land should be prohibited. All land now held by railroads and other corporations in excess of their actual needs, and all lands now owned by aliens, should be reclaimed by the government and held for actual settlers only.

RESOLUTIONS

Whereas, Other questions have been presented for our consideration, we hereby submit the following, not as a part of the platform of the People's party, but as resolutions expressive of the sentiment of this convention.

1. *Resolved,* That we demand a free ballot and a fair count in all elections, and pledge ourselves to secure it to every legal voter without federal intervention, through the adoption by the States of the unperverted Australian or secret ballot system.

2. *Resolved,* That the revenue derived from a graduated income tax should be applied to the reduction of the burden of taxation now resting upon the domestic industries of this country.

3. *Resolved,* That we pledge our support to fair and liberal pensions to ex-Union soldiers and sailors.

4. *Resolved,* That we condemn the fallacy of protecting American labor under the present system, which opens our ports to the pauper and criminal classes of the world, and crowds out our wage-earners; and we denounce the present ineffective laws against contract labor, and demand the further restriction of undesirable immigration.

5. *Resolved,* That we cordially sympathize with the efforts of organized workingmen to shorten the hours of labor, and demand a rigid enforcement of the existing eight-hour law on government work, and ask that a penalty clause be added to the said law.

6. *Resolved,* That we regard the maintenance of a large standing army of mercenaries, known as the Pinkerton system, as a menace to our liberties, and we demand its abolition; and we condemn the recent invasion of the Territory of Wyoming by the hired assassins of plutocracy, assisted by federal officials.

7. *Resolved,* That we commend to the favorable consideration of the people and the reform press the legislative system known as the initiative and referendum.

8. *Resolved,* That we favor a constitutional provision limiting the office of President and Vice-President to one term, and providing for the election of senators of the United States by a direct vote of the people.

9. *Resolved,* That we oppose any subsidy or national aid to any private corporation for any purpose.

10. *Resolved,* That this convention sympathizes with the Knights of Labor and their righteous contest with the tyrannical combine of clothing manufacturers of Rochester, and declares it to be the duty of all who hate tyranny and oppression to refuse to purchase the goods made by said manufacturers, or to patronize any merchants who sell such goods.

BIBLIOGRAPHY

BIBLIOGRAPHY

SOURCES

MANUSCRIPTS AND OFFICIAL PROCEEDINGS

The Donnelly Papers in the possession of the Minnesota Historical Society constitute probably the most extensive collection of Populist material in existence. They include letters to Donnelly from as early as 1856 to as late as 1901. For the most part Donnelly's own letters are not available, but occasionally the first draft of an important letter is preserved, and during the Populist period Donnelly sometimes made use of letter books, several of which are available. He also kept clippings and pamphlets that interested him, and about twenty volumes of scrapbooks filled with such material are available for preservation. The Weller Papers in the possession of the State Historical Society of Wisconsin are of far less consequence. They include, however, many interesting letters to Lemuel H. Weller of Iowa from other and more important third-party leaders. The Allen Papers are still retained by a member of Senator William V. Allen's family, Mrs. W. L. Dowling of Madison, Nebraska. Through the mediation of Miss Mittie Y. Scott permission was secured to examine a large portion of this collection, which is extremely rich in material on the period of the nineties. The Maxwell Correspondence in the possession of the Nebraska Historical Society contains letters to Judge Maxwell from 1853 to 1901. Maxwell's prominent place among the Populists from 1893 on makes this collection of considerable value.

The official proceedings and record books of many national, state, and subordinate farm orders and of a number of Populist organizations are to be found, sometimes in the original manuscript form and sometimes printed, in the various historical society libraries of the states where Populism flourished and in the Library of Congress. Literature of this sort, however, is too widely scattered to be of much use; moreover, nearly all the material it contains is to be found printed at full length in the reform newspapers. Occasional references are made to the original documents in the footnotes, but ordinarily the newspaper accounts are cited.

FEDERAL DOCUMENTS

Much useful material is scattered through the annual reports of the commissioner of agriculture, the yearbooks of the United States Department of Agriculture, the *Statistical Abstract* of the United States for various years, the congressional directories, the *Congressional Globe* and the *Congressional Record*, the *United States Statutes at Large,* the reports of the United States Supreme Court, and the census reports. A few congressional documents of especial importance should also be noted:

Serial
Number

791–801,   Reports of Explorations and Surveys to Ascertain the Most Practicable
758–768   and Economical Route for a Railroad from the Mississippi River to
      the Pacific Ocean, 1855. *House Executive Document* No. 91, 33 Congress, Session 2.

Serial
Number

2134   Report on Cotton Production in the United States, Tenth Census of
the United States, Vol. 6 (Washington, 1884). *House Miscellaneous
Document* No. 42, 47 Congress, Session 2.

2356   Report of the Senate Select Committee on Interstate Commerce, 1886.
2357   *Senate Report* No. 46, 49 Congress, Session 1, Parts 1 and 2. The
Cullom report.

3051   System of Subtreasuries, Report of the House Ways and Means Committee, 1892. *House Report* No. 2143, 52 Congress, Session 1. An
attack on the subtreasury plan.

3074   Wholesale Prices, Wages, and Transportation, Report by Mr. Aldrich
from the Committee on Finance, March 3, 1893. *Senate Report* No.
1394, 52 Congress, Session 2, Parts 1–4.

3173   Report on Real Estate Mortgages in the United States at the Eleventh
Census, 1890. *House Miscellaneous Document* No. 340, 52 Congress,
Session 1, Part 23. A special investigation by George K. Holmes and
John S. Lord.

3290   Condition of Cotton Growers in the United States, the Present Prices
of Cotton, and the Remedy, Report of the United States Senate Committee on Agriculture and Forestry, 1895. *Senate Report* No. 986, 53
Congress, Session 3. An elaborate investigation including much testimony from cotton growers.

7922   The Agricultural Crisis and Its Causes, Report of the Joint Commission of Agricultural Inquiry, 1921. *House Report* No. 408, 67 Congress, Session 1. Report of the investigation made during Harding's
administration.

STATE DOCUMENTS

In the states where Populism was strong — notably Colorado, Kansas, Minnesota, Nebraska, and North Carolina — such official documents as Senate and
House journals, session laws, and the reports of railway commissions, bureaus
of labor statistics, and state boards of agriculture are exceedingly valuable.
The following documents have some particular importance.

Colorado: Governor's message, *House Journal, 1895* (Denver, 1895), 19–64. In
this document Governor Waite reviews the whole history of his adminis-
tration.
Kansas: Governor's message, *House Journal, 1895* (Topeka, 1895), 44–47. Re-
views the history of the legislative war of 1893.
———   *Seventh Biennial Report of the State Board of Agriculture,* 1889–90
(Topeka, 1890). Contains information on the decline of population in Kan-
sas after 1887.
Minnesota: *Governor's Message, 1895* (St. Paul, 1895). Printed separately.
Contains interesting recommendations on the railroad and elevator problems.
———   *Report of the Pine Land Investigating Committee to the Governor*
(St. Paul, 1895). The work of a Minnesota committee headed by Ignatius
Donnelly to uncover frauds.
Nebraska: *Annual Report of the Board of Transportation, 1890* (Omaha,
1890). Full details of efforts at regulation.
———   *Biennial Report of the Attorney-General to the Governor, 1887–1888*
(Omaha, 1888). A good statement of western grievances against the rail-
roads.
———   *Biennial Report of the Bureau of Labor and Industrial Statistics,
1887–1888* (Omaha, 1888); *1893–1894* (Lincoln, 1894). Contain valuable
information on mortgage indebtedness within the state.

—————— *Fourth Annual Report of the President and Secretary of the State Board of Agriculture*, 1873 (Lincoln, 1873). Typical of the early optimistic advertising period.

—————— *Laws, Joint Resolution, and Memorials, 1893* (Lincoln, 1893), 164–348. The famous Newberry Act.

—————— *Report of the Board of Railway Commissioners, 1886* (Omaha, 1887). Contains the record of many typical complaints.

North Carolina: *Annual Report of the Bureau of Labor Statistics, 1887* (Raleigh, 1887). States forcefully some of the problems of southern agriculture.

—————— *Public Documents*, legislative sessions of 1889 to 1893 (Raleigh, 1889, 1891, 1893). Contains considerable material on the political situation precipitated by the Alliance and Populist movements.

COLLECTIONS AND COMPILATIONS

Appletons' Annual Cyclopaedia, 1880–1908. Contains much information of value on the Alliance and Populism in the various states.

Handbook of Facts and Alliance Information, Vol. 1, No. 1, *Library of National Economist Extras*, January, 1891 (Washington, 1890). Contains statistical information not wholly reliable and some well-known facts about early Alliance history.

McPHERSON, EDWARD (ed.), *A Handbook of Politics* (Washington, annually, 1888–92).

POOR, HENRY V. (ed.), *Manual of the Railroads of the United States* (New York, annually, 1868 to date). Railroad statistics.

TAYLOR, HOWARD S. and others, *The Battle of 1900, An Official Handbook for Every American Citizen* (Chicago, 1900). Contains the platforms and arguments of every political party.

Tribune Almanac, 1888–1900. Contains more or less dependable statistics.

AUTOBIOGRAPHIES, MEMOIRS, AND REMINISCENCES

BRYAN, WILLIAM J. and MARY B., *The Memoirs of William Jennings Bryan* (Philadelphia, 1925). Of comparatively little value for Populist affairs.

BUCHANAN, JOHN R., "The Great Railroad Migration into Northern Nebraska," *Proceedings and Collections of the Nebraska State Historical Society*, 15:25–34 (Lincoln, 1907). Buchanan was one of the advertisers for a local Nebraska railroad.

BURNAP, WILLARD A., *What Happened During One Man's Lifetime* (Fergus Falls, Minnesota, 1923). Contains some interesting chapters on the settlement of the West.

ERNST, CARL J., "The Railroad as a Creator of Wealth in the Development of a Community or District," *Nebraska History and Record of Pioneer Days*, (7:18–22, January–March, 1924). The writer records here some of his own experiences as head of the Burlington Land Department from 1876 to 1924.

HOLDREGE, GEORGE W., *The Making of the Burlington*. An address delivered before the Nebraska Historical Society on January 12, 1921. Printed as a pamphlet. Holdrege had a large part in making the Burlington.

LA FOLLETTE, ROBERT M., *Autobiography* (Madison, Wisconsin, 1913). The conditions that produced La Follette were in large part the conditions that produced Populism.

MANNING, JOSEPH C., *The Fadeout of Populism* (New York, 1928). The reminiscences of a prominent southern Populist.

—————— *From Five to Twenty-five* (New York, 1929). Focuses chiefly on the early life of the author.

PETTIGREW, RICHARD F., *Imperial Washington* (Chicago, 1922). A reformer's observations of American public life from 1870 to 1920.

POWDERLY, TERENCE V., *Thirty Years of Labor* (Columbus, 1889). Presents the side of labor.

RIGHTMIRE, W. F., " The Alliance Movement in Kansas — Origin of the People's Party," *Kansas State Historical Society Transactions,* 9:1–8 (Topeka, 1906). A reminiscent statement by a prominent participant.

ROOSEVELT, THEODORE, *Autobiography* (New York, 1890). Roosevelt gave little quarter to the Populists in their time, but he later adopted many of their doctrines.

SEYMOUR, CHARLES (ed.), *The Intimate Papers of Colonel House,* 4 vols. (Boston, 1926–28). Colonel House first became active in Texas politics during Alliance and Populist times.

SHELDON, ADDISON E., " Early Railroad Development of Nebraska," *Nebraska History and Record of Pioneer Days,* 7:16–17 (January–March, 1924). Brief comments by an eyewitness.

THOMPSON, J. M., "The Farmers' Alliance in Nebraska: Something of Its Origin, Growth, and Influence," *Proceedings and Collections of the Nebraska State Historical Society,* 10:199–206 (Lincoln, 1902). The author was an active Nebraska Allianceman.

WITHAM, JAMES W., *Fifty Years on the Firing Line* (Chicago, 1924). The recollections of an ex-agitator for farmers' rights.

VI. MISCELLANEOUS BOOKS AND PAMPHLETS

ALLEN, EMORY A., *Labor and Capital* (Cincinnati, 1891). Contains chapters on each of the prominent farm orders of the period.

The Alliance and Labor Songster (Winfield, Kansas, 1890). A typical Alliance songbook used in Kansas.

ARGUS, MODERN I. (pseud.), *Minor Chronicles of the Goodly Land of Texas Which Pertaineth to the Frauds Practiced in the Management of the Farmers' Alliance Exchange* (Austin, Texas, 1890). An attack on Macune's cooperative venture in Texas.

ASHBY, N. B., *The Riddle of the Sphinx* (Des Moines, Iowa, 1890). An outline of Alliance history and arguments for northern Alliancemen.

BELLAMY, EDWARD, *Looking Backward, 2000–1887* (Boston, 1888). From this source came many of the Populist arguments for government ownership.

BLISS, CHARLES H. (ed.), *The Populist Compendium. References for Reformers* (Auburn, Indiana, 1894). Pro-Populist utterances taken mainly from the writings and speeches of early reformers.

BLOOD, F. G., *Handbook and History of the National Farmers' Alliance and Industrial Union* (Washington, 1893). Useful on the origin and early history of the Southern Alliance.

BRYAN, J. E., *The Farmers' Alliance: Its Origin, Progress and Purposes* (Fayetteville, Arkansas, 1891). Good account of Alliance origin by an Alliance Worker. Alliance arguments well set forth.

BRYAN, WILLIAM J., *The First Battle* (Chicago, 1897). Gives only slight attention to the Populists.

CHAMBERLAIN, HENRY R., *The Farmers' Alliance; What It Aims to Accomplish* (New York, 1891). Expresses the point of view of a newspaper reporter.

CLARK, GORDON, *Handbook of Money (National Watchman Economic Quarterly,* Vol. 2, No. 1, September, 1896, Alexandria, Virginia). History of the money question and arguments from the Populist point of view.

DONNELLY, IGNATIUS, *The American People's Money* (Chicago, 1895). An argument for currency inflation.

———— *Caesar's Column* (Chicago, 1890). A novel, widely circulated, portraying the horrible results of failure to heed the farmer and labor demands.

———— *Facts for the Granges* (n.p., 1873). A copy of this pamphlet is in the library of the Minnesota Historical Society.

———— *The Golden Bottle; or, the Story of Ephraim Benezet of Kansas* (New York, 1892). A novel dealing with free silver and other reforms.

DUNNING, NELSON A., *The Farmers' Alliance History and Agricultural Digest* (Washington, 1891). One of the best of the contemporary Alliance histories.

ELLIOT, J. R., *American Farms; their Condition and Future* (New York, 1890). An interesting survey of the agricultural situation.

GARVIN, WILLIAM L., *History of the Grand State Farmers' Alliance of Texas* (Jonesboro, Texas, 1885). Gives the origins of the Southern Alliance in Texas.

GARVIN, WILLIAM L., and DAWS, S. O., *History of the National Farmers' Alliance and Co-operative Union of America* (Jacksboro, Texas, 1887). Early Alliance history and arguments.

HARVEY, WILLIAM H., *Coin's Financial School* (Chicago, 1894). The most effective of the free-silver tracts.

———— *Coins' Financial School Up-to-date* (Chicago, 1895).

———— *Coin on Money, Trusts, and Imperialism* (Chicago, 1899).

HUDSON, J. K., *Letters to Governor Lewelling* (Topeka, Kansas, 1893). Contains an anti-Populist version of the activities of the Kansas legislature of 1893.

JORY, T. C., *What is Populism? An Exposition of the Principles of the Omaha Platform, Adopted by the People's Party* (Salem, Oregon, 1895). A Populist handbook.

LARRABEE, WILLIAM, *The Railroad Question: A Historical and Practical Treatise on Railroads, and Remedies for their Abuses* (Chicago, 1893). Larrabee was a reform governor of Iowa in the eighties. His book was many times reprinted.

LAUGHLIN, J. LAURENCE, *Facts about Money* (Chicago, 1895). Includes a debate with " Coin " Harvey.

LLOYD, HENRY DEMAREST, *Wealth against Commonwealth* (New York, 1894). A powerful indictment of plutocracy.

MORGAN, W. SCOTT, *History of the Wheel and Alliance, and the Impending Revolution* (Fort Scott, Kansas, 1891). The official Southern Alliance history. This book, which was first published in 1889, went through many editions. Contains all the standard arguments used by the farmers.

OTKEN, CHARLES H., *The Ills of the South; or Related Causes Hostile to the General Prosperity of the Southern People* (New York, 1894). Economic rather than political. Attacks the crop mortgage system.

PEFFER, WILLIAM A., *The Farmer's Side, His Troubles and Their Remedy* (New York, 1891). One of the most widely read statements of the Populist creed.

STREETER, JOHN J., *Populist National Organization United States of America* (Vineland, New Jersey, 1907). A mid-road Populist statement of rules and articles of agreement with reference to the use of the initiative and referendum within the party.

WATSON, THOMAS E., *Life and Speeches of Thomas E. Watson* (Nashville, Tennessee, 1908). A campaign biography.

———— *Not a Revolt, It is a Revolution. The People's Party Campaign Book* (Washington, 1892). A compendium of Populist documents, arguments, and history.

WEAVER, JAMES BAIRD, *A Call to Action* (Des Moines, Iowa, 1892). The standard farmers' arguments by one of their best-known leaders.

CONTEMPORARY ARTICLES

ADAMS, HENRY C., " The Farmer and Railway Legislation," *Century*, 21:780–783 (March, 1892). States clearly the need of railway legislation.

ALLEN, WILLIAM V., " The Populist Program," *Independent*, 52:475–476 (February 22, 1900)

———— " Western Feeling towards the East," *North American Review*, 162:588–593 (May, 1896). Debtor versus creditor.

ATKINSON, EDWARD, "The True Meaning of the Farm Mortgage Statistics," *Forum*, 17:310–325 (May, 1894). Concludes that the western mortgage burden is relatively light.

BEMIS, EDWARD W., "Discontent of the Farmer," *Journal of Political Economy*, 1:193–213 (March, 1893). Appreciates the farmer's point of view.

BUTLER, MARION, "The People's Party," *Forum*, 28:658–662 (February, 1900). Butler was chairman of the Populist national committee and claimed that the party was not disintegrating.

CANFIELD, JAMES H., in "A Bundle of Western Letters," *Review of Reviews*, 10:42–43 (July, 1894). These letters give a good cross section of western opinion on the Populist demands.

———— "Is the West Discontented? A Local Study of Facts," *Forum*, 18:449–461 (December, 1894). Canfield was chancellor of the University of Nebraska. In this article he features the "contented" classes of the West.

CARLISLE, JOHN G., "The Tariff and the Farmer," *Forum*, 8:475–488 (January, 1890). An argument for low tariffs.

CHAMBERLAIN, HENRY R., "The Farmers' Alliance and Other Political Parties," *Chautauquan*, 13:338–342 (June, 1891). A popular article by a newspaper reporter.

CULLOM, SHELBY M., "Protection and the Farmer," *Forum*, 8:136–147 (October, 1889). Tries to show how protection helps the farmer.

DAVIS, C. WOOD, "The Farmer, the Investor and the Railway," *Arena*, 3:291–313 (February, 1891). States effectively the grievances against the roads.

———— "Why the Farmer Is Not Prosperous," *Forum*, 9:231–241 (April, 1890). A contemporary study in prices by a Kansas farmer.

DAVIS, R. MEANS, "The Matter with the Small Farmer," *Forum*, 14:381–389 (November, 1892). Concerned chiefly with the cotton planter.

DIGGS, MRS. ANNIE L., "The Farmers' Alliance and Some of Its Leaders," *Arena*, 5:590–604 (April, 1892). Laudatory sketches with photographs of leading Alliancemen.

———— "The Women in the Alliance Movement," *Arena*, 6:161–179 (July, 1892). Sketches and photographs.

DILLON, SIDNEY, "The West and the Railroads," *North American Review*, 152:443–452 (April, 1891). Dillon was president of the Union Pacific Railway Company.

DODGE, JOHN R., "The Discontent of the Farmer," *Century*, 21:447–456 (January, 1892). Sympathetic.

DREW, FRANK, "The Present Farmers' Movement," *Political Science Quarterly*, 6:282–310 (June, 1891). One of the best of the contemporary evaluations of the pre-Populist movement.

DUNN, J. P., JR., "The Mortgage Evil," *Political Science Quarterly*, 5:65–83 (March, 1890). Concedes the seriousness of the problem.

"The Farmer and the Carpenter," *Nation*, 50:407 (May 22, 1890). An interesting comparison of the lot of the farmer and the laborer by a writer who had little sympathy with either.

FISHER, WILLARD, "'Coin' and his Critics," *Quarterly Journal of Economics*, 10:187–208 (January, 1896). Criticizes severely both sides of the money dispute.

FOWLER, B. O., "Songs of the People," *Arena*, 6:xlvi–l (1892).

FOWLER, W. J., "The Farmers' Alliance: Letter from President Fowler," *Western Rural*, November 20, December 4, 1880.

GARLAND, HAMLIN, "The Alliance Wedge in Congress," *Arena*, 5:447–457 (March, 1892). Sketches of the Populist senators and representatives by a would-be admirer.

GLADDEN, WASHINGTON, " The Embattled Farmers," *Forum,* 10:315–322 (November, 1890). The interesting conclusions of a keen and sympathetic observer.

GLEED, CHARLES S., " The True Significance of the Western Unrest," *Forum,* 16:251–260 (October, 1893). Opposes the idea that the western radicals are socialistic.

GLEED, J. WILLIS, in "A Bundle of Western Letters," *Review of Reviews,* 10:45–46 (July, 1894). Sets forth hopefully the situation in Kansas.

———— " Is New York More Civilized than Kansas? " *Forum,* 17:217–234 (April, 1894). Finds the greater possibilities in Kansas.

———— " Western Lands and Mortgages," *Forum,* 11:468–471 (June, 1891). Regards the prospects of western recovery as bright.

———— " Western Mortgages," *Forum,* 9:93–105 (March, 1890). A fair statement of the existing situation.

GOODLOE, DANIEL R., " Western Farm Mortgages," *Forum,* 10:346–355 (November, 1890). Pessimistic.

GRADY, HENRY W., " Cotton and Its Kingdom," *Harper's New Monthly Magazine,* 63:719–734 (October, 1881). Contains useful material on the break-up of the southern plantations.

GREENE, THOMAS L., " Railroad Stock-Watering," *Political Science Quarterly,* 6:474–492 (September, 1891). Minimizes the evil.

HAGUE, JAMES D., " The Cost of Silver and the Profits of Mining," *Forum,* 15:60–67 (March, 1893). Points out that only the most fortunate producers could mine silver at a profit.

HARGER, CHARLES M., " The Farm Mortgage of To-day," *Review of Reviews,* 33:572–575 (May, 1906). Reviews the history of boom-time mortgages in the West.

———— " New Era in the Middle West," *Harpers' New Monthly Magazine,* 97:276–282 (July, 1898). Describes graphically the collapse of the land boom in the West and the gradual recovery.

HAYNES, FRED E., " The New Sectionalism," *Quarterly Journal of Economics,* 10:269–295 (April, 1896). Points out the significance of the debtor character of West and South.

HOLMES, GEORGE K., " The Peons of the South," *Annals of the American Academy of Political and Social Science,* 4:265–274 (September, 1893). An attack on the evils of the crop-lien system.

———— " A Decade of Mortgages," *Annals of the American Academy of Political and Social Science,* 4:904–918 (May, 1894). Based on the reports in the Census of 1890.

HOWELL, EDWARD B., in " A Bundle of Western Letters," *Review of Reviews,* 10:43 (July, 1894). Claims that western demands are due to the misrule of the East.

LAUGHLIN, J. LAURENCE, " Causes of Agricultural Unrest," *Atlantic Monthly,* 78:577–585 (November, 1896). Finds the root of the trouble in a " boom and wild expansion."

LLOYD, HENRY DEMAREST, " The Populists at St. Louis," *Review of Reviews,* 14:298–303 (September, 1896). Reviews the work of the Populist nominating convention of 1896.

LOUCKS, H. L., " Alliance Business Effort in Dakota," *National Economist,* 1:21 (March 14, 1889). The optimistic observations of an Alliance leader.

MACY, JESSE, in "A Bundle of Western Letters," *Review of Reviews,* 10:43 (July, 1894). States why the Populist movement was not formidable in Iowa.

MAITLAND, WILLIAM, " The Ruin of the American Farmer," *Nineteenth Century,* 32:733–743 (November, 1892). A conservative English view.

MAPPIN, W. F., "Farm Mortgages and the Small Farmer," *Political Science Quarterly*, 4:433–451 (September, 1889). Regards the outlook for the small farmer as encouraging.

MOORE, FREDERICK W., "The Condition of the Southern Farmer," *Yale Review*, 3:56–67 (May, 1894). Points out the usual evils.

MORGAN, JOHN T., "The Danger of the Farmers' Alliance," *Forum*, 12:399–409 (November, 1891). The danger is seen to lie in the politicians who have secured control of the Alliance.

OUSLEY, CLARENCE N., "A Lesson in Co-operation," *Popular Science Monthly*, 36:821–828 (April, 1890). Gives the history of the Texas Farmers' Exchange.

PEFFER, WILLIAM A., "The Farmer's Defensive Movement," *Forum*, 8:464–473 (December, 1889). Senator Peffer was an able defender of the farmers' point of view.

——— "The Farmers' Alliance," *Cosmopolitan*, 10:694–699 (April, 1891).

——— "The Mission of the Populist Party," *North American Review*, 157:665–678 (December, 1893).

——— "The Passing of the People's Party," *North American Review*, 166:12–23 (January, 1898). Predicted that the work attempted by the People's party would yet be done.

——— "People's Party," *Harper's Encyclopaedia of United States History*, Vol. 7. Written during the campaign of 1900 and admits the passing of the party.

POLK, LEONIDAS L., "The Farmers' Discontent," *North American Review*, 153:5–12 (July, 1891). Argues against the overproduction theory.

POST, C. C., "The Sub-Treasury Plan," *Arena*, 5:342–353 (February, 1892). An ardent defense of the subtreasury scheme.

RUSK, JEREMIAH M., "The Duty of the Hour," *North American Review*, 152:423–430 (April, 1891). Rusk was Harrison's secretary of agriculture and urged that the farmers' complaints be given a hearing.

SHAW, ALBERT, "William V. Allen: Populist," *Review of Reviews*, 10:30–42 (July, 1894). A sympathetic sketch.

SMITH, GOLDWIN, "The Brewing of the Storm," *Forum*, 22:436–446 (November, 1896). An anti-Populist view.

SMYTHE, WILLIAM E., in "A Bundle of Western Letters," *Review of Reviews*, 10:44 (July, 1894). Defends western and southern demand for currency expansion.

SNYDER, CARL, "Marion Butler," *Review of Reviews*, 14:429–433 (October, 1896). Reviews Butler's work as chairman of the St. Louis convention.

TRACY, FRANK B., "Menacing Socialism in the Western States," *Forum*, 15:332–342 (May, 1893). The author failed to appreciate the hold that individualism had on the Populist farmer.

——— "Rise and Doom of the Populist Party," *Forum*, 16:240–250 (October, 1893).

VAILE, JOEL F., "Colorado's Experiment with Populism," *Forum*, 18:714–723 (February, 1895). An ill-tempered and hostile account.

"The Vote of the Farm," *Nation*, 50:328–329 (April 24, 1890). An interesting editorial on the farmer and the tariff.

WALKER, C. S., "The Farmers' Alliance," *Andover Review*, 14:127–140 (August, 1890). Favorable.

——— "The Farmers' Movement," *Annals of the American Academy of Political and Social Science*, 4:790–798 (March, 1894). Appreciates the good work that the farmer organizations have done.

WALKER, FRANCIS A., "The Free Coinage of Silver," *Journal of Political Economy*, 1:163–178 (March, 1893). Points out the dangers of silver monometallism.

WARING, GEORGE E., JR., "Secretary Rusk and the Farmers," *North American Review*, 152:751–753 (June, 1891). A low-tariff retort to Secretary Rusk's argument.

WARMAN, CY, "Story of Cripple Creek," *Review of Reviews*, 13:161–166 (February, 1896). The background of the labor outbreak.

WARNER, AMOS G., "Railroad Problems in a Western State," *Political Science Quarterly*, 6:66–89 (March, 1891). A judicial survey of the situation in Nebraska.

WATSON, THOMAS E., "The People's Party's Appeal," *Independent*, 57:829 (October 13, 1904).

———— "Why I Am Still a Populist," *Review of Reviews*, 38:303–306 (September, 1908). Maintains that through Populism the spirit of protest can be kept alive.

WEAVER, JAMES B., "The Threefold Contention of Industry," *Arena*, 5:427–435 (March, 1892). "Land, transportation, and finance."

WELCH, RODNEY, "The Farmers' Changed Condition," *Forum*, 10:689–700 (February, 1891). Points out the difficulty that the farmer has in adapting himself to the changed conditions.

WEST, HENRY LITCHFIELD, "Two Republics or One?" *North American Review*, 162:509–511 (April, 1896). Will the discontented West secede?

WIMAN, ERASTUS, "The Farmer on Top," *North American Review*, 153:12–22 (July, 1891). An optimistic view.

ADVERTISING LITERATURE

Most of the following pamphlets, all to be found in the library of the Nebraska Historical Society, were issued for advertising purposes by railroads operating in Nebraska. They are fairly typical of all such literature.

BURLINGTON PAMPHLETS

B and M Railroad — 750,000 Acres of the Best Lands for Sale Southern Iowa and Southeastern Nebraska (Lincoln, 1878).

The Broken Bow Country in Central and Western Nebraska and How to Get There (Omaha, 1886).

Eastern Colorado. A Brief Description of the New Lands Now Being Opened Up (Lincoln, 1887).

FURNAS, ROBERT W., *An Invitation to South-east Nebraska* (Lincoln, 1887).

North Platte Lands of the Burlington and Missouri River Railroad Company in Nebraska (Lincoln, 1877).

Southwestern Nebraska and North-Western Kansas. A Brief Description of the Country, Its Products and Resources, Together with a Synopsis of the Homestead, Pre-emption, and Timber Culture Laws (Lincoln, 1887).

Views and Descriptions of Burlington and Missouri River Railroad Lands, with Important Information Concerning How to Select and Purchase Farms in Iowa and Nebraska, on Ten Years Credit (Burlington, Iowa, and Lincoln, Nebraska, 1872).

UNION PACIFIC PAMPHLETS

Central Western Nebraska and the Experiences of Its Stockgrowers (Omaha, 1883).

Corn is King (Omaha, 1882).

Guide to the Union Pacific Railroad Lands (Omaha, 1870).

Nebraska and Its Settlers. What They Have Done, and How They Do It; Its Crops and People (Omaha, n d.). Prepared and compiled by J. T. Allan.

Union Pacific Railroad across the Continent West from Omaha, Nebraska, 1865–1870 (New York, 1868). An early bid for settlers.

The State of Nebraska and Its Resources (Lincoln, 1879). An advertising brochure issued by the state showing how lands might be acquired and the rosy prospects of settlers who came to Nebraska.

Statement Brownville, Ft. Kearney and Pacific Railroad Company (Brownville, 1871). The preliminary announcement of a railroad that failed to materialize.

FURNAS, ROBERT W., *Nebraska, Her Resources and Advantages* (New Orleans, 1885).

WEEKLY NEWSPAPERS

Alliance (Lincoln, Nebraska), 1889. Name changed to the *Farmers' Alliance, q. v.* Nebraska Historical Society.

Alliance-Independent (Lincoln, Nebraska), 1892–94. Successor to the *Farmers' Alliance, q. v.* Nebraska Historical Society.

American (Philadelphia), 1895–1900. Edited by Wharton Barker, the middle-of-the-road Populist candidate for the presidency in 1900. Library of Congress.

Anti-Monopolist (St. Paul), 1876–78. An early radical weekly edited by Ignatius Donnelly. Minnesota Historical Society.

Broad-Axe (St. Paul), 1891–96. A free-silver paper. Minnesota Historical Society.

Farm, Stock and Home (Minneapolis), 1890–93. A farm journal that gave only incidental attention to politics. Minnesota Historical Society.

Farmers' Alliance (Lincoln, Nebraska), 1889–92. This weekly was the official organ of the Nebraska State Alliance. Later, under a variety of titles, it was always strongly Populistic. Nebraska Historical Society.

Great West (Minneapolis), 1889–92. A radical paper. Minnesota Historical Society.

Greensboro Patriot, 1890–92 (Greensboro, North Carolina). A conservative southern paper. Files are available in the office of the *Patriot*.

Kansas Weekly Capital and Farm Journal (Topeka), February, 1894. Reports the annual meeting of the Southern Alliance. Library of Congress.

National Economist (Washington, D. C.), 1889–92. Official organ of the Southern Farmers' Alliance. "Devoted to social, financial, and political economy." Library of Congress. Incomplete files owned by the Nebraska Historical Society.

National Watchman (Washington, D. C.), 1900–01. A free-silver paper started in 1892 as an official organ of the People's party. Incomplete files in the Library of Congress.

Nebraska Independent (Lincoln, Nebraska), 1896–99. Successor to the *Wealth-Makers, q. v.* Incomplete files in the library of the Nebraska Historical Society.

Penny Press (Minneapolis), 1894. A journal of politics. Minnesota Historical Society.

People's Party Paper (Atlanta, Georgia), 1891–94. Watson's personal organ. Library of Congress.

Progressive Farmer (Raleigh, North Carolina), 1886–93. A farm journal that championed Alliance and Populist principles. Library of the North Carolina Historical Commission, Raleigh, North Carolina.

Representative (St. Paul), 1893–98. A radical paper edited by Ignatius Donnelly. Minnesota Historical Society.

Southern Mercury (Dallas, Texas), 1892–1901. Official journal of the Farmers' State Alliance of Texas. Exceedingly valuable for southern phases of Populist history. Library of Congress.

Wealth Makers (Lincoln, Nebraska), 1894–96. Successor to the *Alliance-Independent, q. v.* Nebraska Historical Society.
Western Rural (Chicago), 1880–90. Edited by the founder of the Northern Alliance. Incomplete files in the library of the Nebraska Historical Society.

DAILY NEWSPAPERS

Chicago Daily Tribune, 1892–93. Gives considerable attention to Populist news. Library of Congress.
Chicago Dispatch, June, August, 1893. Reports the Antimonopoly convention and the silver congress held in Chicago during June and August, respectively. Library of Congress.
Chicago Evening Post, June, 1893. Reports the Antimonopoly convention. Library of Congress.
Chicago Inter-Ocean, June, 1893. Reports the Antimonopoly convention of that year. Library of Congress.
Chicago Times, August, 1893. Reports the so-called "silver congress" held in Chicago during the early part of August. Library of Congress.
Cincinnati Enquirer, May, 1891; September, 1898; May, 1900. Contains reports of the convention that founded the People's party and of the mid-road conventions of 1898 and 1900. Library of Congress.
Times-Star (Cincinnati), May, 1891. Reports the first People's party convention. Library of Congress.
Duluth Daily News (Duluth, Minnesota), 1892. For local Minnesota politics. Minnesota Historical Society.
Greensboro Daily Record (Greensboro, North Carolina), 1890–92. A conservative Democratic paper. Incomplete files in the Greensboro Public Library.
Kansas City Star (Kansas City, Missouri), September, 1901. Reports the conference of all reform parties held in Kansas City on September 18 and 19, 1901. Library of Congress.
Nebraska State Journal (Lincoln, Nebraska), 1887–1900. A high-church Republican paper, usually favorable to the railroads. Nebraska Historical Society.
Louisville Courier Journal (Louisville, Kentucky), April, 1902. Reports a mid-road conference. Library of Congress.
Memphis Appeal-Avalanche, November, 1892. Reports the annual meeting of the Southern Alliance. Library of Congress.
Commercial Appeal-Avalanche (Memphis), June, 1895. Reports the silver convention held in Memphis during June. Library of Congress.
Minneapolis Tribune, 1893. A conservative Republican paper. Minnesota Historical Society.
Nashville American (Nashville, Tennessee), July, 1897. Reports the proceedings of the Nashville conference of mid-roaders. Library of Congress.
New York Times, December, 1890. Reports the Ocala convention. Library of Congress.
New York Tribune, February, 1892. Reports and criticizes the St. Louis convention. Library of Congress.
Omaha Daily Bee, 1881–96. Only scattering numbers used. A Republican paper with occasional liberal tendencies. Nebraska Historical Society.
Omaha Herald, 1888. A Democratic paper, sometimes favorable to the Alliance. Succeeded by the *World-Herald, q. v.*
Omaha World-Herald, 1889–98. One of the leading Democratic papers in the Middle West. Friendly to Populism and to fusion. Nebraska Historical Society.
News and Observer, 1887–90 (Raleigh, North Carolina). A conservative Democratic paper. Library of the North Carolina Historical Commission.
St. Louis Globe-Democrat, 1889–1908. Useful for reports of the numerous Alliance and Populist conventions held in St. Louis. Library of Congress.

458   THE POPULIST REVOLT

St. Louis Republic, 1889–96. St. Louis was a favorite meeting place for Alliance and Populist conventions. Library of Congress.

St. Paul Globe, 1888–93. Gives considerable space to the farmers' movement in Minnesota. Minnesota Historical Society.

Pioneer Press (St. Paul), 1884–96. A conservative Republican daily. Minnesota Historical Society.

Daily Argus Leader (Sioux Falls, South Dakota), May, 1900. Reports the Fusionist-Populist nominating convention of 1900. Library of Congress.

Illinois State Register (Springfield, Illinois), July, 1904. Reports the Populist nominating conventions of 1904. Library of Congress.

SECONDARY WORKS

Books of General or Local Interest

Allen, Emory A., *Life and Public Service of James Baird Weaver* (Cincinnati, 1892). A campaign biography.

Andrews, E. Benjamin, *An Honest Dollar* (Hartford, 1894). Andrews was a convinced bimetallist.

———— *The United States in Our Own Times* (New York, 1903). Gives full attention to western phases of Populism.

Arnett, Alex M., *The Populist Movement in Georgia* (New York, 1922). A doctoral dissertation of unusual merit.

Banks, Enoch M., *The Economics of Land Tenure in Georgia* (New York, 1905). Valuable on the southern cropping system.

Beard, Charles A., *American Government and Politics*, 4th ed. (New York, 1924). Reveals well the final perseverance of Populist doctrines.

Beard, Charles A. and Mary R., *The Rise of American Civilization*, 2 vols. (New York, 1930). Thoroughly appreciative of the causes of agricultural discontent but short on details of Alliance and Populist history.

Bogart, Ernest L., *An Economic History of the United States* (New York, 1922). Useful for general background.

Boyle, James E., *The Financial History of Kansas* (Madison, Wisconsin, 1908). A useful doctoral dissertation.

Brewton, William W., *The Life of Thomas E. Watson* (Atlanta, 1926). Poorly done, but draws upon manuscript material not yet generally available.

Brooks, Robert P., *The Agrarian Revolution in Georgia, 1865–1912* (Madison, Wisconsin, 1914). Perhaps the best study of the economic background of Populism in the South.

Browne, Waldo Ralph, *Altgeld of Illinois: A Record of His Life and Work* (New York, 1924). Attempts to dispel the popular prejudice against Altgeld.

Bryce, James, *The Study of American History* (New York, 1922). Contains sagacious observations on the significance of the West in American history.

Buck, Solon J., *The Agrarian Crusade* (New Haven, 1920). A brief but informing essay on the various farmers' movements since the Civil War.

———— *The Granger Movement: A Study of Agricultural Organization and Its Political, Economic, and Social Manifestations, 1870–1880* (Cambridge, Massachusetts, 1913). A pioneer work in the history of American agrarianism.

Commons, John R., and others, *History of Labor in the United States*, 2 vols. (New York, 1918). Gives only incidental attention to Populism.

Connor, Robert D. W., and Poe, Clarence, *The Life and Speeches of Charles Brantley Aycock* (Garden City, 1912). Aycock ushered in the " new day " as governor of North Carolina after Populism had run its course.

CROLY, HERBERT D., *Marcus Alonzo Hanna; His Life and Work* (New York, 1912). A brilliant interpretation of the period as well as of the man.

EVANS, HARRY C., *The Pioneers and Politics of Davis County, Iowa* (Bloomfield, Iowa, 1929). Contains an excellent sketch of James B. Weaver.

FINE, NATHAN, *Labor and Farmer Parties in the United States, 1828–1928* (New York, 1928). More labor than farmer.

FLEMING, WALTER L., *The Sequel of Appomattox* (New Haven, 1919). Excellent on the condition of southern agriculture following the Civil War.

FOSSUM, PAUL R., *The Agrarian Movement in North Dakota* (Baltimore, 1925). Written from the point of view of a conservative economist.

GARLAND, HAMLIN, *A Son of the Middle Border* (New York, 1917). The author catches the spirit of the Populist West.

GRESHAM, MATILDA, *Life of Walter Quintin Gresham*, 2 vols. (Chicago, 1919). Throws a little light on Gresham's flirtations with the Populists.

HAMILTON, J. G. DE ROULHAC, *North Carolina since 1860* (*History of North Carolina*, ed. by Hamilton, Vol. 3, New York, 1919). An exceptionally good state history.

HAMMOND, MATTHEW B., *The Cotton Industry: An Essay in American Economic History* (Ithaca, New York, 1897). The standard authority on the subject.

HANEY, LEWIS HENRY, *A Congressional History of Railways in the United States, 1850–1887* (Madison, Wisconsin, 1910). Excellent on the limited field covered.

HAYNES, FRED E., *James Baird Weaver* (Iowa City, Iowa, 1919). A good political biography.

——— *Social Politics in the United States* (Boston, 1894). Gives relatively scant attention to Populism.

——— *Third Party Movements since the Civil War, with Special Reference to Iowa* (Iowa City, Iowa, 1916). A good mosaic pieced together from contemporary writings.

HEPBURN, ALONZO B., *A History of Currency in the United States* (New York. 1915). A useful compendium.

HIBBARD, BENJAMIN H., *The History of Agriculture in Dane County, Wisconsin* (Madison, Wisconsin, 1904). Dane County was typical in many ways of the whole West.

——— *A History of the Public Land Policies* (New York, 1924). Useful particularly on later phases of public land policies.

HICKS, JOHN D., *The Constitutions of the Northwest States* (Lincoln, Nebraska, 1923). A study of the constitution-making of North and South Dakota, Montana, Wyoming, Idaho, and Montana in the pre-Populist period.

HIGGINS, ELIZABETH, *Out of the West* (New York, 1902). A novel with the hard times in the West as a setting.

HOLDSWORTH, JOHN T., *Money and Banking* (New York, 1924). Useful on certain historical aspects of the money question.

HOLT, WILLIAM S., *The Federal Farm Loan Bureau, Its History, Activities, and Organization* (Baltimore, 1924). Adequate.

JONES, ELIOT, *The Trust Problem in the United States* (New York, 1921). A good survey.

KELSEY, RAYNER W., *Farm Relief and Its Antecedents* (Haverford, Pennsylvania, 1929). Contains an excellent historical sketch.

KEMMERER, EDWIN WALTER, *The A B C of the Federal Reserve System*, 5th ed. (Princeton, 1926). An untechnical presentation of the subject.

LAUCK, W. JETT, *The Causes of the Panic of 1893* (Boston, 1907). Lauck holds that the money situation was primarily responsible for the panic of 1893.

LAUGHLIN, J. LAURENCE, *The History of Bimetallism in the United States* (New York, 1892). The standard conservative work on the subject. Went through many editions.

LINGLEY, CHARLES R., *Since the Civil War*, rev. ed. (New York, 1926). Gives a good account of the monetary and financial problems that helped to produce Populism.

LOCKLIN, DAVID PHILIP, *Railroad Regulation since 1920* (New York, 1928). Useful on the latest phase of the railroad problem.

McMURRAY, DONALD R., *Coxey's Army: A Study of the Industrial Army Movement of 1894* (Boston, 1929). A detailed story of the various marches on Washington.

McVEY, FRANK L., *The Populist Movement* (New York, 1896). A contemporary study, strongly anti-Populist.

MERCER, ASA SHINN, and BOOTS, JOHN MERCER, *Powder River Invasion. War on the Rustlers in 1892* (n. p., 1923). Revision of a book earlier published as *The Banditti of the Plains*. Contains an account of the Johnson County War in Wyoming.

MITCHELL, WESLEY C., *Business Cycles* (Berkeley, California, 1913). Scholarly and suggestive.

MOODY, JOHN, *The Masters of Capital* (New Haven, 1920). A popular statement of some of the problems caused by "big business."

MORTON, J. STERLING, and WATKINS, ALBERT, *Illustrated History of Nebraska* (Lincoln, Nebraska, 1905-13). One of the best of the traditional type of biographic state histories.

MUZZEY, DAVID S., *The United States of America*, 2 vols. (Boston, 1924). The second volume contains admirable chapters on "The Revolt of the West" and "The Progressive Movement" that followed it.

MYRICK, HERBERT, *The Federal Farm Loan System* (New York, 1916). Contains text of the Federal Farm Loan Act.

NEVINS, ALLAN, *The Emergence of Modern America, 1865-1878* (New York, 1927). Contains some useful material on the economic and social background of Populism.

NOYES, ALEXANDER D., *Forty Years of American Finance* (New York, 1909). A useful popular treatise.

OBERHOLTZER, ELLIS P., *The Referendum in America together with Some Chapters on the Initiative and the Recall* (New York, 1912). An authoritative work.

OSGOOD, ERNEST S., *The Day of the Cattleman* (The University of Minnesota Press, 1929). A carefully documented study of the northern cattle range.

OSTROGORSKI, MOISEI IAKOVLEVICH, *Democracy and the Organization of Political Parties*, 2 vols. (New York, 1902). Takes cognizance of the People's party.

PAXSON, FREDERIC L., *History of the American Frontier, 1763-1893* (Boston, 1924). Good on the frontier background of Populism.

——— *The Last American Frontier* (New York, 1910). An understanding view of the conquest of the trans-Mississippi West.

——— *Recent History of the United States* (Boston, 1928). Devotes considerable space to the Populist movement and its consequences.

——— *When the West is Gone* (Boston, 1930). Concedes fully the significance of the Populist revolt.

PECK, HARRY THURSTON, *Twenty Years of the Republic, 1885-1905* (New York, 1906). A useful popular history.

RASTALL, BENJAMIN M., *The Labor History of the Cripple Creek District: A Study in Industrial Evolution* (Madison, Wisconsin, 1908). Governor Waite's activities on behalf of the strikers are here fairly set forth.

RIEGEL, ROBERT E., *The Story of the Western Railroads* (New York, 1926). A satisfactory general account of railway expansion into the trans-Mississippi West.

RIPLEY, WILLIAM Z., *Railroads; Rates and Regulation* (New York, 1912).
A standard authority.

SANBORN, JOHN B., *Congressional Grants of Land in Aid of Railways* (Madison,
Wisconsin, 1899). Useful for the narrow field of railway history it covers.

SEAGER, HENRY R., and GULICK, CHARLES A., *Trust and Corporation Problems*
(New York, 1929). Some of the problems described would have been
familiar enough to the Populists.

SELIGMAN, EDWIN R. A., *The Economics of Farm Relief* (New York, 1929).
A recent survey of the agricultural situation.

SHIPPEE, LESTER B., *Recent American History* (New York, 1924). A textbook
by an author who appreciates the significance of the Populist movement.

SIMKINS, FRANCIS B., *The Tillman Movement in South Carolina* (Durham,
North Carolina, 1926). An admirable study of the South Carolina version
of Populism.

SMALLEY, EUGENE V., *A History of the Republican Party . . . of Minnesota*
(St. Paul, 1896). A good political history.

STANWOOD, EDWARD, *A History of the Presidency*, 2 vols. (Boston, 1912). Contains nearly all the Populist platforms and some third-party history.

SULLIVAN, MARK, *The Turn of the Century, 1900–1904. Our Times, The United
States, 1900–1925,* Vol. 1 (New York, 1926). Contains a discerning review
of the Populist period.

TAUSSIG, FRANK W., *The Silver Situation in the United States* (New York,
1892). An intelligent contemporary view, which went through many editions.

THOMPSON, HOLLAND, *The New South* (New Haven, 1919). Good on southern
development since Reconstruction.

TURNER, FREDERICK J., *The Frontier in American History* (New York, 1921).
Several of the essays here collected were written with a view to explaining
the Populist movement.

VANDERBLUE, HOMER B., and BURGESS, KENNETH F., *Railroads. Rates —
Service — Management.* (New York, 1923). Contains a valuable historical
sketch.

VAN HISE, CHARLES R., *The Conservation of National Resources in the United
States* (New York, 1910). A pioneer work of enduring value on the subject.

WATKINS, MYRON W., *Industrial Combinations and Public Policy* (New York,
1927). A recent view.

WHITE, HORACE, *Money and Banking* (Boston, 1914). Contains the essentials
on the money question.

WHITE, WILLIAM ALLEN, *A Certain Rich Man* (New York, 1909). A novel
that includes much that is true about the history of Kansas and the
pioneer West.

WIEST, EDWARD, *Agricultural Organization in the United States* (Lexington,
Kentucky, 1923). A good resumé of the various farmers' movements.

WILLIS, H. PARKER, *The Federal Reserve System, Legislation, Organization, and
Operation* (New York, 1923).

WILLIS, H. PARKER, and STEINER, WILLIAM H., *Federal Reserve Banking Practice*
(New York, 1926). A scholarly study.

WIPRUD, A. CLARENCE, *The Federal Farm Loan System in Operation* (New
York, 1921). Good on the organization and early history of the system.

ARTICLES

BARNHART, JOHN D., " Rainfall and the Populist Party in Nebraska," *American
Political Science Review,* 19:527–540 (August, 1925). Presents an interesting parallel.

BARR, ELIZABETH N., " The Populist Uprising," *A Standard History of Kansas
and Kansans,* ed. by William E. Connelly, 2:1115–1195 (Chicago, 1918).
A strongly pro-Populist study.

462   THE POPULIST REVOLT

BECKER, CARL, "Kansas," *Essays in American History Dedicated to Frederick Jackson Turner* (New York, 1910), 85–111. A brilliant study of Kansas variations on the typical western characteristics.

BUCK, SOLON J., "Independent Parties in the Western States, 1873–1876," *Essays in American History Dedicated to Frederick Jackson Turner* (New York, 1910), 137–164. These parties foreshadow Populism.

CLARK, FREDERICK C., "State Railroad Commissions, and How They May Be Made Effective," *Publications of the American Economic Association*, Vol. 6, No. 6, pp. 473–583 (November, 1891). Assembles much useful material on early railway regulation.

COUTTS, W. A., "Agricultural Depression in the United States. Its Causes and Remedies," *Publications of the Michigan Political Science Association*, 2:213–308 (April, 1897). A prize essay written by a law student in the University of Michigan.

DELAP, SIMEON A., "The Populist Party in North Carolina," *Historical Papers Published by the Trinity College Historical Society* (Durham, North Carolina, 1922), 14:40–74. Superficial, but contains some useful information.

DEY, PETER A., "Railroad Legislation in Iowa," *Iowa Historical Record*, 9:540–566 (October, 1893). Merely a review of early railroad legislation in the state.

DIXON, FRANK H., "Railroad Control in Nebraska," *Political Science Quarterly*, 13:617–647 (December, 1898). A fair and accurate account.

EMERICK, C. F., "An Analysis of Agricultural Discontent in the United States," *Political Science Quarterly*, 11:433–463, 601–639; 12:93–127 (September, December, 1896; March, 1897). Finds abundant reasons for the unrest of the nineties.

FARMER, HALLIE, "The Economic Background of Frontier Populism," *Mississippi Valley Historical Review*, 10:406–427 (March, 1924). A scholarly and illuminating study.

———— "The Railroads and Frontier Populism," *Mississippi Valley Historical Review*, 13:387–397 (December, 1926). Stresses the importance of railroad oppression as a cause of agrarian revolt.

GLYNN, HERBERT L., "The Urban Real Estate Boom in Nebraska during the Eighties," *Nebraska Law Bulletin*, 6:455–481; 7:228–254 (May, November, 1928). Useful especially on the subject of the small town booms.

HAMMOND, MATTHEW B., "The Southern Farmer and the Cotton Question," *Political Science Quarterly*, 12:450–475 (September, 1897). Sets forth much the same type of information as appears in *The Cotton Industry* by the same author.

HEDGES, JAMES B., "Colonization Work of the Northern Pacific Railroad," *Mississippi Valley Historical Review*, 13:311–342 (December, 1926). An excellent piece of work.

HICKS, JOHN D., "The Birth of the Populist Party," *Minnesota History*, 9:219–247 (September, 1928).

———— "The Farmers' Alliance in North Carolina," *North Carolina Historical Review*, 2:162–187 (April, 1925).

———— "The Origin and Early History of the Farmers' Alliance in Minnesota," *Mississippi Valley Historical Review*, 9:203–226 (December, 1922).

———— "The People's Party in Minnesota," *Minnesota History Bulletin*, 5:531–560 (November, 1924).

———— "The Political Career of Ignatius Donnelly," *Mississippi Valley Historical Review*, 8:80–132 (June–September, 1921).

———— "The Sub-Treasury: A Forgotten Plan for the Relief of Agriculture," *Mississippi Valley Historical Review*, 15:355–373 (December, 1928).

HICKS, JOHN D., and BARNHART, JOHN D., "The Farmers' Alliance," *North Carolina Historical Review*, 6:254–280 (July, 1929). Covers the subject taken up in Chapter IV.

KENDRICK, BENJAMIN B., "Agrarian Discontent in the South: 1880–1900," *Annual Report of the American Historical Association, 1920* (Washington, 1925), 267–272. Sets forth the need of further research along this line.

KLOTSCHE, JOHANNES, "The Political Career of Samuel Maxwell," *Nebraska Law Bulletin*, 6:439–454 (May, 1928). Maxwell turned Populist in the nineties after serving for years as a Republican judge.

LEWINSON, PAUL, "The Negro in the White Class and Party Struggle," *Southwestern Political and Social Science Quarterly*, 8:358–382 (March, 1928). An adequate and scholarly account.

LIBBY, ORIN G., "A Study of the Greenback Movement, 1876–1884," *Transactions of the Wisconsin Academy*, 12:530–543 (Madison, Wisconsin, 1898). Shows the reasons back of the western demand for money inflation.

McMURRAY, DONALD R., "The Industrial Armies and the Commonwealth," *Mississippi Valley Historical Review*, 10:215–252 (December, 1923). Later expanded into a book, *Coxey's Army*, q. v.

MILLER, RAYMOND C., "The Economic Background of Populism in Kansas," *Mississippi Valley Historical Review*, 11:469–489 (March, 1925). A scholarly monograph based on exhaustive research.

MORRIS, RALPH C., "The Notion of a Great American Desert East of the Rockies," *Mississippi Valley Historical Review*, 13:190–200 (September, 1926). Shows the slow evaporation of a popular myth.

NEWCOMB, H. T., "The Progress of Federal Railway Rates," *Political Science Quarterly*, 11:201–221 (June, 1896). Reviews the workings of the Interstate Commerce Act to date.

NIXON, HERMAN C., "The Cleavage within the Farmers' Alliance Movement," *Mississippi Valley Historical Review*, 15:22–33 (June, 1928). Cottonseed oil versus butter and lard.

——— "The Economic Basis of the Populist Movement in Iowa," *Iowa Journal of History and Politics*, 21:373–396 (July, 1923). A good preliminary survey.

——— "The Populist Movement in Iowa," *Iowa Journal of History and Politics*, 24:3–107 (January, 1926). An excellent study, limited in importance somewhat, however, by the fact that Populism was relatively weak in Iowa.

PAXSON, FREDERIC L., "The Pacific Railroads and the Disappearance of the Frontier in America," *Annual Report of the American Historical Association, 1907*, 1:105–118 (Washington, 1908). An early study on the significance of the disappearance of free lands.

REFSELL, OSCAR N., "The Farmers' Elevator Movement," *Journal of Political Economy*, 22:872–895 (November, 1914). A good discussion of the elevator situation at the close of the Populist period.

SCHMIDT, LOUIS B., "Some Significant Aspects of the Agrarian Revolution in the United States," *Iowa Journal of History and Politics*, 28:371–395 (July, 1920). Contains some thoughtful observations on the service rendered by the farmers' organizations.

SIMKINS, FRANCIS B., "The Problems of South Carolina Agriculture after the Civil War," *North Carolina Historical Review*, 7:46–77 (January, 1930). An excellent piece of research.

"Some Lost Towns of Kansas," *Kansas Historical Collections*, 12:426–490 (Topeka, 1912). Records the disappearance of some of the "boom" towns of the eighties.

STEWART, ERNEST D., "The Populist Party in Indiana," *Indiana Magazine of History*, 14:332–367; 15:53–74 (December, 1918; March, 1919). A good local study.

VEBLEN, THORSTEIN, "The Price of Wheat since 1867," *Journal of Political Economy*, 1:68–103 (December, 1892). Concludes that with some exceptions wheat-growing is still generally profitable.

WELLBORN, FRED, "The Influence of the Silver-Republican Senators, 1889–1891," *Mississippi Valley Historical Review*, 14:462–480 (March, 1928). On the "deal" between the free-silver and the anti-Force Bill senators.

WHITE, MELVIN J., "Populism in Louisiana during the Nineties," *Mississippi Valley Historical Review*, 5:1–19 (June, 1918). A brief political survey.

WHITE, WILLIAM ALLEN, "The End of an Epoch," *Scribners' Magazine*, 79:561–570 (June, 1926). An illuminating backward glance by the man who wrote "What's the Matter with Kansas?"

WOODBURN, JAMES A., "Western Radicalism in American Politics," *Mississippi Valley Historical Review*, 13:143–168 (September, 1926). Shows the similarity between Populist and other western third-party demands.

UNPUBLISHED MONOGRAPHS

ANDERSON, HELEN, "The Influence of Railway Advertising upon the Settlement of Nebraska." Master's thesis, University of Nebraska, 1926.

BARNHART, JOHN D., "The History of the Farmers' Alliance and of the People's Party in Nebraska." Doctoral dissertation, Harvard University, 1929. Of great merit. Soon to be published.

BOELL, JESSE E., "William Jennings Bryan before 1896." Master's thesis, University of Nebraska, 1929.

HARMER, MARIE U., "The Life of Charles H. Van Wyck." Master's thesis, University of Nebraska, 1929. Van Wyck was prominent in Populist circles.

JONES, VIRGINIA BOWEN, "The Influence of the Railroads of Nebraska on Nebraska State Politics." Master's thesis, University of Nebraska, 1927.

KEEFER, E. ELOISE, "The Chicago, Burlington and Quincy in the Early History of Nebraska, 1860–1885." Master's thesis, University of Nebraska, 1929.

KUESTER, FRIEDA C., "The Farmers' Alliance in Nebraska." Master's thesis, University of Nebraska, 1927.

LINDSAY, CHARLES, "The Political Antecedents of the Progressive Movement." Master's thesis, University of Nebraska, 1925.

MILLER, RAYMOND C., "The Populist Party in Kansas." Doctoral dissertation, Chicago University, 1928. Especially convincing on the economic background that produced Populism. Soon to be published.

ROWLEY, R. D., "The Judicial Career of Samuel Maxwell." Master's thesis, University of Nebraska, 1928.

SCOTT, MITTIE Y., "The Life and Political Career of William Vincent Allen." Master's thesis, University of Nebraska, 1927.

STORMS, HELEN, "The Nebraska State Election of 1890." Master's thesis, University of Nebraska, 1924.

ZIMMERMAN, WILLIAM F., "Legislative History of Nebraska Populism, 1890–1895." Master's thesis, University of Nebraska, 1926.

INDEX

INDEX

Foght, H. W., 30

Fowler, B. O., editor of *Arena*, 350

Free lands, disappearance of, 72

Free railroad passes, 70, 419

Free silver, in campaign of *1892*, 264–268; Waite's plan to legalize, 292; opposed by Cleveland, 311; arguments for, 314–320, 333; conquers Populist party, 318, 343–346; conquers Democratic party, 342, 356; mid-road attitude towards, 382

Fusion, between Populists and Republicans, 245–247, 329–331, 369–372, 391–393; between Populists and Democrats, 255–257, 261, 265–267, 327, 394; opposition to, 326, 347, 358, 363–366, 380–387; experiment with, in North Carolina, 377, 391–393

Garland, Hamlin, 4

Gary, Martin, of South Carolina, 53

Gay, Mrs. Bettie, of Texas, 166

George, Milton, of Illinois, 98–100

Georgia, mortgages in, 85; Alliance exchange in, 137; political activities of Alliance in, 176; election of *1892* in, 246, 251; of *1894*, 330, 334; of *1896*, 371, 377; of *1898*, 393

Gold reserve, declines, 307; maintained by purchases, 321

Gold standard, 388

"Good-bye, My Party, Good-bye," 168

Gould, Jay, 68

Grady, Henry W., of Georgia, 37, 49

Grain, grading and marketing of, 76–78, 287–289

Granger movement, 96

Great American Desert, legend of, 5

Greenback party, 87, 96

Gresham, Judge Walter I., of Illinois, 233

Gulf and Interstate Railway, 284

Hallowell, James R., of Kansas, 162

Hard times, 31–35, 309

Harvey, William H., 340

Hepburn Act, 419

Hill, James J., 62

Hogg, H. H., of Texas, 177, 249

Holcomb, Silas A., of Nebraska, 285, 328

Homestead Act, 8–11

Homestead Strike, 323

Howard, M. W., of Alabama, 400

Humphrey, R. M., 114

Idaho, election of *1892* in, 265; of *1894*, 333; of *1896*, 377

Independent voting, encouraged by Populism, 409

Indianapolis, meeting of Southern Alliance at, 221

Industrial armies, 323

Industrial development, in South, 50

Initiative and referendum, favored by Populists, 406–408

Interest charges, 82–84, 89

International bimetallism, 320

International monetary conference, *1867*, 304

Interstate Commerce Act, 418

Iowa, population, 16, 20; business agency in, 134; political activities of Alliance in, 150; election of *1894* in, 333

Jones, Evan, of Texas, 113

Jute-bagging trust, boycotted, 140

Kansas, beginnings, 5; population, 16, 20; railroad advertising, 17, 20; mortgages, 23, 84; real estate boom, 25, 28; loss of population, 32–34; production of corn, 58, 60; taxation, 86; drought, 100; Farmers' Alliance in, 100, 103; Southern Alliance in, 121; Union Labor party in, 155; People's party organized, 156; campaign of *1890* in, 167–170, 179; Alliance legislation, 182; election of *1892* in, 256, 261, 268; first Populist government, 275; legislative war in, 276–281; election of *1894* in, 328, 333; of *1896*, 377; of *1898*, 395

Kansas City conference, *1901*, 402

Knights of Labor, fraternize with Alliance, 103, 115, 123

Kolb, R. F., of Alabama, 177, 249, 330, 335

Kyle, James H., of South Dakota, 181, 236

Labor demands, 420

Land, grants of, to railroads, 3, 71; speculation in, 24–29; disappearance of free land, 72

Larrabee, William, of Iowa, 148, 150

Lease, Mrs. Mary E., of Kansas, in campaign of *1890*, 159; of *1892*, 244, 266; quarrels with Governor Lewelling, 281; in campaign of *1893*, 286; evaluates Populist contribution, 421